Digital Electronics: Principles and Theories

Digital Electronics: Principles and Theories

Rachell Hawkins

NY RESEARCH PRESS
New York

Published by NY Research Press
118-35 Queens Blvd., Suite 400,
Forest Hills, NY 11375, USA
www.nyresearchpress.com

Digital Electronics: Principles and Theories
Rachell Hawkins

International Standard Book Number: 978-1-63238-703-5 (Hardback)

Cataloging-in-Publication Data

Digital electronics : principles and theories / Rachell Hawkins.
p. cm.
Includes bibliographical references and index.
ISBN 978-1-63238-703-5
1. Digital electronics. 2. Electronic systems. 3. Electronics. I. Hawkins, Rachell.
TK7868.D5 D54 2019
621.381--dc23

Printed in China.

Contents

Preface

Digital electronics is the branch of physics that deals with electronic circuits that use digital signals. In digital circuits, signals can be transmitted without degradation. A digital circuit is constructed using logic gates that are created from electrically controlled switches such as transistors. An integrated circuit consists of multiple transistors on a single silicon chip, and designed using electronic design automation software. Digital circuit complexity can be minimized using algorithms such as Espresso heuristic logic minimizer. Algorithms for the optimization of large logic systems employ binary decision diagrams or algebraic manipulations. This book is a valuable compilation of topics, ranging from the basic to the most complex theories and principles in the field of digital electronics. The various aspects of digital electronics along with technological progress that have future implications have also been glanced at. In this book, constant effort has been made to make the understanding of the difficult concepts of digital electronics as easy and informative as possible, for the readers.

A foreword of all chapters of the book is provided below:

Chapter 1- In a digital signal, data is represented in terms of a sequence of discrete values. Electronics that use digital signal to operate are known as digital electronics This is an introductory chapter, which will introduce briefly all the significant principles of digital signals and digital electronics;

Chapter 2- In electronics, a logic gate is a device that performs a logical operation on binary inputs producing a single binary output. These are generally constructed using diodes or transistors, vacuum tubes, relays, etc. This chapter has been carefully written to provide an easy understanding of the different types of logic gates, such as NOT, AND, OR, NAND, NOR, XOR, etc. It also delves into the fundamental concepts of diode logic, resistor-transistor logic, diode-transistor logic and direct-coupled transistor logic, among others;

Chapter 3- In order to completely understand the use of number systems in digital electronics, it is vital to understand the fundamentals of binary code, Boolean algebra, bitwise operation and binary-coded decimal. The following chapter discusses in detail about the different number systems and in particular, about the binary number system and its use in digital electronic;

Chapter 4- Circuits that operate using digital signals are called digital circuits. These operate using logic gates that are printed on integrated circuits. This chapter closely examines some of the devices used in digital circuits, such as binary adder, subtractor, multiplier, encoder, decoder, shift register, etc. The principles of combinational logic circuits, sequential logic circuits, asynchronous and synchronous sequential circuits, etc. which are crucial to the understanding of digital circuit operations have also been carefully dealt in this chapter.

At the end, I would like to thank all the people associated with this book devoting their precious time and providing their valuable contributions to this book. I would also like to express my gratitude to my fellow colleagues who encouraged me throughout the process.

Rachell Hawkins

1

Digital Signal and Digital Electronics

In a digital signal, data is represented in terms of a sequence of discrete values. Electronics that use digital signal to operate are known as digital electronics. This is an introductory chapter, which will introduce briefly all the significant principles of digital signals and digital electronics.

Digital Signal

A digital signal - a must for computer processing - is described as using binary (0s and 1s), and therefore, cannot take on any fractional values. As illustrated in the graphic below, digital signals retain a uniform structure, providing a constant and consistent signal. Because of the inherent reliability of the digital signal, technology using it is rapidly replacing a large percentage of analog applications and devices. For example, the wristwatch, showing the time of day, with its minute, hour, and sweeping second hands, is being replaced by the digital watch, which offers the time of day and other information using a numerical display. A typical digital signal is represented below. Note the equally dispersed 1s and 0s.

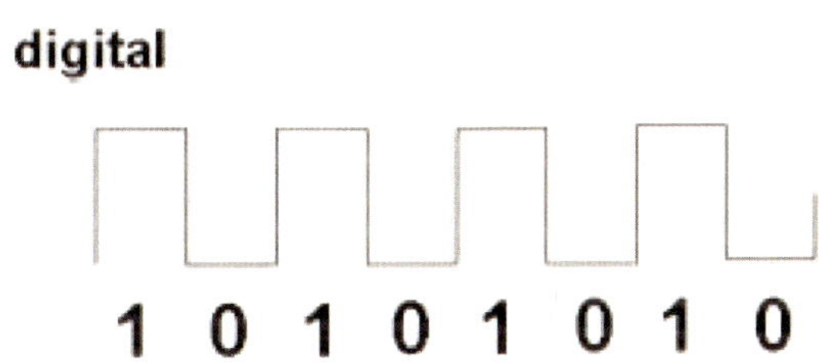

Digital signal with binary

Digital signals must have a finite set of possible values. The number of values in the set can be anywhere between two and a-very-large-number-that's-not-infinity. Most commonly digital signals will be one of two values – like either 0V or 5V. Timing graphs of these signals look like square waves.

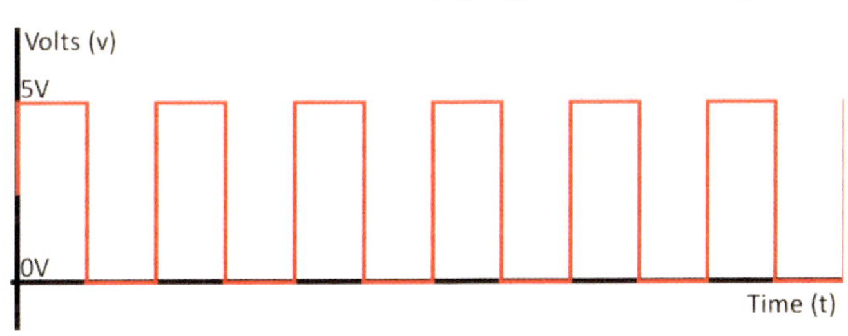

Or a digital signal might be a discrete representation of an analog waveform. Viewed from afar, the wave function below may seem smooth and analog, but when you look closely there are tiny discrete steps as the signal tries to approximate values:

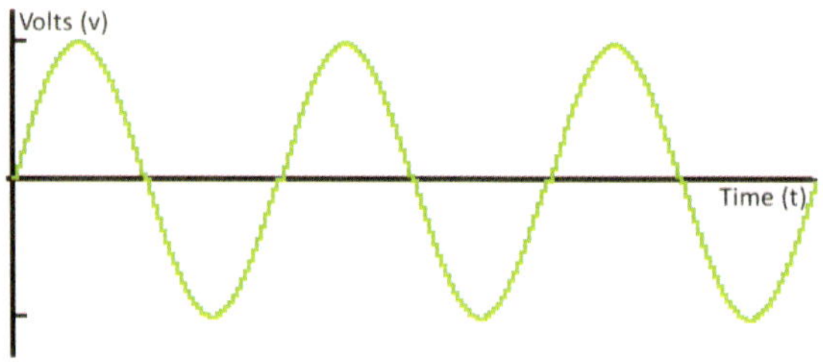

That's the big difference between analog and digital waves. Analog waves are smooth and continuous, digital waves are stepping, square, and discrete.

Examples of Digital Signals

Not all audio and video signals are analog. Standardized signals like HDMI for video (and audio) and MIDI, I^2C, or AC'97 for audio are all digitally transmitted.

Most communication between integrated circuits is digital. Interfaces like serial, I^2C, and SPI all transmit data via a coded sequence of square waves.

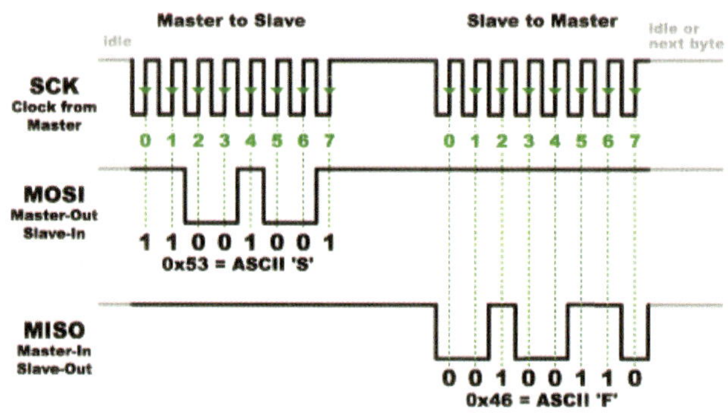

Serial peripheral interface (SPI) uses many digital signals to transmit data between devices.

Characteristics of Digital Signals

1. Bit interval

It is the time required to send one single bit

2. Bit rate

- It refers to the number of bit intervals in one second.
- Therefore bit rate is the number of bits sent in one second as shown in fig.
- Bit rate is expressed in bits per second (bps).
- Other units used to express bit rate are Kbps, Mbps and Gbps.

  1 kilobit per second (Kbps) = 1,000 bits per second

  1 Megabit per second (Mbps) = 1,000,000 bits per second

  1 Gigabit per second (Gbps) = 1,000,000,000 bits per second

Advantages of Digital Signals

Digital Data - Digital transmission certainly has the advantage where binary computer data is being transmitted. The equipment required to convert digital data to analog format and transmitting the digital bit streams over an analog network can be expensive, susceptible to failure, and can create errors in the information.

Compression - Digital data can be compressed relatively easily, thereby increasing the efficiency of transmission. As a result, substantial volumes of voice, data, video and image information can be transmitted using relatively little raw bandwidth.

Security - Digital systems offer better security. While analog systems offer some measure of security through the scrambling of several frequencies. Scrambling is fairly simple to defeat. Digital information, on the other hand, can be encrypted to create the appearance of a single, pseudorandom bit stream. Thereby, the true meaning of individual bits, sets of bits, or the total bit stream cannot be determined without having the key to unlock the encryption algorithm employed.

Quality - Digital transmission offers improved error performance (quality) as compared to analog. This is due to the devices that boost the signal at periodic intervals in the transmission system in order to overcome the effects of attenuation. Additionally, digital networks deal more effectively with noise, which always is present in transmission networks.

Cost - The cost of the computer components required in digital conversion and transmission has dropped considerably, while the ruggedness and reliability of those components has increased over the years.

Upgradeability - Since digital networks are comprised of computer (digital) components, they are relatively easy to upgrade. Such upgrading can increase bandwidth, reduces the incidence of error and enhance functional value. Some upgrading can be effected remotely over a network, eliminating the need to dispatch expensive technicians for that purpose.

Management - digital networks can be managed much more easily and effectively due to the fact that such networks consist of computerized components. Such components can sense their own level of performance, isolate and diagnose failures, initiate alarms, respond to queries, and respond to commands to correct any failure. Further, the cost of these components continues to drop.

Digital Electronics

Digital Electronics is the sub-branch of electronics which deals with digital signals for processing and controlling various systems and sub-systems. In various applications like sensors and actuators, usage of digital electronics is increasing extensively.

The fundamentals and implementation of digital electronics are essential to understanding the design and working of consumer/industrial electronics, communications, embedded systems, computers, security and military equipment.

Devices used in applications such as these are constantly decreasing in size and employing more complex technology. It is therefore essential for engineers and students to understand the fundamentals, implementation and application principles of digital electronics, devices and integrated circuits. This is so that they can use the most appropriate and effective technique to suit their technical need.

Digital Electronics includes:

- Information on number systems, binary codes, digital arithmetic, logic gates and families, and Boolean algebra;
- An in-depth look at multiplexers, de-multiplexers, devices for arithmetic operations, flip-flops and related devices, counters and registers, and data conversion circuits;
- Up-to-date coverage of recent application fields, such as programmable logic devices, microprocessors, microcontrollers, digital troubleshooting and digital instrumentation.

Introduction to Digital Electronics

Digital electronics deals with the electronic manipulation of numbers, or with the manipulation of varying quantities by means of numbers. Because it is convenient to do so, today's digital systems deal only with the numbers 'zero' and 'one', because they can be represented easily by 'off and 'on' within a circuit. This is not the limitation it might seem, for the binary system of counting can be used to represent any number that we can represent with the usual denary (0 to 9) system that we use in everyday life.

First, compare the way a digital system might be used to multiply two numbers together compared with a linear, or 'analogue', system. Let us imagine that we want to multiply 7 by 6 and arrive at the answer: 42. In an analogue system we could start with a voltage of exactly 7 V. If this were fed into an op-amp with a gain of exactly 6, the output voltage could be measured accurately and it would be found to be 42 V. This kind of 'computer' is quite feasible, but has limitations. Now suppose we want to multiply 7234 by 27 300. This is a different problem. Clearly it would now be impractical to use 1 V to represent 1, so we must use, say, 1 mV to represent 1. The input to the amplifier is thus 7.234 V. There now comes the problem of the amplifier to give a gain of 27 300-a massive design problem in itself. If it could be made, and made to work, the output would be 197 488.2 V, or nearly 200 kV! Even representing the input as 1 J.J.V per 1, the output level would be almost 200 V -not really a practical solution, even supposing we could measure 200 V to the nearest microvolt.

A digital system, on the other hand, would treat the problem in a different way. Back to 7 times 6. Seven, in binary notation, is 111, and six is 110-see Figure for a table giving comparable denary and binary numbers. We can represent a 0 by 0 V, and a 1 by, say,+ 5 V. Both six and seven could be represented using only three wires at one of two different potentials. The answer, 42, is represented in binary as 101010, which needs six wires. To increase the capacity of the system to calculate 7234 x 27 300 simply needs more wires; 7234 in binary is 1110001000010 and 27 300 in binary is 110101010100100. Hardly a readily recognisable or convenient number for us, but it is still a way of representing a big number without having to use electronic systems needing more than a few volts. Also, the accuracy of the system in terms of reading voltage levels is not required to be high; all that is needed is the ability to differentiate in the circuits between 0 V and + 5 V. The solution to the problem, 197 488 200, is actually a rather long binary number, 1011110001010110111001001110,

28 binary digits, or 'bits', as the computer people call them. Although binary numbers are long, the difficulty of manipulating them is nothing compared with the problems associated with the equivalent analogue systems. Rather than use 28 wires, a computer or calculator would use a register holding the 28 'ons' and 'offs' in bistables, and the digits would be processed in sequence rather than altogether, but such details are part of computer science, not mastering electronics! Digital systems clearly have big advantages.

Figure some denary numbers with their binary equivalents

Denary	Binary
0	0
1	1
2	10
3	11
4	100
5	101
6	110
7	111
8	1000
9	1001
10	1010
11	1011
42	101010
100	1100100

A circuit involved in storing and manipulating binary numbers is rather complex, but in much the same way as a knitted sweater is complex, containing as it does a very large number of similar units. Digital systems remained the province of the specialist until the advent of microelectronics.

A digital circuit is a circuit where the signal must be one of two discrete levels. Each level is interpreted as one of two different states (for example, on/off, 0/1, true/false). Digital circuits use transistors to create logic gates in order to perform Boolean logic. This logic is the foundation of digital electronics and computer processing. Digital circuits are less susceptible to noise or degradation in quality than analog circuits. It is also easier to perform error detection and correction with digital signals. To automate the process of designing digital circuits, engineers use electronic design automation (EDA) tools, a type of software that optimizes the logic in a digital circuit.

Digital System

The term digital system refers to elements such as hardware, software and networks and their use. There may be many different components that make up one system; for example, a computer has a central processing unit, a hard disk, keyboard, mouse, screen etc. A peripheral device is a digital component that can be connected to a digital system, for example, a digital camera or printer.

Digital hardware and software *components* (internal and external) used to transform *data* into a *digital solution*. When *digital systems* are connected, they form a network. For example:

- A smartphone is a digital system that has software (apps, an operating system), input *components* (for example, touch screen, keyboard, camera and microphone), output *components* (for example, screen and speakers), memory *components* (for example, silicon chips, solid state drives), communication *components* (for example, SIM card, wi-fi, bluetooth or mobile network antennas), and a processor made up of one or more silicon chips.
- A desktop computer with specific software and hardware *components* for dairy farming. The computer is connected via cables to milking *equipment* and via wi-fi to sensors that read tags on the cows. Through these hardware *components* the software records how much milk each cow provides. Such systems can also algorithmically control attaching milking *equipment* to each cow, providing feed and opening gates.

Advantages of Digital Systems

- Easier Designing: The Digital systems can be easily designed as they involve digital signals. These signals do not require exact value at a particular time but it consists of range of particular values of voltage. Thus, it comprises of basically two values 0 and 1 I.e high or low.
- Noise Immune: Digital systems are noise immune because digital signal consists of range of particular values. Thus, when noise is introduced in the medium and digital signal and analog signal both passes through it. The analog signal will be affected more because it varies continuous with time so it is difficult to identify that noise has destroyed which value of voltage. While in case of digital system, noise effect the particular range of the signal thus, it is clear to identify the particular range of filtering is also easy in case of digital signals.
- Information Storage is Simpler: The storage of information in digital systems is easy. It can be stored by latching thus, the it can be stored for a long period of time.
- High Accuracy And Precision: The digital signal offers high accuracy and precision. This is because the processing of digital signal is done through the switching circuit. While in case of analog signals the processing and its output is highly dependant on circuit components. The accuracy obtained in analog circuits is restricted to 3-4 digits while in case of digital signals the accuracy is far more than the analog circuit.

- Programmable: The digital systems are easily programmable but analog system becomes complex when excessive programming of components is done.

Disadvantages of Digital Systems

Everything comes up with pros and cons. Similarly digital systems too hold certain disadvantages which are as follows:-

- Expensive: Digital systems are expensive because it involves switching elements.
- Analog nature of Real World Entities: We need to convert the digital output in analog form because all the real world entities are analog.

Digital to Analog and Analog to Digital Conversion

These types of conversion circuits are required so that we can convert analog signal into digital and then process the digital signal using digital systems and convert back the digital signal into analog form so that it is can be understood by real world.

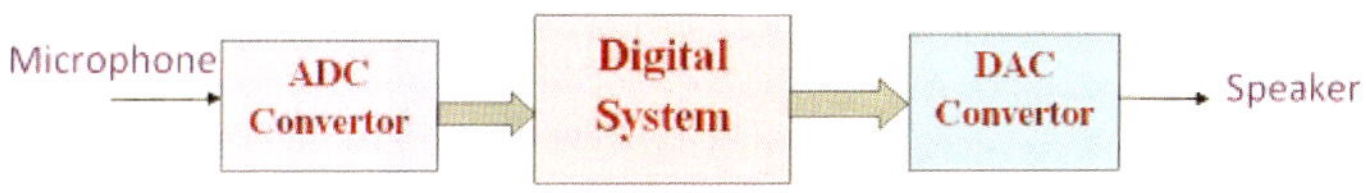

Conversion System

In this way we can convert one form of signal into another form and process it digitally.

References

- What-are-digital-and-analog-signals-definition-lesson-quiz : study.com, Retrieved 19 July 2018
- Tutorials, analog-vs-digital: learn.sparkfun.com, Retrieved 29 June 2018
- Digital-signal, computer networking notes, communication-networks: ecomputernotes.com, Retrieved 19 July 2018
- Digital-systems, teachers: digitaltechnologieshub.edu.au, Retrieved 9 July 2018
- Digital-electronics: electronicscoach.com, Retrieved 29 May 2018

Logic Gates

In electronics, a logic gate is a device that performs a logical operation on binary inputs producing a single binary output. These are generally constructed using diodes or transistors, vacuum tubes, relays, etc. This chapter has been carefully written to provide an easy understanding of the different types of logic gates, such as NOT, AND, OR, NAND, NOR, XOR, etc. It also delves into the fundamental concepts of diode logic, resistor-transistor logic, diode-transistor logic and direct-coupled transistor logic, among others.

Nowadays, computers have become an integral part of life as they perform many tasks and operations in quite a short span of time. One of the most important functions of the CPU in a computer is to perform logical operations by utilizing hardware like Integrated Circuits, electronic circuits and software technologies. But, how these hardware and software perform such operations is a mysterious puzzle. In order to have a better understanding of such a complex issue, we must have to acquaint ourselves with the term Boolean Logic, developed by George Boole. For a simple operation, computers utilize binary digits rather than digital digits. All the operations are carried out by the Basic Logic gates.

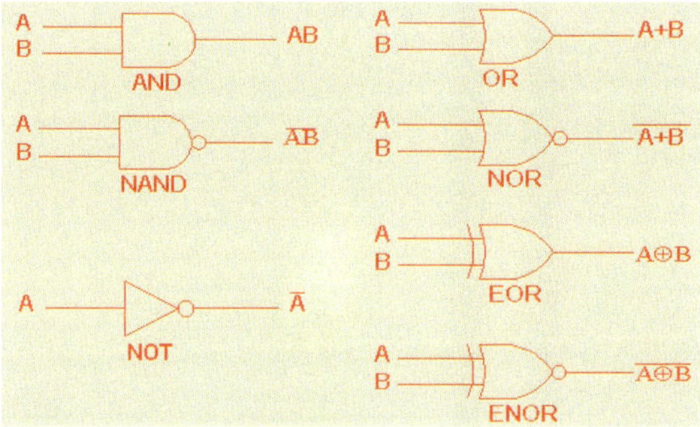

Basic Logic Gates

A logic gate is a basic building block of a digital circuit that has two inputs and one output.The relationship between the i/p and the o/p is based on a certain logic. These gates are implemented using electronic switches like transistors, diodes. But, in practice basic logic gates are built using CMOS technology, FETS and MOSFET(Metal Oxide Semiconductor FET)s. Logic gates are used in microprocessors, microcontrollers, embedded system applications and in electronic and electrical project circuits.

Digital Logic Gates can be made from discrete components such as Resistors, Transistors and Diodes to form RTL (resistor-transistor logic) or DTL (diode-transistor logic) circuits, but today's modern digital 74xxx series integrated circuits are manufactured using TTL (transistor-transistor logic) based on NPN bipolar transistor technology or the much faster and low power CMOS based MOSFET transistor logic used in the 74Cxxx, 74HCxxx, 74ACxxx and the 4000 series logic chips.

The eight most "standard" individual Digital Logic Gates are summarised below along with their corresponding truth tables.

Standard Logic Gates

The Logic AND Gate

Symbol	Truth Table		
	B	A	Q
	0	0	0
	0	1	0
	1	0	0
2-input AND Digital Logic Gate	1	1	1
Boolean Expression Q = A.B	Read as A AND B gives Q		

The Logic OR Gate

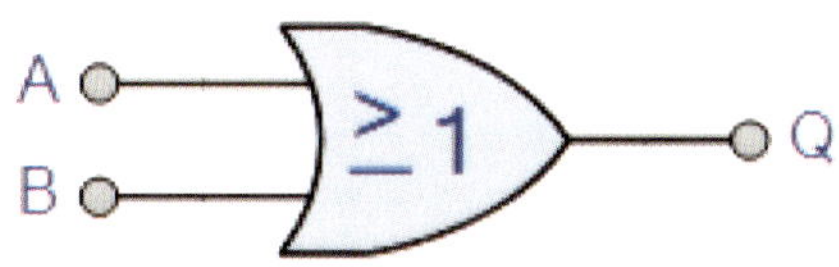

Symbol	Truth Table		
	B	A	Q
	0	0	0
	0	1	1
	1	0	1
	1	1	1
Boolean Expression Q = A + B	Read as A OR B gives Q		

Inverting Logic Gates

The Logic NAND Gate

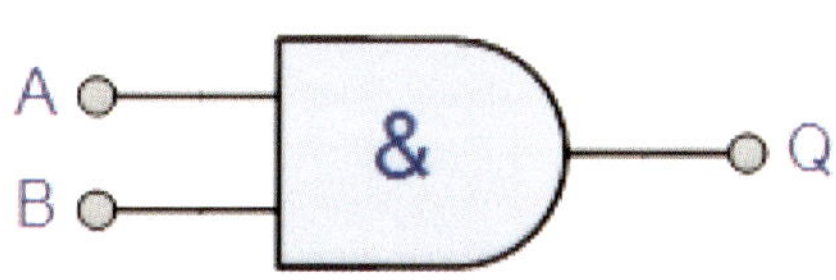

Symbol	Truth Table		
	B	A	Q
	0	0	1
	0	1	1
	1	0	1
	1	1	0
Boolean Expression $Q = \overline{A.B}$	Read as A AND B gives NOT Q		

The Logic NOR Gate

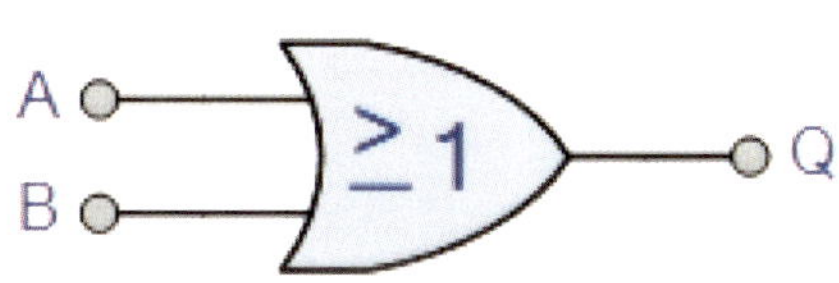

Symbol	Truth Table		
	B	A	Q
	0	0	1
	0	1	0
	1	0	0
	1	1	0
Boolean Expression $Q = \overline{A+B}$	Read as A OR B gives NOT Q		

Exclusive Logic Gates

The Logic Exclusive-OR Gate (Ex-OR)

<table>
<tr><th>Symbol</th><th colspan="3">Truth Table</th></tr>
<tr><td rowspan="5">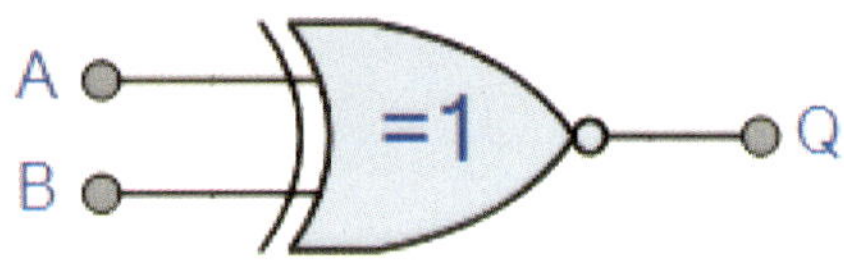</td><td>B</td><td>A</td><td>Q</td></tr>
<tr><td>0</td><td>0</td><td>0</td></tr>
<tr><td>0</td><td>1</td><td>1</td></tr>
<tr><td>1</td><td>0</td><td>1</td></tr>
<tr><td>1</td><td>1</td><td>0</td></tr>
<tr><td>Boolean Expression Q = A ⊕ B</td><td colspan="3">Read as A OR B but not BOTH gives Q (odd)</td></tr>
</table>

The Logic Exclusive-NOR Gate (Ex-NOR)

<table>
<tr><th>Symbol</th><th colspan="3">Truth Table</th></tr>
<tr><td rowspan="5">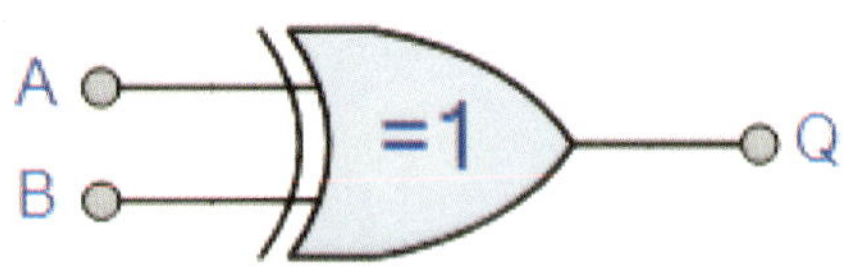</td><td>B</td><td>A</td><td>Q</td></tr>
<tr><td>0</td><td>0</td><td>1</td></tr>
<tr><td>0</td><td>1</td><td>0</td></tr>
<tr><td>1</td><td>0</td><td>0</td></tr>
<tr><td>1</td><td>1</td><td>1</td></tr>
<tr><td>Boolean Expression Q = $\overline{A \oplus B}$</td><td colspan="3">Read if A AND B the SAME gives Q (even)</td></tr>
</table>

Single Input Logic Gates

The Hex Buffer

<table>
<tr><th>Symbol</th><th colspan="2">Truth Table</th></tr>
<tr><td rowspan="3"></td><td>A</td><td>Q</td></tr>
<tr><td>0</td><td>0</td></tr>
<tr><td>1</td><td>1</td></tr>
<tr><td>Boolean Expression Q = A</td><td colspan="2">Read as A gives Q</td></tr>
</table>

The NOT gate (Inverter)

<table>
<tr><th>Symbol</th><th colspan="2">Truth Table</th></tr>
<tr><td rowspan="3"></td><td>A</td><td>Q</td></tr>
<tr><td>0</td><td>1</td></tr>
<tr><td>1</td><td>0</td></tr>
<tr><td>Boolean Expression Q = not A or $\overline{A}$</td><td colspan="2">Read as inverse of A gives Q</td></tr>
</table>

The operation of the above Digital Logic Gates and their Boolean expressions can be summarised into a single truth table as shown below. This truth table shows the relationship between each output of the main digital logic gates for each possible input combination.

Digital Logic Gate Truth Table Summary

The following logic gates truth table compares the logical functions of the 2-input logic gates detailed above.

Inputs		Truth Table Outputs For Each Gate					
B	A	AND	NAND	OR	NOR	EX-OR	EX-NOR
0	0	0	1	0	1	0	1
0	1	0	1	1	0	1	0
1	0	0	1	1	0	1	0
1	1	1	0	1	0	0	1

Truth Table Output for Single-input Gates		
A	NOT	Buffer
0	1	0
1	0	1

Pull-up and Pull-down Resistors

One final point to remember, when connecting together digital logic gates to produce logic circuits, any "unused" inputs to the gates must be connected directly to either a logic level "1" or a logic level "0" by means of a suitable "Pull-up" or "Pull-down" resistor (for example 1kΩ resistor) to produce a fixed logic signal. This will prevent the unused input to the gate from "floating" about and producing false switching of the gate and circuit.

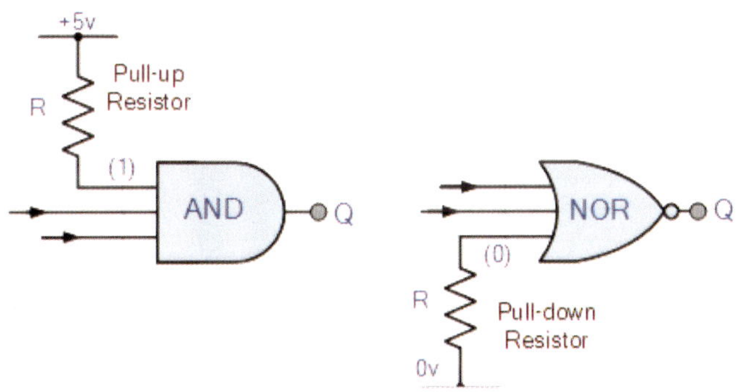

As well as using pull-up or pull-down resistors to prevent unused logic gates from floating about, spare inputs to gates and latches can also be connected together or connected to left-over or spare gates within a single IC package as shown.

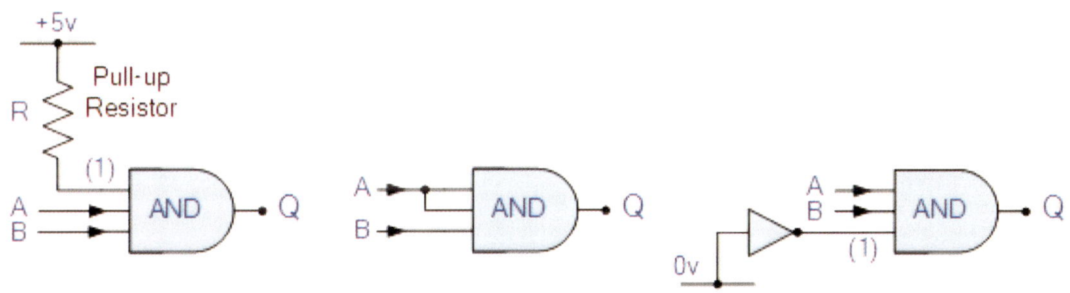

Digital logic gates may have more than one input, (A, B, C, etc.) but generally only have one digital output, (Q). Individual logic gates can be connected together to form combinational or sequential circuits, or larger logic gate functions.

Standard commercially available digital logic gates are available in two basic families or forms, TTL which stands for *Transistor-Transistor Logic* such as the 7400 series, and CMOS which stands for *Complementary Metal-Oxide-Silicon* which is the 4000 series of chips. This notation of TTL or CMOS refers to the logic technology used to manufacture the integrated circuit, (IC) or a "chip" as it is more commonly called.

Digital Logic Gate

Generally speaking, TTL logic IC's use NPN and PNP type Bipolar Junction Transistors while CMOS logic IC's use complementary MOSFET or JFET type Field Effect Transistors for both their input and output circuitry.

As well as TTL and CMOS technology, simple digital logic gates can also be made by connecting together diodes, transistors and resistors to produce RTL, Resistor-Transistor logic gates, DTL, Diode-Transistor logic gates or ECL, Emitter-Coupled logic gates but these are less common now compared to the popular CMOS family.

Integrated Circuits or IC's as they are more commonly called, can be grouped together into families according to the number of transistors or "gates" that they contain. For example, a simple AND gate my contain only a few individual transistors, were as a more complex microprocessor may contain many thousands of individual transistor gates. Integrated circuits are categorised according to the number of logic gates or the complexity of the circuits within a single chip with the general classification for the number of individual gates given as:

Classification of Integrated Circuits

- Small Scale Integration or (SSI) – Contain up to 10 transistors or a few gates within a single package such as AND, OR, NOT gates.
- Medium Scale Integration or (MSI) – between 10 and 100 transistors or tens of gates within a single package and perform digital operations such as adders, decoders, counters, flip-flops and multiplexers.
- Large Scale Integration or (LSI) – between 100 and 1,000 transistors or hundreds of gates and perform specific digital operations such as I/O chips, memory, arithmetic and logic units.
- Very-Large Scale Integration or (VLSI) – between 1,000 and 10,000 transistors or

thousands of gates and perform computational operations such as processors, large memory arrays and programmable logic devices.

- Super-Large Scale Integration or (SLSI) – between 10,000 and 100,000 transistors within a single package and perform computational operations such as microprocessor chips, micro-controllers, basic PICs and calculators.
- Ultra-Large Scale Integration or (ULSI) – more than 1 million transistors – the big boys that are used in computers CPUs, GPUs, video processors, micro-controllers, FPGAs and complex PICs.

While the "ultra large scale" ULSI classification is less well used, another level of integration which represents the complexity of the Integrated Circuit is known as the System-on-Chip or (SOC) for short. Here the individual components such as the microprocessor, memory, peripherals, I/O logic etc, are all produced on a single piece of silicon and which represents a whole electronic system within one single chip, literally putting the word "integrated" into integrated circuit.

These complete integrated chips which can contain up to 100 million individual silicon-CMOS transistor gates within one single package are generally used in mobile phones, digital cameras, micro-controllers, PIC's and robotic type applications.

Moore's Law

In 1965, Gordon Moore co-founder of the Intel corporation predicted that "*The number of transistors and resistors on a single chip will double every 18 months*" regarding the development of semiconductor gate technology. When Gordon Moore made his famous comment way back in 1965 there were approximately only 60 individual transistor gates on a single silicon chip or die.

The worlds first microprocessor in 1971 was the Intel 4004 that had a 4-bit data bus and contained about 2,300 transistors on a single chip, operating at about 600kHz. Today, the Intel Corporation have placed a staggering 1.2 Billion individual transistor gates onto its new Quad-core i7-2700K Sandy Bridge 64-bit microprocessor chip operating at nearly 4GHz, and the on-chip transistor count is still rising, as newer faster microprocessors and micro-controllers are developed.

De Morgan Equivalent Symbols

By use of De Morgan's laws, an *AND* function is identical to an *OR* function with negated inputs and outputs. Likewise, an *OR* function is identical to an *AND* function with negated inputs and outputs. A NAND gate is equivalent to an OR gate with negated inputs, and a NOR gate is equivalent to an AND gate with negated inputs.

This leads to an alternative set of symbols for basic gates that use the opposite core symbol (*AND* or *OR*) but with the inputs and outputs negated. Use of these alternative symbols can make logic circuit diagrams much clearer and help to show accidental connection of an active high output to an active low input or vice versa. Any connection that has logic negations at both ends can be replaced by a negationless connection and a suitable change of gate or vice versa. Any connection that has a negation at one end and no negation at the other can be made easier to interpret by instead using the De Morgan equivalent symbol at either of the two ends. When negation or polarity indicators on both ends of a connection match, there is no logic negation in that path (effectively, bubbles "cancel"), making it easier to follow logic states from one symbol to the next. This is commonly seen in real logic diagrams – thus the reader must not get into the habit of associating the shapes exclusively as OR or AND shapes, but also take into account the bubbles at both inputs and outputs in order to determine the "true" logic function indicated.

A De Morgan symbol can show more clearly a gate's primary logical purpose and the polarity of its nodes that are considered in the "signaled" (active, on) state. Consider the simplified case where a two-input NAND gate is used to drive a motor when either of its inputs are brought low by a switch. The "signaled" state (motor on) occurs when either one OR the other switch is on. Unlike a regular NAND symbol, which suggests AND logic, the De Morgan version, a two negative-input OR gate, correctly shows that OR is of interest. The regular NAND symbol has a bubble at the output and none at the inputs (the opposite of the states that will turn the motor on), but the De Morgan symbol shows both inputs and output in the polarity that will drive the motor.

De Morgan's theorem is most commonly used to implement logic gates as combinations of only NAND gates, or as combinations of only NOR gates, for economic reasons.

Data Storage

Logic gates can also be used to store data. A storage element can be constructed by connecting several gates in a "latch" circuit. More complicated designs that use clock signals and that change only on a rising or falling edge of the clock are called edge-triggered "flip-flops". Formally, a flip-flop is called a bistable circuit, because it has two stable states which it can maintain indefinitely. The combination of multiple flip-flops in parallel, to store a multiple-bit value, is known as a register. When using any of these gate setups the overall system has memory; it is then called a sequential logic system since its output can be influenced by its previous state(s), i.e. by the *sequence* of input states. In contrast, the output from combinational logic is purely a combination of its present inputs, unaffected by the previous input and output states.

These logic circuits are known as computer memory. They vary in performance, based on factors of speed, complexity, and reliability of storage, and many different types of designs are used based on the application.

Three-state Logic Gates

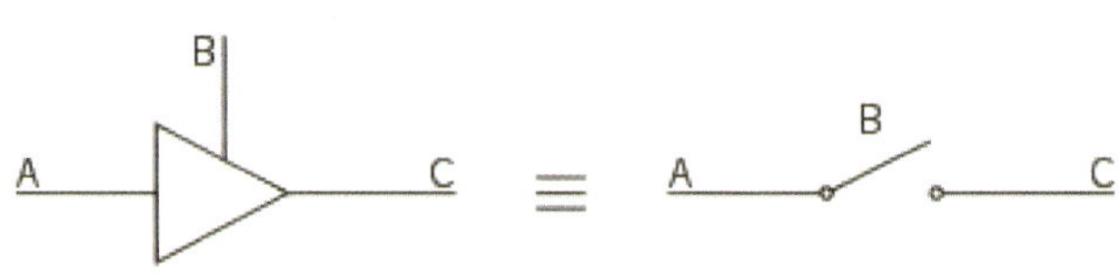

A tristate buffer can be thought of as a switch. If *B* is on, the switch is closed. If B is off, the switch is open.

A three-state logic gate is a type of logic gate that can have three different outputs: high (H), low (L) and high-impedance (Z). The high-impedance state plays no role in the logic, which is strictly binary. These devices are used on buses of the CPU to allow multiple chips to send data. A group of three-states driving a line with a suitable control circuit is basically equivalent to a multiplexer, which may be physically distributed over separate devices or plug-in cards.

In electronics, a high output would mean the output is sourcing current from the positive power terminal (positive voltage). A low output would mean the output is sinking current to the negative power terminal (zero voltage). High impedance would mean that the output is effectively disconnected from the circuit.

Implementations

Since the 1990s, most logic gates are made in CMOS (complementary metal oxide semiconductor) technology that uses both NMOS and PMOS transistors. Often millions of logic gates are packaged in a single integrated circuit.

There are several logic families with different characteristics (power consumption, speed, cost, size) such as: RDL (resistor–diode logic), RTL (resistor-transistor logic), DTL (diode–transistor logic), TTL (transistor–transistor logic) and CMOS. There are also sub-variants, e.g. standard CMOS logic vs. advanced types using still CMOS technology, but with some optimizations for avoiding loss of speed due to slower PMOS transistors.

Non-electronic implementations are varied, though few of them are used in practical applications. Many early electromechanical digital computers, such as the Harvard Mark I, were built from relay logic gates, using electro-mechanical relays. Logic gates can be made using pneumatic devices, such as the Sorteberg relay or mechanical logic gates, including on a molecular scale. Logic gates have been made out of DNA and used to create a computer called MAYA (MAYA-II). Logic gates can be made from quantum mechanical effects (though quantum computing usually diverges from boolean design). Photonic logic gates use nonlinear optical effects.

In principle any method that leads to a gate that is functionally complete (for example, either a NOR or a NAND gate) can be used to make any kind of digital logic circuit. Note that the use of 3-state logic for bus systems is not needed, and can be replaced by digital multiplexers, which can be built using only simple logic gates (such as NAND gates, NOR gates, or AND and OR gates).

NOT Gate

The Logic NOT Gate is the most basic of all the logical gates and is often referred to as an Inverting Buffer or simply an Inverter.

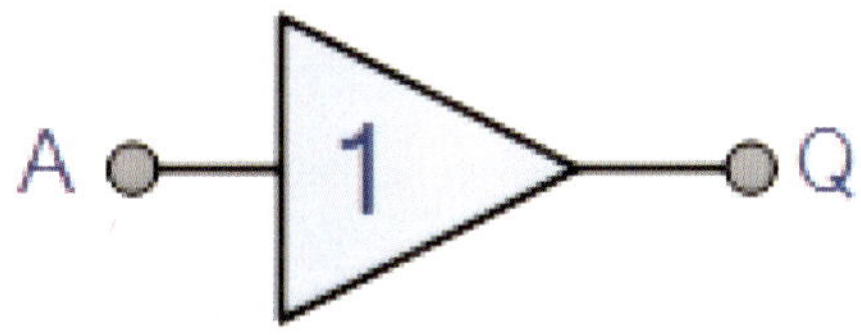

It is a single input device which has an output level that is normally at logic level "1" and goes "LOW" to a logic level "0" when its single input is at logic level "1", in other words it "inverts" (complements) its input signal. The output from a NOT gate only returns "HIGH" again when its input is at logic level "0" giving us the Boolean expression of: $\overline{A} = Q$.

Then we can define the operation of a single input digital logic NOT gate as being:

"If A is NOT true, then Q is true".

Transistor NOT Gate

A simple 2-input logic NOT gate can be constructed using a RTL Resistor-transistor switches as shown below with the input connected directly to the transistor base. The transistor must be saturated "ON" for an inverted output "OFF" at Q.

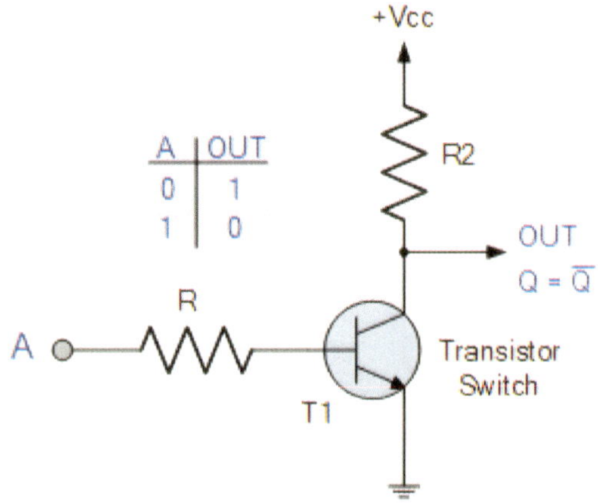

Logic NOT Gates are available using digital circuits to produce the desired logical function. The standard NOT gate is given a symbol whose shape is of a triangle pointing to the right with a circle at its end. This circle is known as an "inversion bubble" and is used in NOT, NAND and NOR symbols at their output to represent the logical operation of the NOT function. This bubble denotes a signal inversion (complementation) of the signal and can be present on either or both the output and the input terminals.

The Logic NOT Gate Truth Table

Symbol	Truth Table	
 Inverter or NOT Gate	A	Q
	0	1
	1	0
Boolean Expression Q = not A or $\overline{A}$	Read as inverse of A gives Q	

Logic NOT gates provide the complement of their input signal and are so called because when their input signal is "HIGH" their output state will NOT be "HIGH". Likewise, when their input signal is "LOW" their output state will NOT be "LOW". As they are single input devices, logic NOT gates are not normally classed as "decision" making devices or even as a gate, such as the AND or OR gates which have two or more logic inputs. Commercial available NOT gates IC's are available in either 4 or 6 individual gates within a single IC package.

The "bubble" (o) present at the end of the NOT gate symbol above denotes a signal inversion (complementation) of the output signal. But this bubble can also be present at the gates input to indicate an active-LOW input. This inversion of the input signal is not restricted to the NOT gate only but can be used on any digital circuit or gate as shown with the operation of inversion being exactly the same whether on the input or output terminal. The easiest way is to think of the bubble as simply an inverter.

Signal Inversion using Active-low Input Bubble

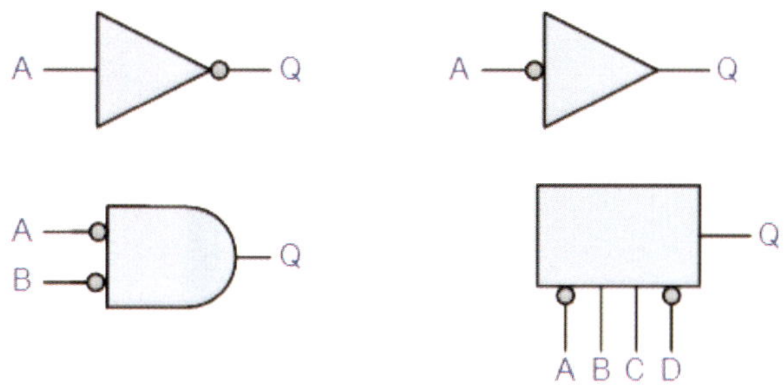

Bubble Notation for Input Inversion

Explanation of Not Gate with Light Switch Circuit

The NOT gate can be easily understood by using a LED (light emitted Diode) circuit. This is also called Light switch circuit. In this circuit, NOT gate functions like an electronic switch. When it got high input, the LED connected at the output will be OFF, as the output of NOT gate becomes 0. In the same way, when the logic gate is connected with LOW input, the LED will be ON, as the output becomes 1. The Light switching circuit with NOT gate is shown below.

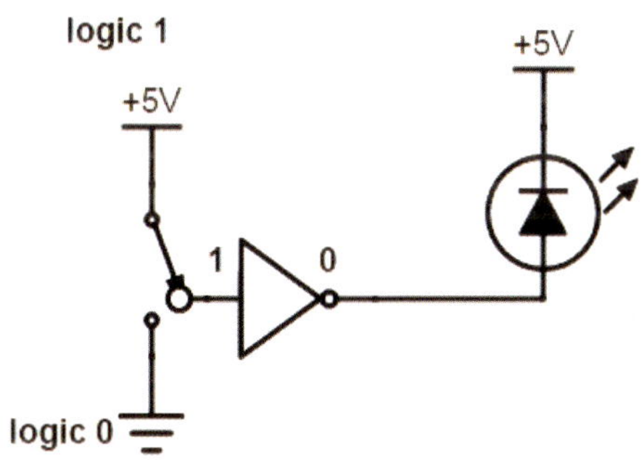

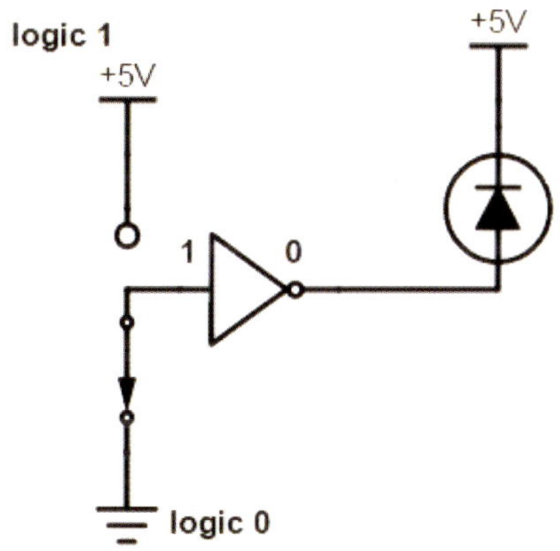

Here we connect an alterable switch with the NOT gate and the output of NOT gate is connected to a LED. An LED is an electronic device which will ON and off when it receives high voltage and low voltage respectively. When the switch is connected to +5 V, the switch is on position so the LED emits light. When the NOT gate is connected to Ground, the LED will OFF so it doesn't emit any light.

Pulsed Operation

When we apply a pulse signal to the NOT gate, it will ON and OFF for HIGH and LOW levels respectively. This means,

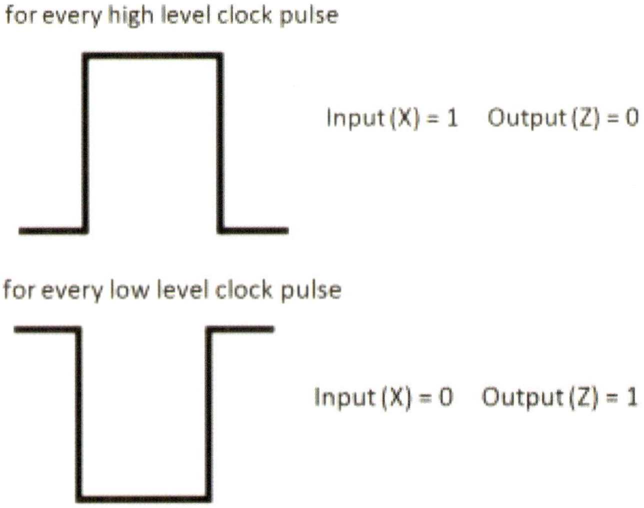

For a sequence of clock pulses, the input and outputs of NOT gate will be

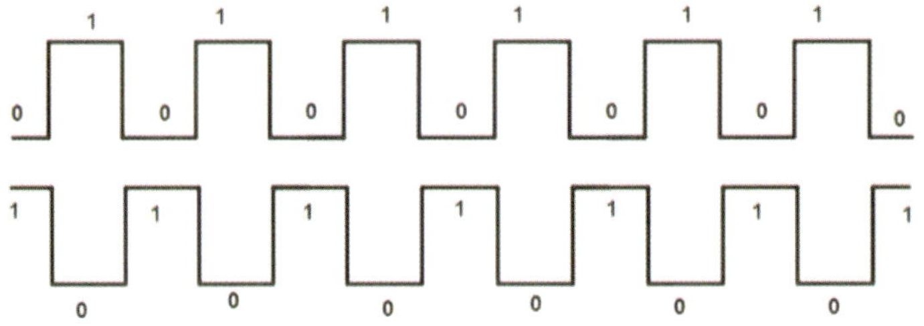

If we observe the above diagram, we can observe how the input will invert for every HIGH and LOW pulses of a clock cycle.

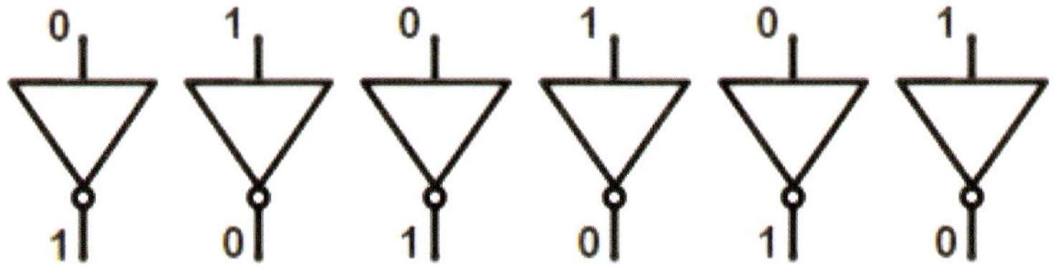

A very simple inverter can also be made using just a single stage transistor switching circuit as shown.

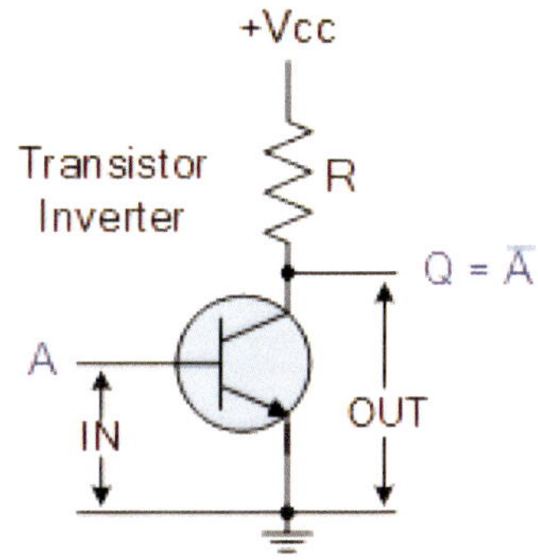

When the transistors base input at "A" is high, the transistor conducts and collector current flows producing a voltage drop across the resistor R thereby connecting the output point at "Q" to ground thus resulting in a zero voltage output at "Q".

Likewise, when the transistors base input at "A" is low (0v), the transistor now switches "OFF" and no collector current flows through the resistor resulting in an output voltage at "Q" high at a value near to +Vcc.

Then, with an input voltage at "A" HIGH, the output at "Q" will be LOW and an input voltage at "A" LOW the resulting output voltage at "Q" is HIGH producing the complement or inversion of the input signal.

Hex Schmitt Inverters

A standard Inverter or Logic NOT Gate, is usually made up from transistor switching circuits that do not switch from one state to the next instantly, there will always be some delay in the switching action.

Also as a transistor is a basic current amplifier, it can also operate in a linear mode and any small variation to its input level will cause a variation to its output level or may even switch "ON" and "OFF" several times if there is any noise present in the circuit. One way to overcome these problems is to use a Schmitt Inverter or Hex Inverter.

We know from the previous pages that all digital gates use only two logic voltage states and that these are generally referred to as Logic "1" and Logic "0" any TTL voltage input between 2.0v and 5v is recognised as a logic "1" and any voltage input below 0.8v is recognised as a logic "0" respectively.

A Schmitt Inverter is designed to operate or switch state when its input signal goes above an "Upper Threshold Voltage" or UTV limit in which case the output changes and goes "LOW", and will remain in that state until the input signal falls below the "Lower Threshold Voltage" or LTV level in which case the output signal goes "HIGH". In other words a Schmitt Inverter has some form of Hysteresis built into its switching circuit.

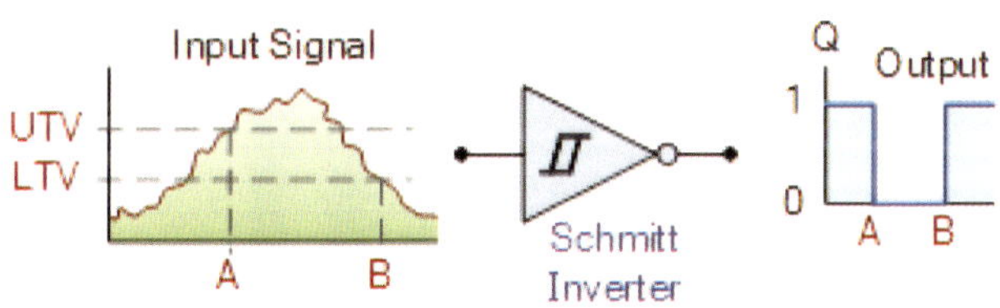

This switching action between an upper and lower threshold limit provides a much cleaner and faster "ON/OFF" switching output signal and makes the Schmitt inverter ideal for switching any slow-rising or slow-falling input signal and as such we can use a Schmitt trigger to convert these analogue signals into digital signals as shown.

Schmitt Inverter

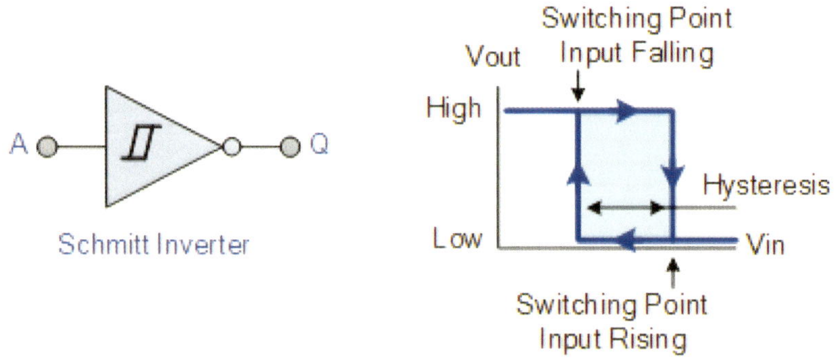

A very useful application of Schmitt inverters is when they are used as oscillators or sine-to-square wave converters for use as square wave clock signals.

Schmitt NOT Gate Inverter Oscillator

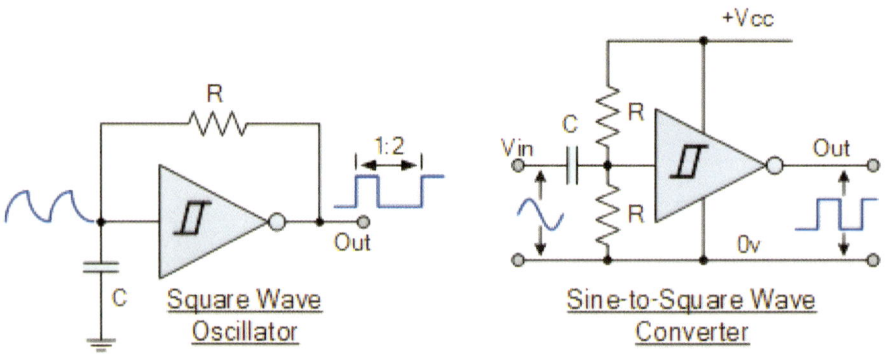

The first circuit shows a very simple low power RC type oscillator using a Schmitt inverter to generate a square wave output waveform. Initially the capacitor C is fully discharged so the input to the inverter is "LOW" resulting in an inverted output which is "HIGH". As the output from the inverter is fed back to its input and the capacitor via the resistor R the capacitor begins to charge up.

When the capacitors charging voltage reaches the upper threshold limit of the inverter, the inverter changes state, the output becomes "LOW" and the capacitor begins to discharge through the resistor until it reaches the lower threshold level were the inverter changes state again. This switching back and forth by the inverter produces a square wave output signal with a 33% duty cycle and whose frequency is given as: $f = 680/RC$.

The second circuit converts a sine wave input (or any oscillating input for that matter) into a square wave output. The input to the inverter is connected to the junction of the potential divider network which is used to set the quiescent point of the circuit. The input capacitor blocks any DC component present in the input signal only allowing the sine wave signal to pass.

As this signal passes the upper and lower threshold points of the inverter the output also changes from "HIGH" to "LOW" and so on producing a square wave output waveform. This circuit produces an output pulse on the positive rising edge of the input waveform, but by connecting a second Schmitt inverter to the output of the first, the basic circuit can be modified to produce an output pulse on the negative falling edge of the input signal.

Commonly available logic NOT gate and Inverter IC's include:

TTL Logic NOT Gates

- 74LS04 Hex Inverting NOT Gate
- 74LS14 Hex Schmitt Inverting NOT Gate
- 74LS1004 Hex Inverting Drivers

CMOS Logic NOT Gates

- CD4009 Hex Inverting NOT Gate
- CD4069 Hex Inverting NOT Gate

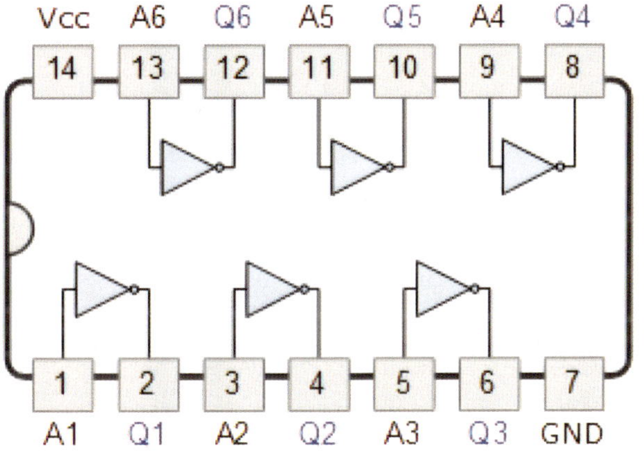

7404 NOT Gate or Inverter

AND Gate

A Logic AND Gate is a type of digital logic gate whose output goes HIGH to a logic level 1 when all of its inputs are HIGH

The output state of a "Logic AND Gate" only returns "LOW" again when ANY of its inputs are at a logic level "0". In other words for a logic AND gate, any LOW input will give a LOW output.

The logic or Boolean expression given for a digital logic AND gate is that for *Logical Multiplication* which is denoted by a single dot or full stop symbol, (.) giving us the Boolean expression of: A.B = Q.

Then we can define the operation of a 2-input logic AND gate as being:

"If both A and B are true, then Q is true"

AND gate can be designed by using two simple diodes. The circuit driving voltage V is applied to the parallely connected diodes and the output is collected as the voltage drop at the diodes. In logic gates, the terms high voltage level means +5 V and low logic level means 0 V or ground.

When one of the inputs of the AND gate is connected to logic HIGH and other is connected to logic LOW then the diodes are at reverse bias condition and no voltage drop at the output. So the output is measured as LOW. If the two inputs are connected to LOW level input, then also the diodes will turn to reverse biased condition and allows no current. So again the output is measured as 0.

But when the two inputs (two diodes) are connected to a HIGH voltage level, then the two diodes are in forward biased condition (diodes switches are ON) so the output of the AND gate is HIGH, and measured as logic 1.

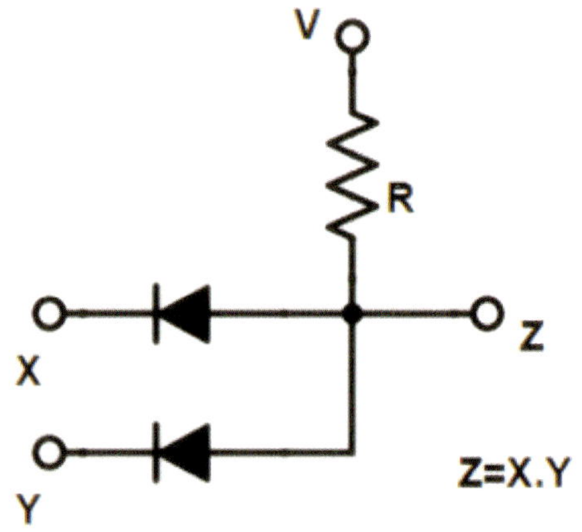

Explanation of AND Gate with Light Switch Circuit

The AND gate switching circuit will have two inputs with two manually toggled switches. Let the two switches be A and B, then the we can explain the switching operation of AND gate as

- When both switches A and B are open (switches are supplied with low level input signal) i.e. A=0, B=0, then the bulb will not glow.
- When switch A is close (supplied with high level input signal) and B is open (supplied with low level input signal) i.e. A=1, B=0, then the bulb will not glow.
- When switch A is open (supplied with low level input signal) and B is close (supplied with high level input signal) i.e. A=0, B=1, then the bulb will not glow.
- When both switches A and B are close (switches are supplied with high level input signal) i.e. A=1, B=1, then the bulb will glow.

This operation of AND gate as light switching circuit is described in below pictures

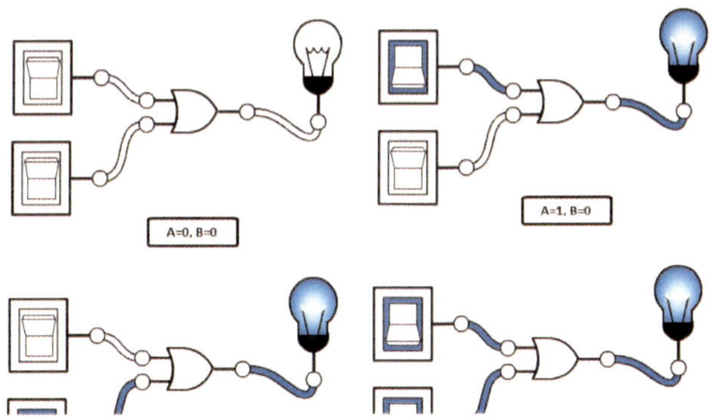

Pulsed Operation

If we apply two different clock signals the inputs of AND logic gate X and Y, then if we observe the output, it will be as shown below (X, Y are inputs and Z is output)

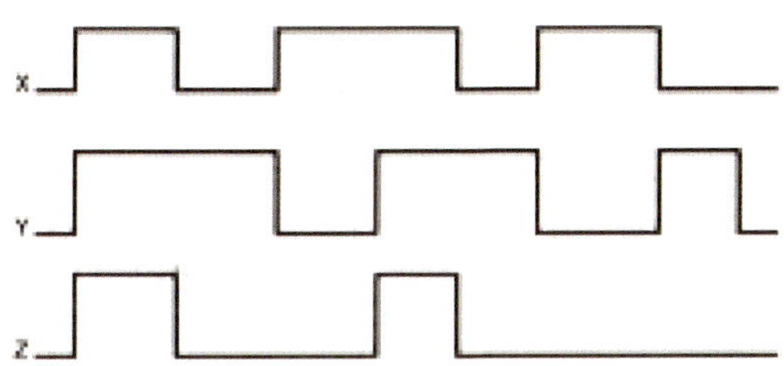

When both inputs are high, the output of the AND gate is also high and when either of the inputs is low, then the output comes to low level. At the end of the clock pulses in the above diagram, the output is at low level as both the inputs are low.

2-input Transistor AND Gate

A simple 2-input logic AND gate can be constructed using RTL Resistor-transistor switches connected together as shown below with the inputs connected directly to the transistor bases. Both transistors must be saturated "ON" for an output at Q.

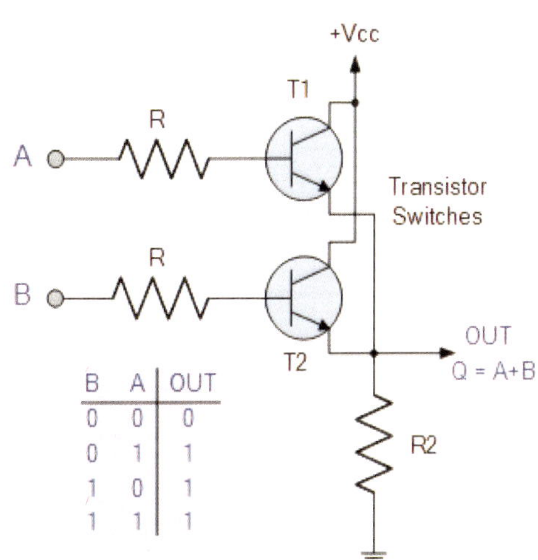

Logic AND Gates are available using digital circuits to produce the desired logical function and is given a symbol whose shape represents the logical operation of the AND gate.

Digital Logic "AND" Gate Types

The 2-input Logic AND Gate

Symbol

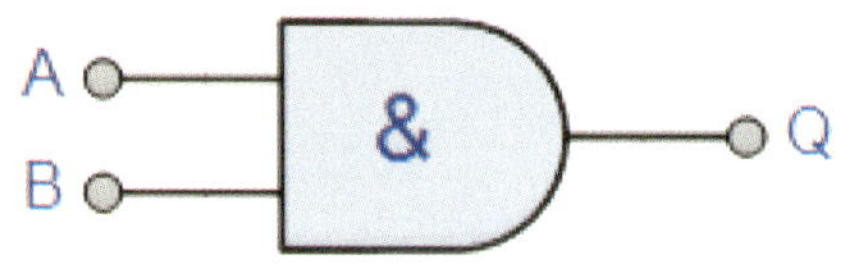

2-input AND Gate

Truth Table

B	A	Q
0	0	0
0	1	0
1	0	0
1	1	1

Boolean Expression Q = A.B	Read as A AND B gives Q

The 3-input Logic AND Gate

Symbol

3-input AND Gate

Truth Table

C	B	A	Q
0	0	0	0
0	0	1	0
0	1	0	0
0	1	1	0
1	0	0	0
1	0	1	0
1	1	0	0
1	1	1	1

Boolean Expression Q = A.B.C	Read as A AND B AND C gives Q

Because the Boolean expression for the logic AND function is defined as (.), which is a binary operation, AND gates can be cascaded together to form any number of individual inputs. However, commercial available AND gate IC's are only available in standard 2, 3, or 4-input packages. If additional inputs are required, then standard AND gates will need to be cascaded together to obtain the required input value, for example.

Multi-input AND Gate

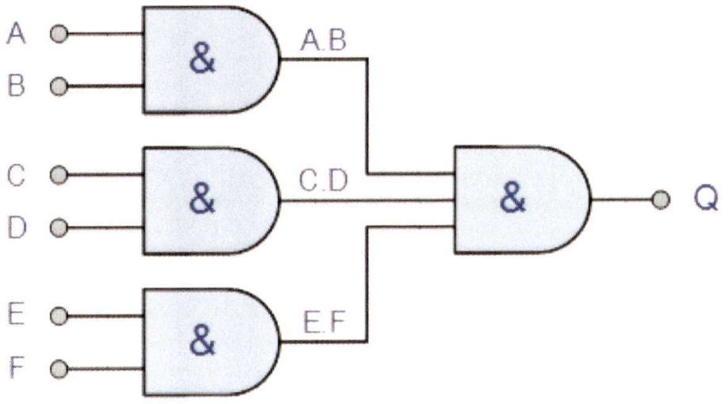

6-input "AND" Function

The Boolean Expression for this 6-input AND gate will therefore be:

Q = (A.B).(C.D).(E.F)

In other words;

A and B and C and D and E and F gives Q

If the number of inputs required is an odd number of inputs any "unused" inputs can be held HIGH by connecting them directly to the power supply using suitable "Pull-up" resistors.

Commonly available digital logic AND gate IC's include:

TTL Logic AND Gates

- 74LS08 Quad 2-input
- 74LS11 Triple 3-input
- 74LS21 Dual 4-input

CMOS Logic AND Gates

- CD4081 Quad 2-input
- CD4073 Triple 3-input
- CD4082 Dual 4-input

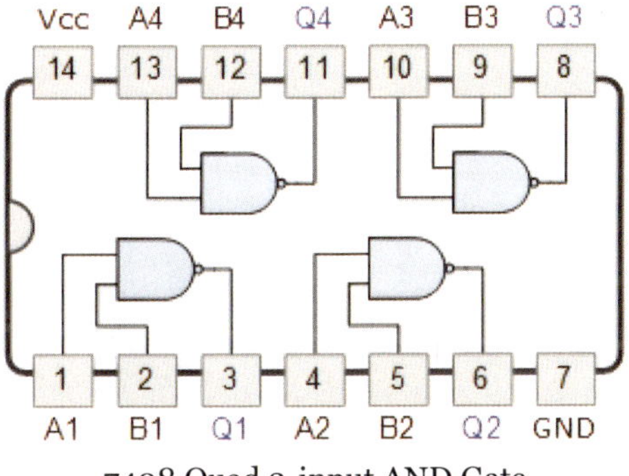

7408 Quad 2-input AND Gate

OR Gate

A Logic OR Gate is a type of digital logic gate whose output goes HIGH to a logic level 1 when one or more of its inputs are HIGH

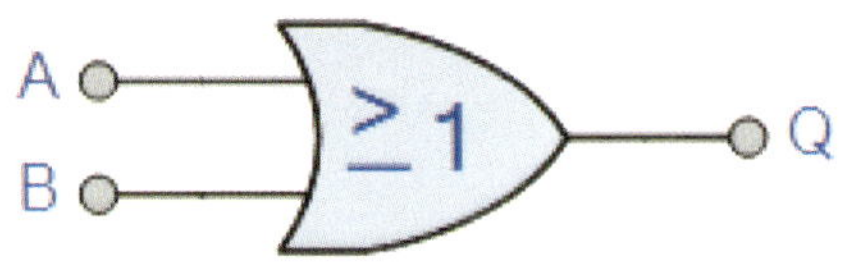

The output, Q of a "Logic OR Gate" only returns "LOW" again when ALL of its inputs are at a logic level "0". In other words for a logic OR gate, any "HIGH" input will give a "HIGH", logic level "1" output.

The logic or Boolean expression given for a digital logic OR gate is that for *Logical Addition* which is denoted by a plus sign, (+) giving us the Boolean expression of: $A+B = Q$.

Thus a logic OR gate can be correctly described as an "Inclusive OR gate" because the output is true when both of its inputs are true (HIGH). Then we can define the operation of a 2-input logic OR gate as being:

"If either A or B is true, then Q is true"

The circuit uses two diodes at input side. In all logic circuits, +5 volts is represented as HIGH level logic and 0 volts or ground is represented as the LOW level logic. The two logic levels are represented as binary numbers 0 and 1.

Here 0 means low (switch is at OFF position) and 1 means high (switch is at ON position). The OR gate produces low output only when both the inputs are low, in all other conditions or for all other combinations of inputs , the output of OR gate will be high.

If we connect HIGH logic level voltage i.e. +5 volts to any one of the input, then the diode becomes forward biased and that makes the output to be high. Or if we connect high level voltage to both inputs, then both the inputs will be high making the diodes to be in forward bias condition.

Then the OR circuit output will be HIGH. If the two inputs of the OR gate are low, then the inputs diodes become reverse biased, allowing no current to flow in the circuit. So the output will be LOW (0).

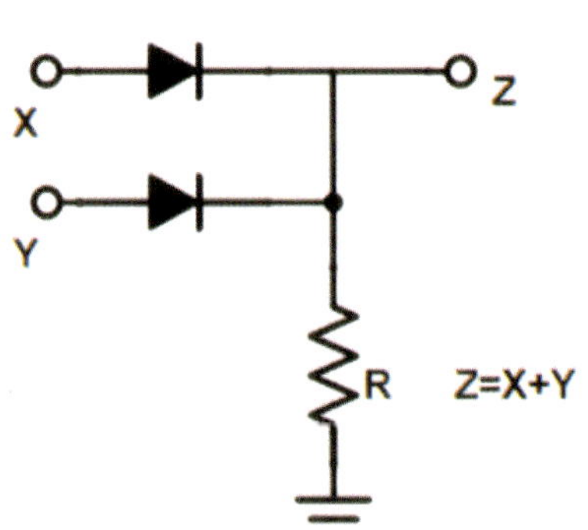

Explanation of OR Gate with Light Switch Circuit

The OR gate switching circuit will have two inputs connected to two manually toggled switches. Let the two switches be A and B, then the we can explain the switching operation of OR gate as

- When both switches A and B are open (switches are supplied with low level input signal) i.e. A=0, B=0, then the bulb will not glow.
- When switch A is close (supplied with high level input signal) and B is open (supplied with low level input signal) i.e. A=1, B=0, then the bulb will glow.
- When switch A is open (supplied with low level input signal) and B is close (supplied with high level input signal) i.e. A=0, B=1, then the bulb will glow.
- When both switches A and B are close (switches are supplied with high level input signal) i.e. A=1, B=1, then the bulb will glow.

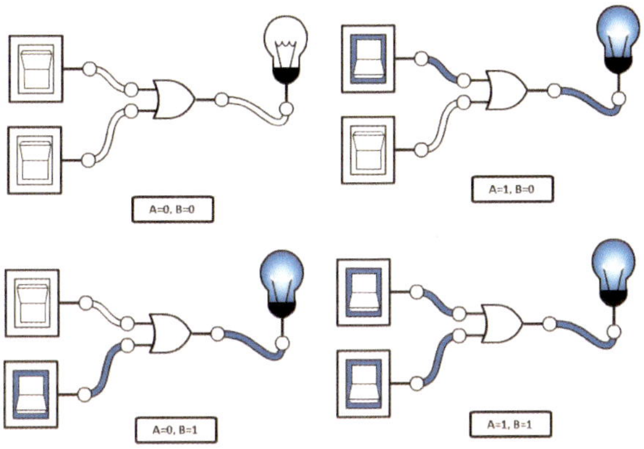

Pulsed Operation

Let's apply two different clock signals to both the inputs of OR logic gate A and B, then if we observe the output of the OR gate it will be as shown below.

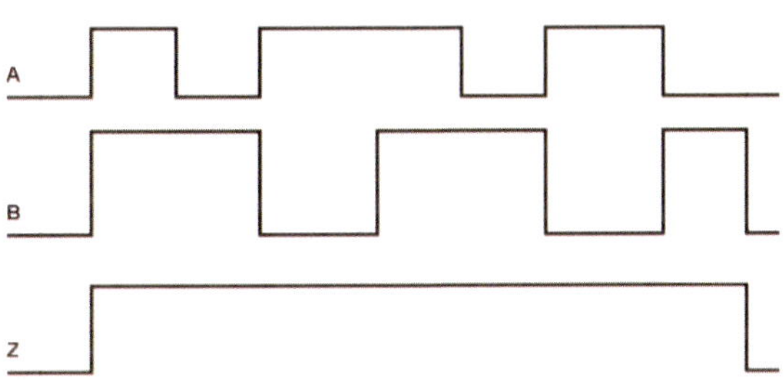

When the two inputs are high, then the output is also high. And further even when either of the inputs is high, then the output of OR gates continued to be high state. When both the inputs of the OR gate are at low level, then the output also comes to low level.

2-input Transistor OR Gate

A simple 2-input inclusive OR gate can be constructed using RTL Resistor-transistor switches connected together as shown below with the inputs connected directly to the transistor bases. Either transistor must be saturated "ON" for an output at Q.

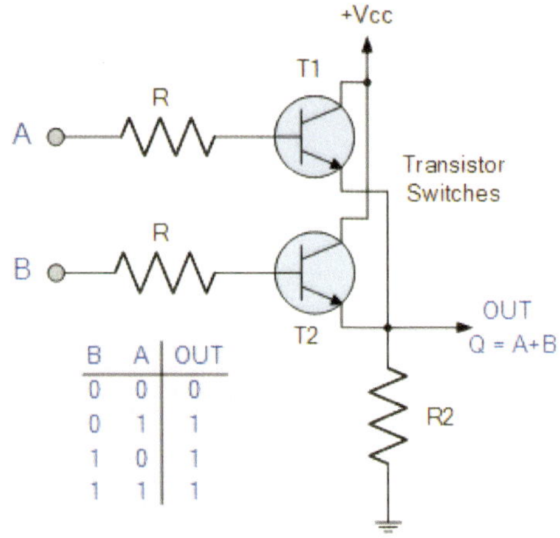

Logic OR Gates are available using digital circuits to produce the desired logical function and is given a symbol whose shape represents the logical operation of the OR gate.

Digital Logic "OR" Gate Types

The 2-input Logic OR Gate

<table>
<tr><th>Symbol</th><th colspan="3">Truth Table</th></tr>
<tr><td rowspan="5">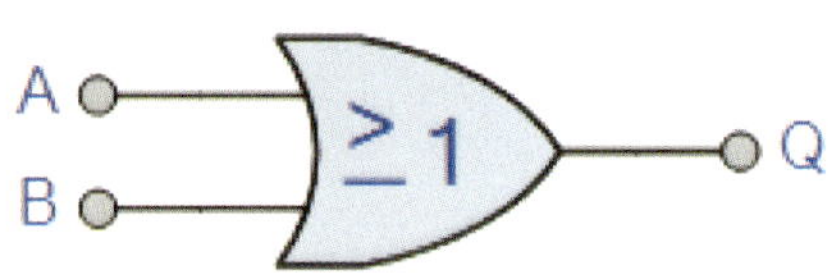
2-input OR Gate</td><td>B</td><td>A</td><td>Q</td></tr>
<tr><td>0</td><td>0</td><td>0</td></tr>
<tr><td>0</td><td>1</td><td>1</td></tr>
<tr><td>1</td><td>0</td><td>1</td></tr>
<tr><td>1</td><td>1</td><td>1</td></tr>
<tr><td>Boolean Expression Q = A+B</td><td colspan="3">Read as A OR B gives Q</td></tr>
</table>

The 3-input Logic OR Gate

<table>
<tr><th>Symbol</th><th colspan="4">Truth Table</th></tr>
<tr><td rowspan="9">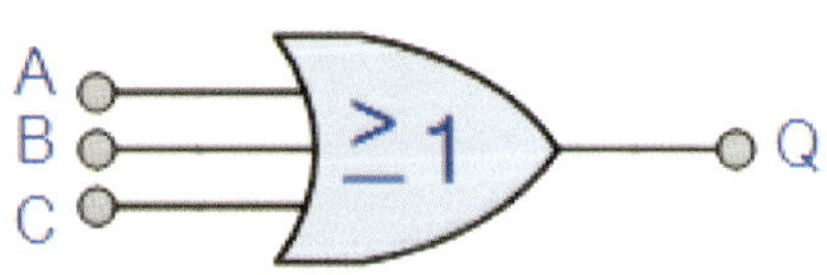
3-input OR Gate</td><td>C</td><td>B</td><td>A</td><td>Q</td></tr>
<tr><td>0</td><td>0</td><td>0</td><td>0</td></tr>
<tr><td>0</td><td>0</td><td>1</td><td>1</td></tr>
<tr><td>0</td><td>1</td><td>0</td><td>1</td></tr>
<tr><td>0</td><td>1</td><td>1</td><td>1</td></tr>
<tr><td>1</td><td>0</td><td>0</td><td>1</td></tr>
<tr><td>1</td><td>0</td><td>1</td><td>1</td></tr>
<tr><td>1</td><td>1</td><td>0</td><td>1</td></tr>
<tr><td>1</td><td>1</td><td>1</td><td>1</td></tr>
<tr><td>Boolean Expression Q = A+B+C</td><td colspan="4">Read as A OR B OR C gives Q</td></tr>
</table>

Like the AND gate, the OR function can have any number of individual inputs. However, commercial available OR gates are available in 2, 3, or 4 inputs types. Additional inputs will require gates to be cascaded together for example.

Multi-input OR Gate

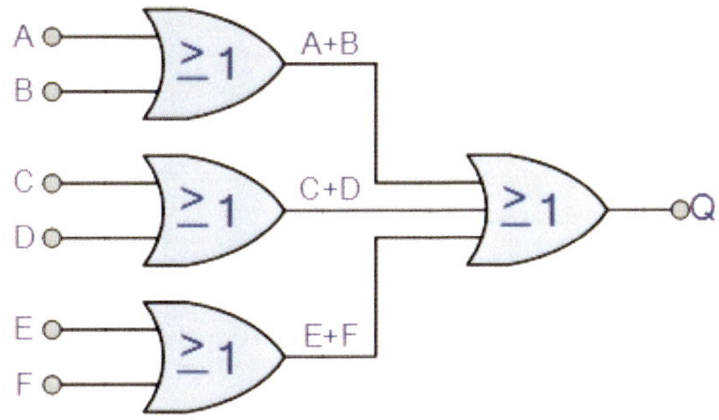

6-input "OR" Function

The Boolean Expression for this 6-input OR gate will therefore be:

Q = (A+B)+(C+D)+(E+F)

In other words:

A or B or C or D or E or F gives Q

If the number of inputs required is an odd number of inputs any "unused" inputs can be held LOW by connecting them directly to ground using suitable "Pull-down" resistors.

Commonly available digital logic OR gate IC's include:

TTL Logic OR Gates

- 74LS32 Quad 2-input

CMOS Logic OR Gates

- CD4071 Quad 2-input
- CD4075 Triple 3-input
- CD4072 Dual 4-input

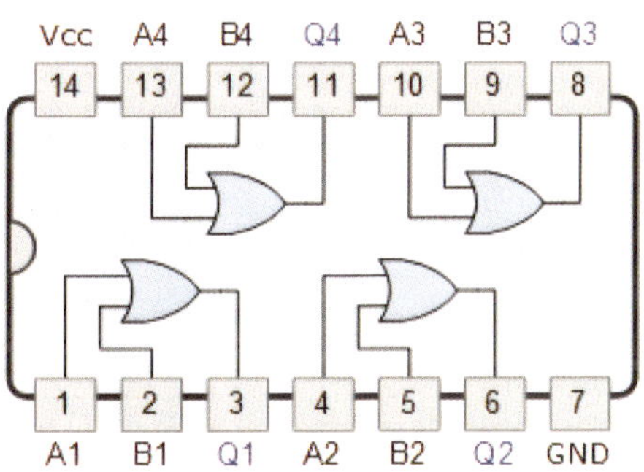

7432 Quad 2-input Logic OR Gate

NAND Gate

The Logic NAND Gate is a combination of a digital logic AND gate and a NOT gate connected together in series.

The NAND (Not – AND) gate has an output that is normally at logic level "1" and only goes "LOW" to logic level "0" when ALL of its inputs are at logic level "1". The Logic NAND Gate is the reverse or "Complementary" form of the AND gate.

Logic NAND Gate Equivalence

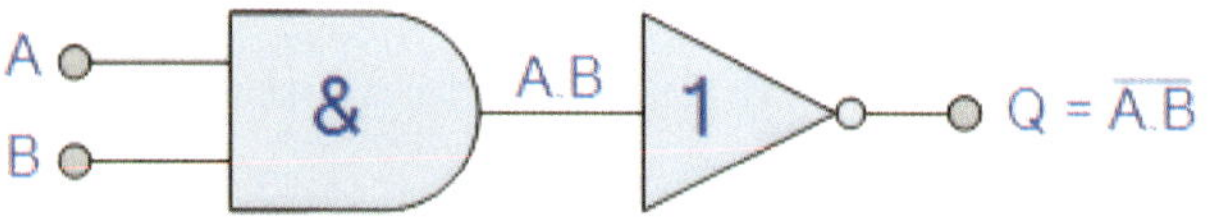

2-input "AND" gate plus a "NOT" gate

The logic or Boolean expression given for a logic NAND gate is that for Logical Addition, which is the opposite to the AND gate, and which it performs on the complements of the inputs. The Boolean expression for a logic NAND gate is denoted by a single dot or full stop symbol, (.) with a line or Overline, ($\overline{\quad}$) over the expression to signify the NOT or logical negation of the NAND gate giving us the Boolean expression of: $\overline{A.B}=Q$.

Then we can define the operation of a 2-input digital logic NAND gate as being:

> "If either A or B are NOT true, then Q is true"

It has the capability to perform the operations of 3 logic gates such as OR, AND gate and NOT gate. The outputs of AND gate and NAND gates are inverse to each other. One should know the fact that AND gate is constructed from the AND gate.

It has 2 inputs X, Y and single output, Z. The equivalent circuit of NAND Gate using diodes and transistors is shown below.

The inputs to the NAND gate is applied through the diodes and these diodes are connected to BJT.

When the inputs of both diodes of NAND gate are connected to a high voltage level i.e. +5 V (logic HIGH), then the 2 diodes will be reverse biased. Since the diodes don't conduct, they both will be in OFF state. At this condition , transistor Q1 is capable for driving the supply voltage from the resistor.

This makes the transistor to be in ON state, so the voltage at the output Vce (Sat) will be 0. In the same way, if we apply the low voltage supply to both the diodes (logic low) i.e. 0, then no current flows through the transistor, so it will become OFF. This makes the output to become HIGH.

The circuit and logic symbol for NAND gate is given below.

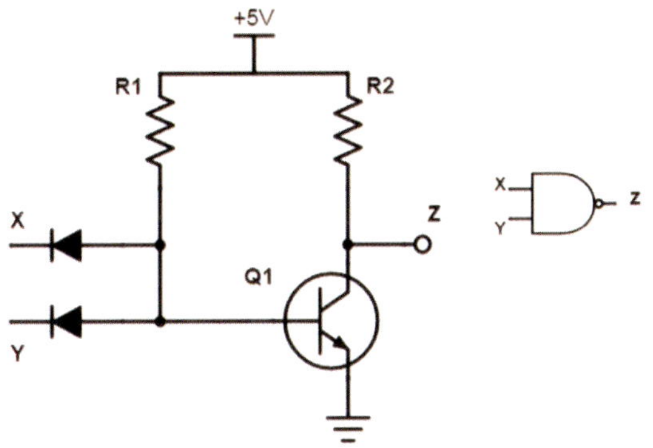

Pulsed Operation

If we apply 2 different clock signals as the inputs of NAND gate X and Y, then the output of the NAND gate is shown below (X, Y are inputs and Z is output)

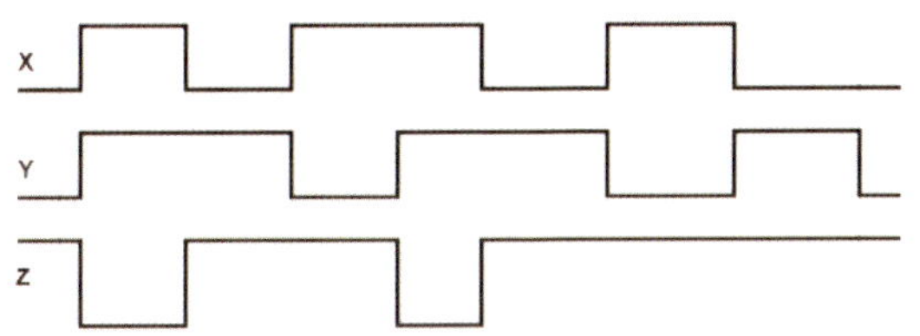

When both inputs are high, then the output of the NAND gate will be LOW and when either of the inputs is low, then the output goes to HIGH level. At the end of the clock pulses in the above, the output is at high level as one of the inputs is low.

Transistor NAND Gate

A simple 2-input logic NAND gate can be constructed using RTL Resistor-transistor switches connected together as shown below with the inputs connected directly to the transistor bases. Either transistor must be cut-off "OFF" for an output at Q.

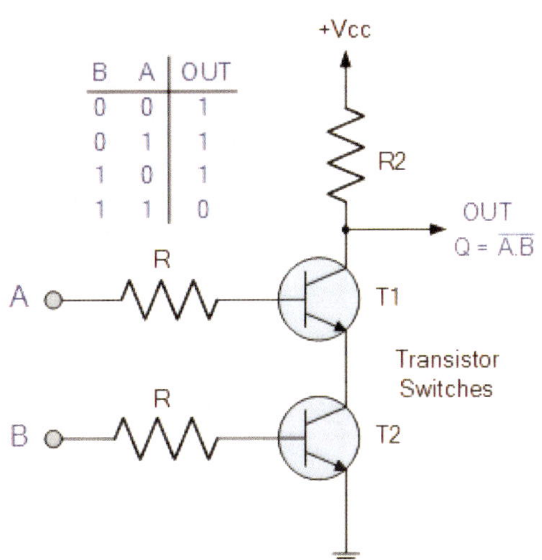

Logic NAND Gates are available using digital circuits to produce the desired logical function and is given a symbol whose shape is that of a standard AND gate with a circle, sometimes called an "inversion bubble" at its output to represent the NOT gate symbol with the logical operation of the NAND gate given as.

The Digital Logic "NAND" Gate

2-input Logic NAND Gate

Symbol

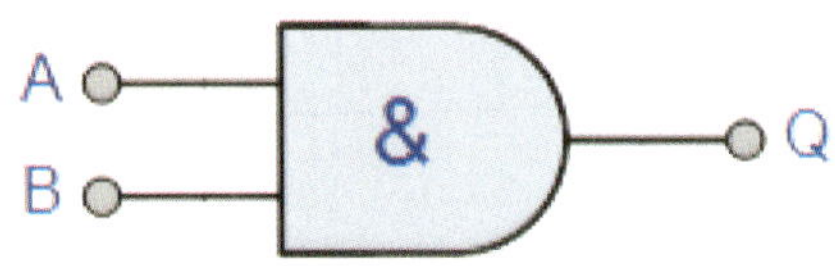

2-input NAND Gate

Truth Table

B	A	Q
0	0	1
0	1	1
1	0	1
1	1	0

Boolean Expression $Q = \overline{A.B}$

Read as A AND B gives NOT Q

3-input Logic NAND Gate

Symbol

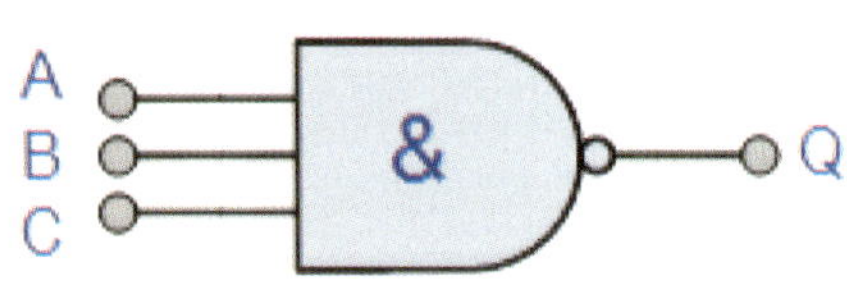

3-input NAND Gate

Truth Table

C	B	A	Q
0	0	0	1
0	0	1	1
0	1	0	1
0	1	1	1
1	0	0	1
1	0	1	1
1	1	0	1
1	1	1	0

Boolean Expression $Q = \overline{A.B.C}$

Read as A AND B AND C gives NOT Q

As with the AND function, the NAND function can also have any number of individual inputs and commercial available NAND Gate IC's are available in standard 2, 3, or 4 input types. If additional inputs are required, then the standard NAND gates can be cascaded together to provide more inputs for example.

A 4-input NAND Function

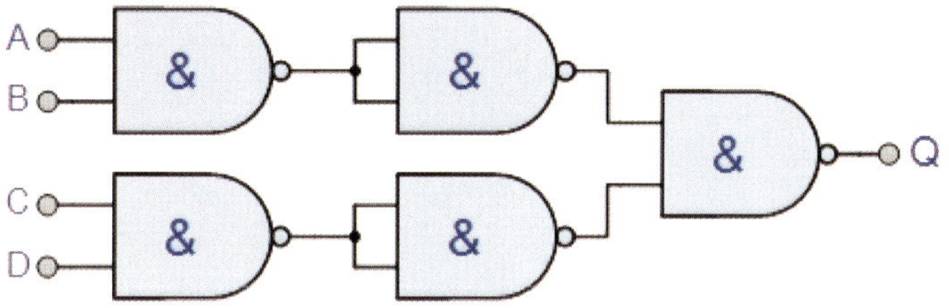

The Boolean Expression for this 4-input logic NAND gate will therefore be: $Q = \overline{A.B.C.D}$

If the number of inputs required is an odd number of inputs any "unused" inputs can be held HIGH by connecting them directly to the power supply using suitable "Pull-up" resistors.

The Logic NAND Gate function is sometimes known as the Sheffer Stroke Function and is denoted by a vertical bar or upwards arrow operator, for example, A NAND B = A|B or A↑B.

The "Universal" NAND Gate

The Logic NAND Gate is generally classed as a "Universal" gate because it is one of the most commonly used logic gate types. NAND gates can also be used to produce any other type of logic gate function, and in practice the NAND gate forms the basis of most practical logic circuits.

By connecting them together in various combinations the three basic gate types of AND, OR and NOT function can be formed using only NAND's, for example.

Various Logic Gates using only NAND Gates

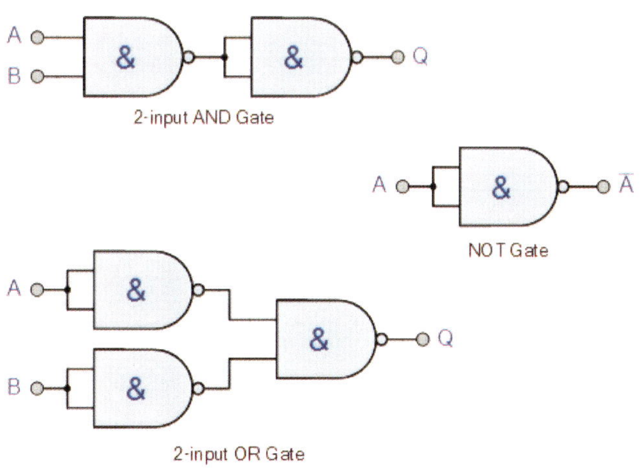

As well as the three common types above, Ex-Or, Ex-Nor and standard NOR gates can be formed using just individual NAND gates.

Commonly available digital logic NAND gate IC's include:

TTL Logic NAND Gates

- 74LS00 Quad 2-input

- 74LS10 Triple 3-input
- 74LS20 Dual 4-input
- 74LS30 Single 8-input

CMOS Logic NAND Gates

- CD4011 Quad 2-input
- CD4023 Triple 3-input
- CD4012 Dual 4-input

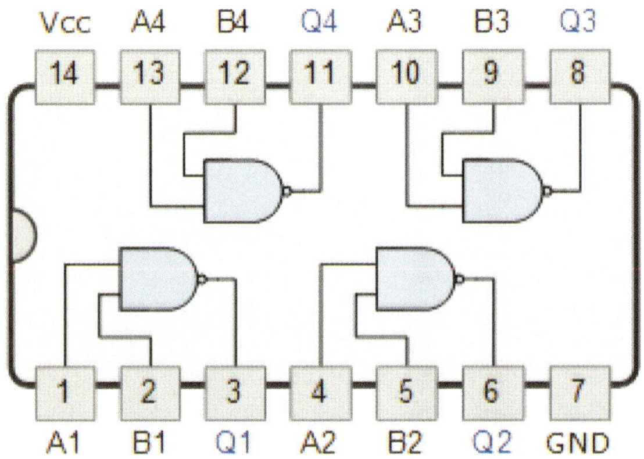

7400 Quad 2-input Logic NAND Gate

NOR Gate

The Logic NOR Gate gate is a combination of the digital logic OR gate and an inverter or NOT gate connected together in series.

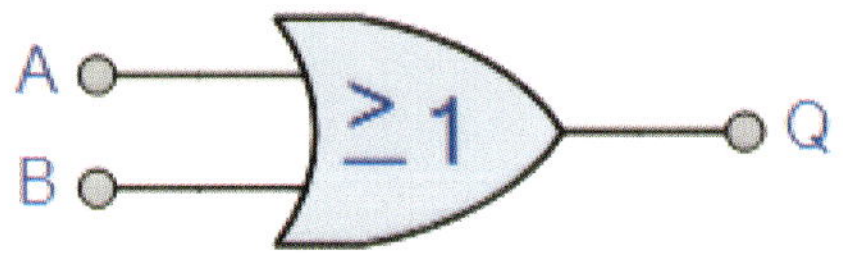

The inclusive NOR (Not – OR) gate has an output that is normally at logic level "1" and only goes "LOW" to logic level "0" when ANY of its inputs are at logic level "1". The Logic NOR Gate is the reverse or "Complementary" form of the inclusive OR gate.

Logic NOR Gate Equivalent

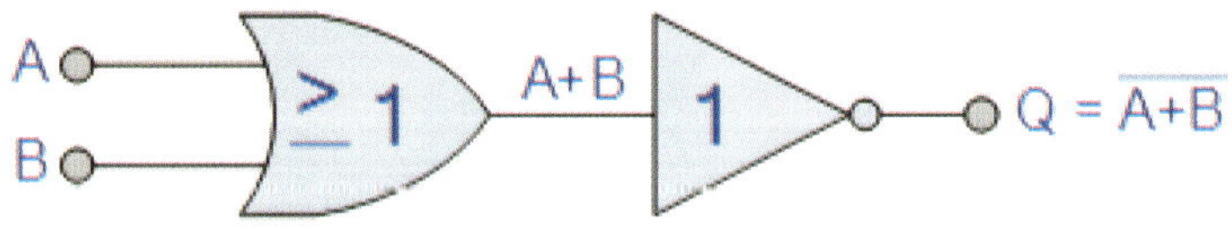

2-input "OR" gate plus a "NOT" gate

The logic or Boolean expression given for a logic NOR gate is that for Logical Multiplicationwhich it performs on the complements of the inputs. The Boolean expression for a logic NOR gate is denoted by a plus sign, (+) with a line or Overline, ($\overline{\ \ }$) over the expression to signify the NOT or logical negation of the NOR gate giving us the Boolean expression of: $\overline{A+B} = Q$.

Then we can define the operation of a 2-input digital logic NOR gate as being:

"If both A and B are NOT true, then Q is true"

Pulsed Operation

If we apply 2 different clock signals as the inputs of NOR gate X and Y, then the output of the NOR gate will be as shown below (X, Y are inputs and Z is output)

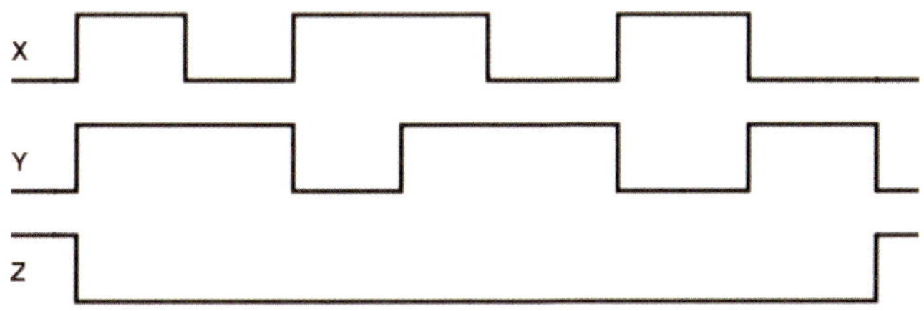

If we observe the above clock signals, we can say that when both inputs are high, then the output of the NOR gate is LOW and the LOW output state continued for every input logic low level. At the end of the clock pulses, the output becomes high as both the inputs of NOR gate the inputs are low.

Transistor NOR Gate

A simple 2-input logic NOR gate can be constructed using RTL Resistor-transistor switches connected together as shown below with the inputs connected directly to the transistor bases. Both transistors must be cut-off "OFF" for an output at Q.

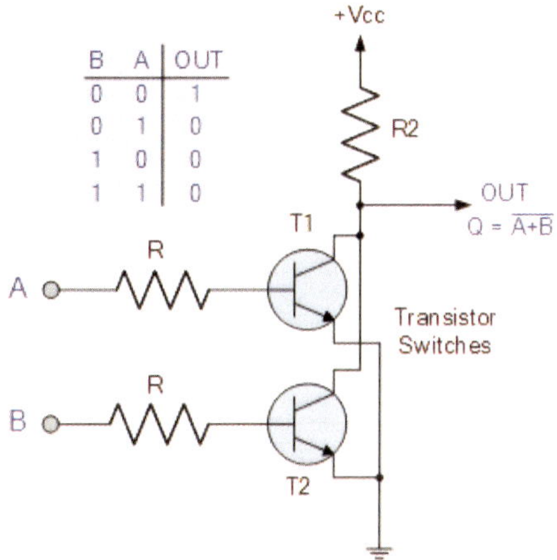

Logic NOR Gates are available using digital circuits to produce the desired logical function and is given a symbol whose shape is that of a standard OR gate with a circle, sometimes called an "inversion bubble" at its output to represent the NOT gate symbol with the logical operation of the NOR gate given as.

The Digital Logic "NOR" Gate

2-input NOR Gate

Symbol	Truth Table		
	B	A	Q
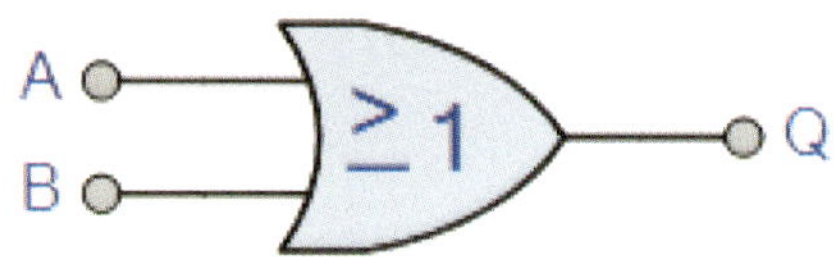	0	0	1
	0	1	0
2-input NOR Gate	1	0	0
	1	1	0
Boolean Expression Q = A+B	Read as A OR B gives NOT Q		

3-input NOR Gate

Symbol	Truth Table			
	C	B	A	Q
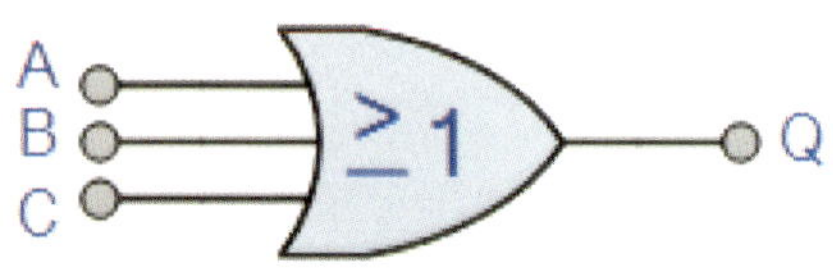	0	0	0	1
	0	0	1	0
	0	1	0	0
	0	1	1	0
3-input NOR Gate	1	0	0	0
	1	0	1	0
	1	1	0	0
	1	1	1	0
Boolean Expression Q = $\overline{A+B+C}$	Read as A OR B OR C gives NOT Q			

As with the OR function, the NOR function can also have any number of individual inputs and commercial available NOR Gate IC's are available in standard 2, 3, or 4 input types. If additional inputs are required, then the standard NOR gates can be cascaded together to provide more inputs for example.

4-input NOR Function

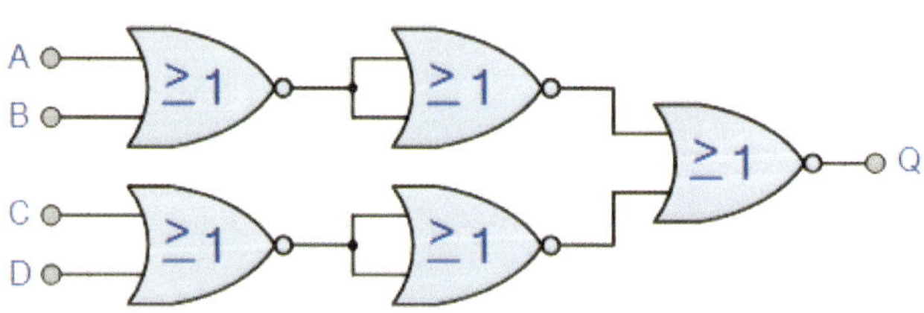

The Boolean Expression for this 4-input NOR gate will therefore be: Q = $\overline{A+B+C+D}$

If the number of inputs required is an odd number of inputs any "unused" inputs can be held LOW by connecting them directly to ground using suitable "Pull-down" resistors.

The Logic NOR Gate function is sometimes known as the Pierce Function and is denoted by a downwards arrow operator as shown, A↓B.

The "Universal" NOR Gate

Like the NAND gate in the last topic, the NOR gate can also be classed as a "Universal" type gate. NOR gates can be used to produce any other type of logic gate function just like the NAND gate and by connecting them together in various combinations the three basic gate types of AND, OR and NOT function can be formed using only NOR's, for example.

Various Logic Gates using only NOR Gates

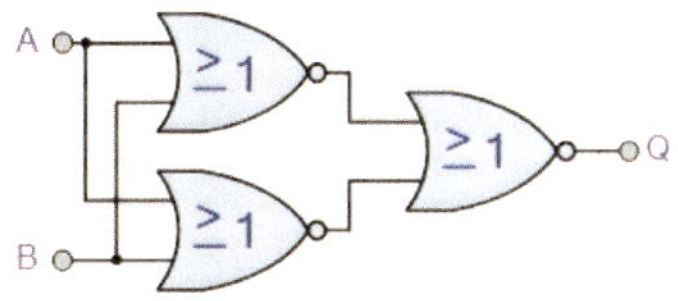

2-input OR Gate

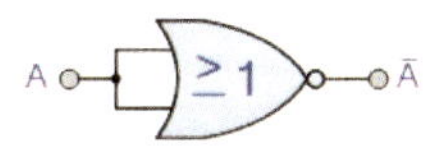

NOT Gate

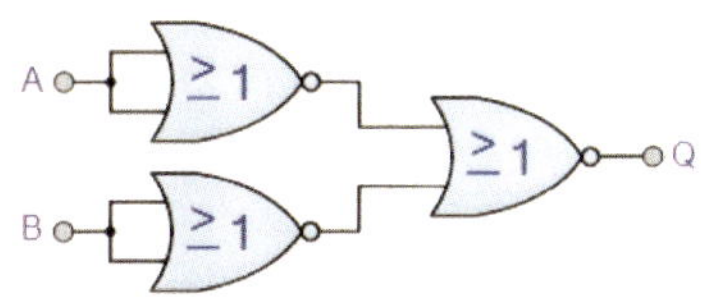

2-input AND Gate

As well as the three common types above, Ex-Or, Ex-Nor and standard NOR gates can also be formed using just individual NOR gates.

Commonly available digital logic NOR gate IC's include:

TTL Logic NOR Gates

- 74LS02 Quad 2-input
- 74LS27 Triple 3-input
- 74LS260 Dual 4-input

CMOS Logic NOR Gates

- CD4001 Quad 2-input
- CD4025 Triple 3-input
- CD4002 Dual 4-input

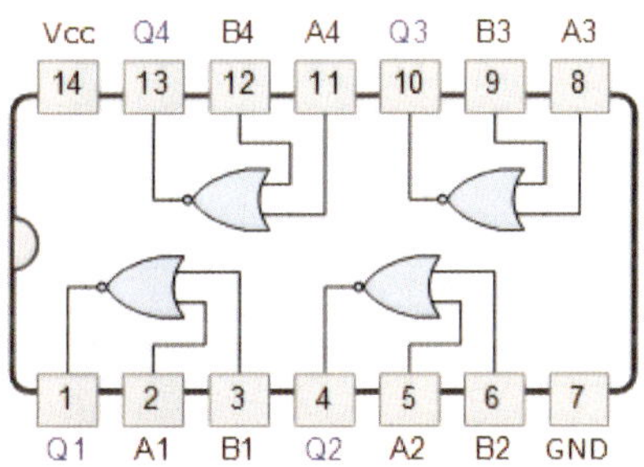

7402 Quad 2-input NOR Gate

XOR Gate

The Exclusive-OR logic function is a very useful circuit that can be used in many different types of computational circuits.

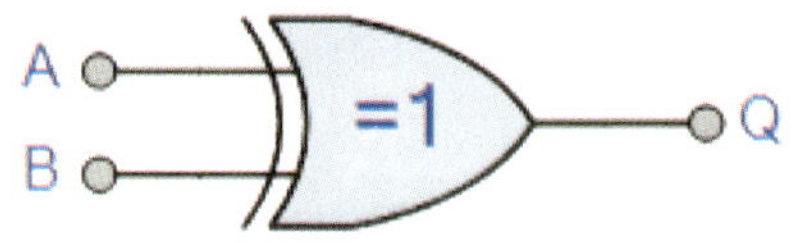

There are two other types of digital logic gates which although they are not a basic gate in their own right as they are constructed by combining together other logic gates, their output Boolean function is important enough to be considered as complete logic gates. These two "hybrid" logic gates are called the Exclusive-OR (Ex-OR) Gate and its complement the Exclusive-NOR (Ex-NOR) Gate.

Previously, we saw that for a 2-input OR gate, if A = "1", OR B = "1", OR BOTH A + B = "1"then the output from the digital gate must also be at a logic level "1" and because of this, this type of logic gate is known as an Inclusive-OR function. The gate gets its name from the fact that it *includes* the case of Q = "1" when both A and B = "1".

If however, an logic output "1" is obtained when ONLY A = "1" or when ONLY B = "1" but NOT both together at the same time, giving the binary inputs of "01" or "10", then the output will be "1". This type of gate is known as an Exclusive-OR function or more commonly an Ex-Or function for short. This is because its boolean expression excludes the "OR BOTH" case of Q = "1" when both A and B = "1".

In other words the output of an Exclusive-OR gate ONLY goes "HIGH" when its two input terminals are at "DIFFERENT" logic levels with respect to each other.

An odd number of logic "1's" on its inputs gives a logic "1" at the output. These two inputs can be at logic level "1" or at logic level "0" giving us the Boolean expression of: $Q = (A \oplus B) = \overline{A}.B + \overline{A}.B$.

The Exclusive-OR Gate function, or Ex-OR for short, is achieved by combining standard logic gates together to form more complex gate functions that are used extensively in building arithmetic logic circuits, computational logic comparators and error detection circuits.

The two-input "Exclusive-OR" gate is basically a modulo two adder, since it gives the sum of two binary numbers and as a result are more complex in design than other basic types of logic gate.

Pulsed Operation

The pulsed operation of 2 input XOR gate is shown below.

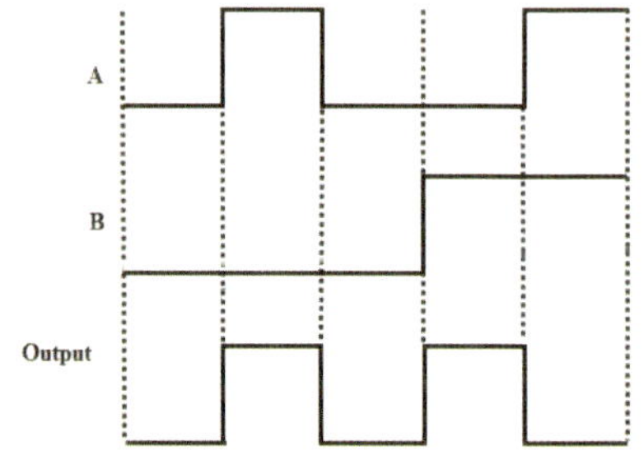

The Digital Logic "Exclusive-OR" Gate

2-input Ex-OR Gate

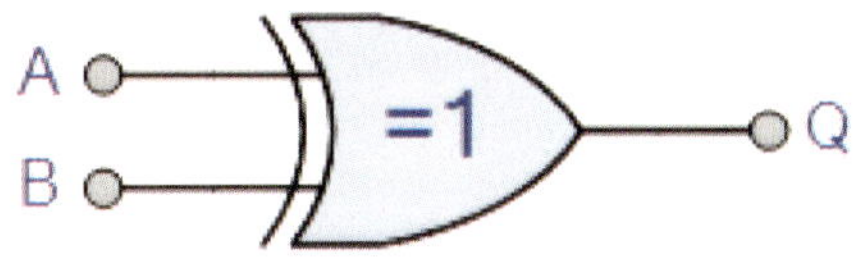

2-input Ex-OR Gate

B	A	Q
0	0	0
0	1	1
1	0	1
1	1	0

Boolean Expression $Q = A \oplus B$

A OR B but NOT BOTH gives Q

Giving the Boolean expression of: $Q = A\overline{} + \overline{A}\,B$

The truth table above shows that the output of an Exclusive-OR gate ONLY goes "HIGH" when both of its two input terminals are at "DIFFERENT" logic levels with respect to each other. If these two inputs, A and B are both at logic level "1" or both at logic level "0" the output is a "0" making the gate an "odd but not the even gate". In other words, the output is "1" when there are an odd number of 1's in the inputs.

This ability of the *Exclusive-OR gate* to compare two logic levels and produce an output value dependent upon the input condition is very useful in computational logic circuits as it gives us the following Boolean expression of:

$Q = (A \oplus B) = \overline{A}.B + A.\overline{B}$

The logic function implemented by a 2-input Ex-OR is given as either: "A OR B but NOT both" will give an output at Q. In general, an Ex-OR gate will give an output value of logic "1" ONLY when there are an ODD number of 1's on the inputs to the gate, if the two numbers are equal, the output is "0".

Then an Ex-OR function with more than two inputs is called an "odd function" or modulo-2-sum (Mod-2-SUM), not an Ex-OR. This description can be expanded to apply to any number of individual inputs as shown below for a 3-input Ex-OR gate.

3-input Ex-OR Gate

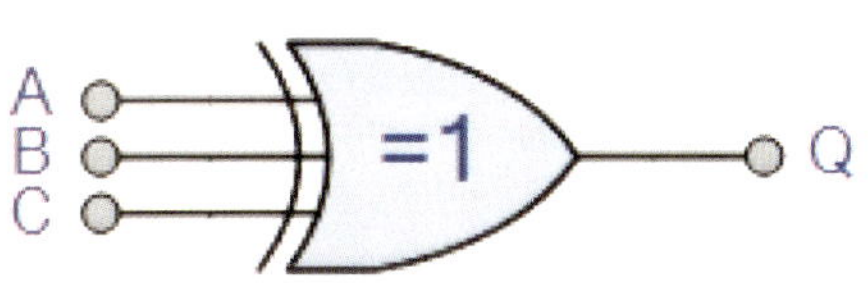

3-input Ex-OR Gate

C	B	A	Q
0	0	0	0
0	0	1	1
0	1	0	1
0	1	1	0
1	0	0	1
1	0	1	0
1	1	0	0
1	1	1	1

Boolean Expression $Q = A \oplus B \oplus C$

"Any ODD Number of Inputs" gives Q

Giving the Boolean expression of: $Q = A\overline{B}\,\overline{C} + \overline{A}B\overline{C} + \overline{A}\,\overline{B}C + ABC$

The symbol used to denote an Exclusive-OR odd function is slightly different to that for the standard Inclusive-OR Gate. The logic or Boolean expression given for a logic OR gate is that of logical addition which is denoted by a standard plus sign.

The symbol used to describe the Boolean expression for an Exclusive-OR function is a plus sign, (+) within a circle (O). This exclusive-OR symbol also represents the mathematical "direct sum of sub-objects" expression, with the resulting symbol for an *Exclusive-OR* function being given as: ($\oplus$).

We said previously that the Ex-OR function is not a basic logic gate but a combination of different logic gates connected together. Using the 2-input truth table above, we can expand the Ex-OR function to: $(A+B).\overline{(A.B)}$ which means that we can realise this new expression using the following individual gates.

Ex-OR Gate Equivalent Circuit

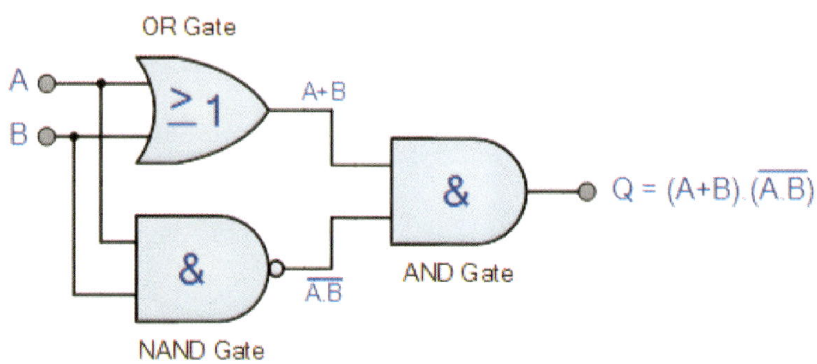

One of the main disadvantages of implementing the Ex-OR function above is that it contains three different types logic gates OR, NAND and finally AND within its design. One easier way of producing the Ex-OR function from a single gate is to use our old favourite the NAND gate as shown below.

Ex-OR Function Realisation using NAND Gates

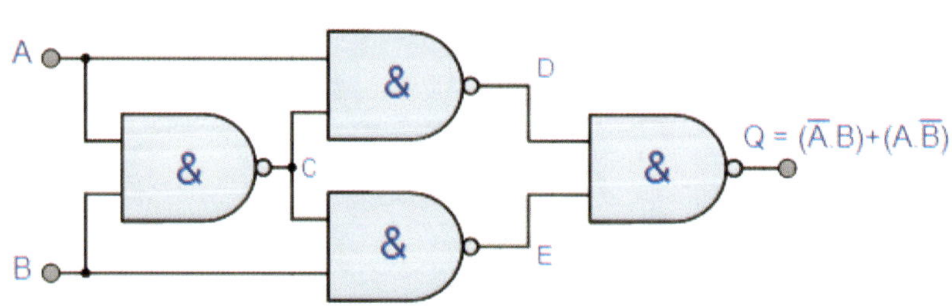

NAND Gate Realisation

Exclusive-OR Gates are used mainly to build circuits that perform arithmetic operations and calculations especially Adders and Half-Adders as they can provide a "carry-bit" function or as a controlled inverter, where one input passes the binary data and the other input is supplied with a control signal.

Commonly available digital logic Exclusive-OR gate IC's include:

TTL Logic Ex-OR Gates

- 74LS86 Quad 2-input

CMOS Logic Ex-OR Gates

- CD4030 Quad 2-input

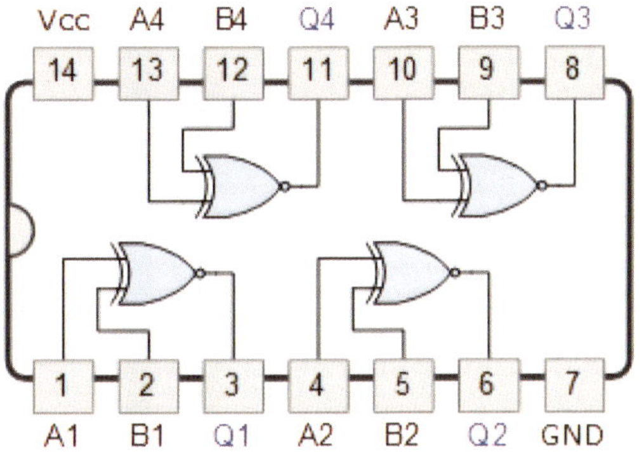

7486 Quad 2-input Exclusive-OR Gate

The Exclusive-OR logic function is a very useful circuit that can be used in many different types of computational circuits. Although not a basic logic gate in its own right, its usefulness and versatility has turned it into a standard logical function complete with its own Boolean expression, operator and symbol. The Exclusive-OR Gate is widely available as a standard quad two-input 74LS86 TTL gate or the 4030B CMOS package.

Ex-OR Gate Applications

X-OR logic gate is used in many applications. Some of them are explained below.

Uses in Addition

We can design single bit adder, which will add two bits and produce a single bit output. The single bit Adder designed by using XOR gate is shown below.

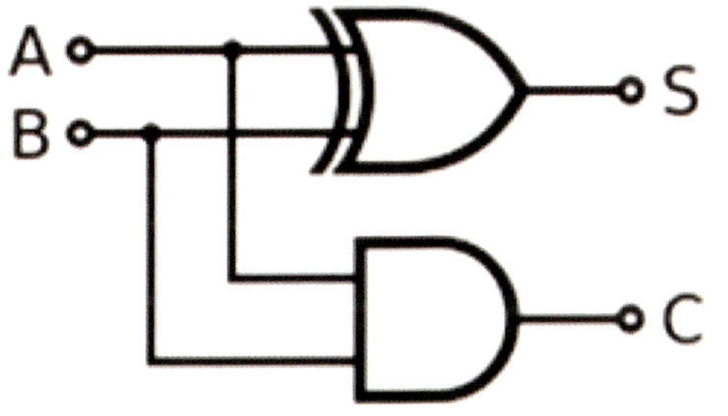

For example, if we add two bits 1 and 1 in binary addition, we get the answer 10 and in decimal addition method we get 2. The main principle of half adders is that the trailing sum is achieved by the output of XOR gate and the carry bit is calculated by AND gate.

We can cascade many single bit adder circuits to form n-bit adder circuit, to calculate the sum of longer binary numbers.

Pseudo-random Number Generation

Linear shift registers are also called as Pseudo Random Number Generators (PNRs). In order to generate the random numbers, we arrange the XOR logic gate in a specific order by forming a linear feedback shift register.

Correlation and Sequence Detection

XOR gate is capable of producing a low level input i.e. 0 when all of its inputs are high or low. When we are searching a particular bit sequence in a long data sequence, we use XOR gates to find the required data bit sequence. The accuracy of finding the required string of data bits in the target sequence is determined by calculating the number of 0's obtained. In many communication devices such as decoders and CDMA receivers, we use co-relaters, which are used to extract the parity of a specific Pseudo Random Number sequence in a group of PRN sequence.

XNOR Gate

The Exclusive-NOR Gate function is a digital logic gate that is the reverse or complementary form of the Exclusive-OR function.

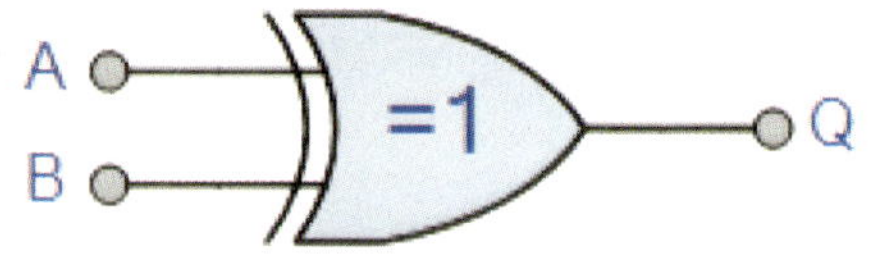

The "Exclusive-NOR Gate" is a combination of the Exclusive-OR gate and the NOT gate but has a truth table similar to the standard NOR gate in that it has an output that is normally at logic level "1" and goes "LOW" to logic level "0" when ANY of its inputs are at logic level "1".

However, an output "1" is only obtained if BOTH of its inputs are at the same logic level, either binary "1" or "0". For example, "00" or "11". This input combination would then give us the Boolean expression of: $Q = (\overline{A \oplus B}) = \overline{A.B} + A.B$

Then the output of a digital logic Exclusive-NOR gate ONLY goes "HIGH" when its two input terminals, A and B are at the "SAME" logic level which can be either at a logic level "1" or at a logic level "0". In other words, an even number of logic "1's" on its inputs gives a logic "1" at the output, otherwise is at logic level "0".

Then this type of gate gives and output "1" when its inputs are "logically equal" or "equivalent" to each other, which is why an Exclusive-NOR gate is sometimes called an Equivalence Gate.

The logic symbol for an Exclusive-NOR gate is simply an Exclusive-OR gate with a circle or "inversion bubble", (o) at its output to represent the NOT function. Then the Logic Exclusive-NOR Gate is the reverse or "Complementary" form of the Exclusive-OR gate, ($A \oplus B$).

Ex-NOR Gate Equivalent

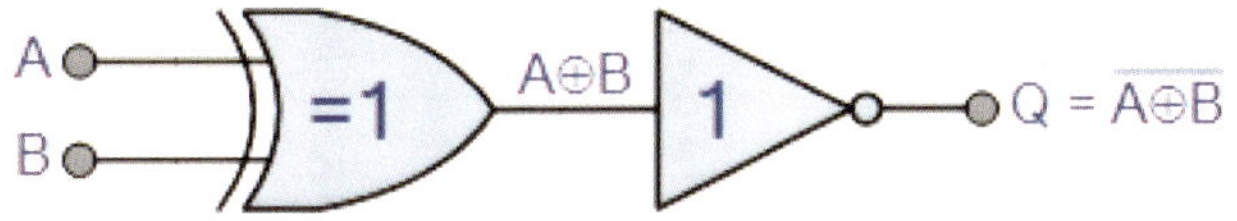

2-input "Ex-OR" gate plus a "NOT" gate

The Exclusive-NOR Gate function is achieved by combining standard gates together to form more complex gate functions.

Ex – NOR Gate Pulsed Operation

The pulsed operation of 2 input XOR gate is shown below.

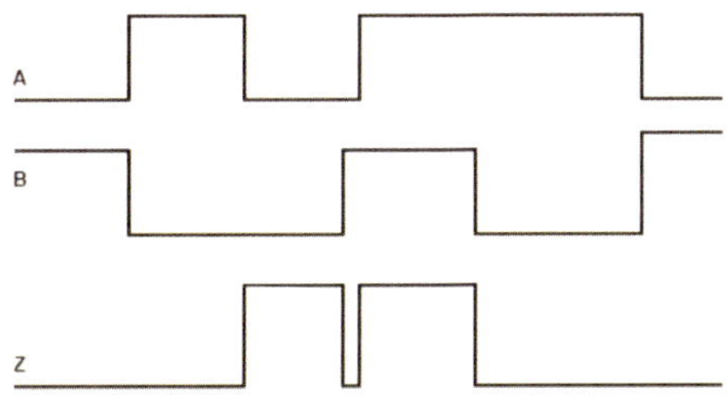

Initially when one of the two inputs of the XNOR gate is low, so the output is at low level. When both the input changes to low level, the XNOR output will rise to HIGH level and again when one of the inputs comes to low level, the output also comes to low level.

The Digital Logic "Ex-NOR" Gate

2-input Ex-NOR Gate

<table>
<tr><th>Symbol</th><th colspan="3">Truth Table</th></tr>
<tr><td rowspan="5">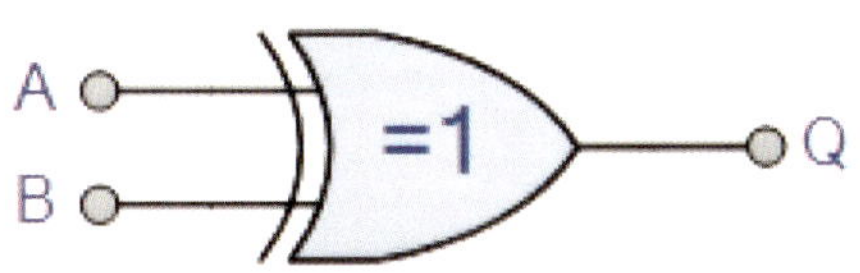
2-input Ex-NOR Gate</td><td>B</td><td>A</td><td>Q</td></tr>
<tr><td>0</td><td>0</td><td>1</td></tr>
<tr><td>0</td><td>1</td><td>0</td></tr>
<tr><td>1</td><td>0</td><td>0</td></tr>
<tr><td>1</td><td>1</td><td>1</td></tr>
<tr><td>Boolean Expression Q = $\overline{A \oplus B}$</td><td colspan="3">Read if A AND B the SAME gives Q</td></tr>
</table>

Giving the Boolean expression of: Q = $\overline{AB} + AB$

The logic function implemented by a 2-input Ex-NOR gate is given as "when both A AND B are the SAME" will give an output at Q. In general, an Exclusive-NOR gate will give an output value of logic "1" ONLY when there are an EVEN number of 1's on the inputs to the gate (the inverse of the Ex-OR gate) except when all its inputs are "LOW".

Then an Ex-NOR function with more than two inputs is called an "even function" or modulo-2-sum (Mod-2-SUM), not an Ex-NOR. This description can be expanded to apply to any number of individual inputs as shown below for a 3-input Exclusive-NOR gate.

3-input Ex-NOR Gate

Symbol	Truth Table			
	C	B	A	Q
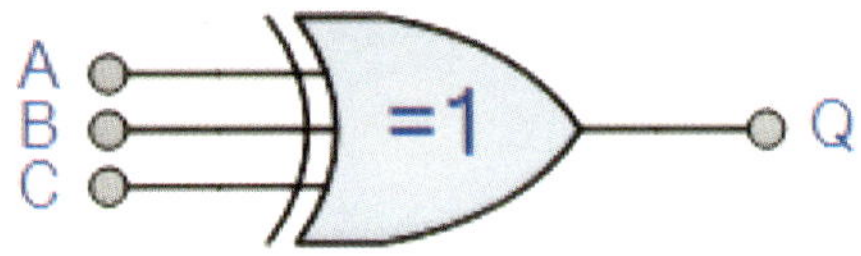 3-input Ex-NOR Gate	0	0	0	1
	0	0	1	0
	0	1	0	0
	0	1	1	1
	1	0	0	0
	1	0	1	1
	1	1	0	1
	1	1	1	0
Boolean Expression $Q = \overline{A \oplus B \oplus BC}$	Read as "any EVEN number of Inputs" gives Q			

Giving the Boolean expression of: Q = ABC + ABC + ABC + ABC

We said previously that the Ex-NOR function is a combination of different basic logic gates Ex-OR and a NOT gate, and by using the 2-input truth table above, we can expand the Ex-NOR function to: $Q = \overline{A \oplus B} = (A.B) + (\overline{A}.\overline{B})$ which means we can realise this new expression using the following individual gates.

Ex-NOR Gate Equivalent Circuit

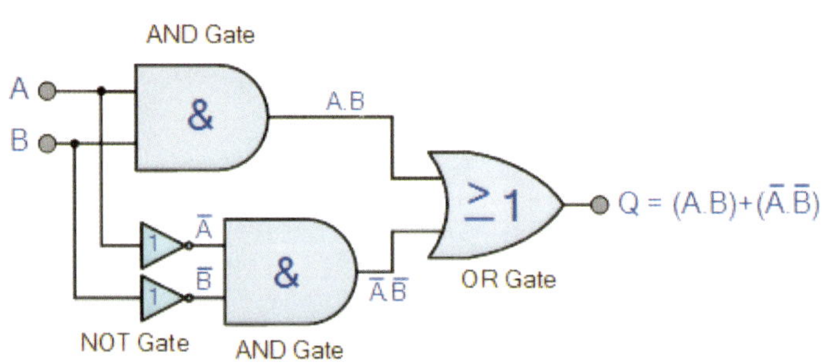

One of the main disadvantages of implementing the Ex-NOR function above is that it contains three different types logic gates the AND, NOT and finally an OR gate within its basic design. One easier way of producing the Ex-NOR function from a single gate type is to use NAND gates as shown below.

Ex-NOR Function Realisation using NAND Gates

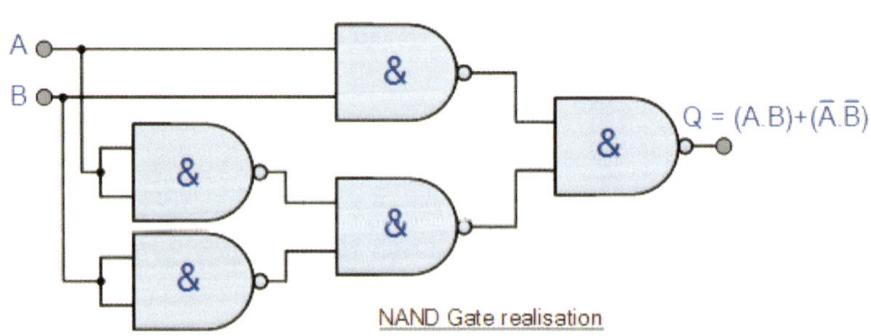

Ex-NOR gates are used mainly in electronic circuits that perform arithmetic operations and data checking such as Adders, Subtractors or Parity Checkers, etc. As the Ex-NOR gate gives an output of logic level "1" whenever its two inputs are equal it can be used to compare the magnitude of two binary digits or numbers and so Ex-NOR gates are used in Digital Comparator circuits.

Commonly available digital logic Exclusive-NOR gate IC's include:

TTL Logic Ex-NOR Gates

- 74LS266 Quad 2-input

CMOS Logic Ex-NOR Gates

- CD4077 Quad 2-input

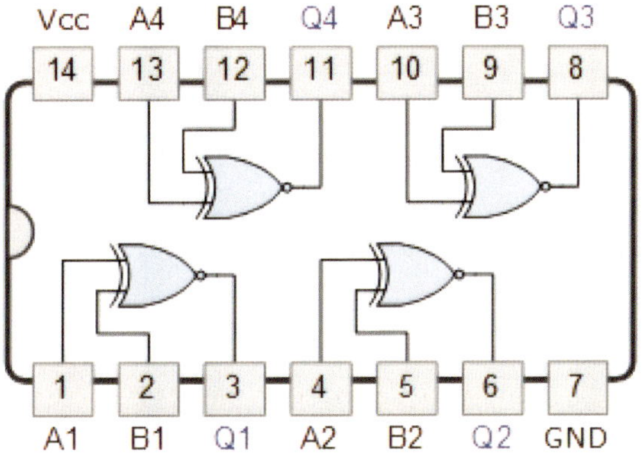

74266 Quad 2-input Ex-NOR Gate

Ex-NOR Gate Applications

- The XNOR logic gates are used in error detecting circuits which are to detect Odd parity or even parity bits in digital data transmission circuits.
- XNOR gate is mainly used in arithmetic and encryption circuits. This process is the combinational operation of the XOR and XNOR gates, by using 6 transistors for low power application.

This is also used as Heat exchanger tank, which will ring the alarm when the water temperature level goes up or down, to the pre-set level. The Heat exchanger circuit is explained below.

Construction

This circuit is used to warm the water. It has two outlets and two float indicators. The float indicators are connected as the inputs of XNOR gate. The output of the XNOR gate is connected to the alarm circuit, which used to alert when the water level changes in the glass beaker. There is a heating element at the bottom of the beaker to heat the certain amount of water.

When the water reaches high level, the excess water will overflow from the upper outlet. When the water comes to low level, then the heating element situated at the bottom is not covered by water so it will burn out. As we said earlier, it alerts the when the water level increases or decreases to certain level. To do this, the float switches are to be placed at high level and low level positions of water, inside the tank.

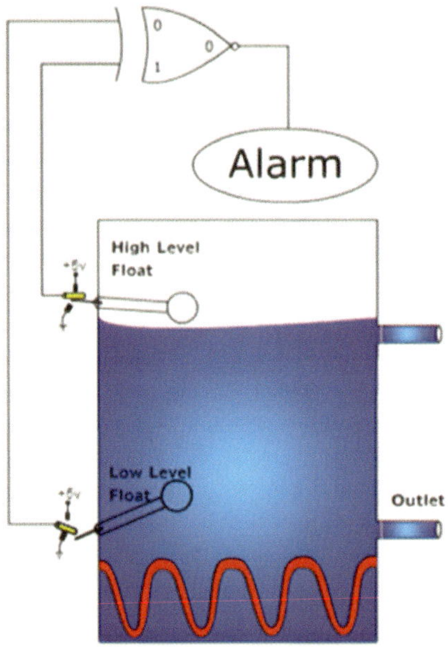

When a float indicator gets in contact with water level, it is pushed upwards. This will cause the float's leaver to disconnect from the armature. At this situation, the XNOR gate's input is 0. If the float indicator is not in contact with the water level, then the leaver is connected to +5 V supply, making the armature active. At this situation, the XNOR gate's input will be HIGH, which is equal to 1. When the 2 leavers are connected to the +5 volts supply or to ground voltage, then the 2 inputs of the XNOR gate is 1, and thus the output of the XNOR gate is high, so that it activates the alarm and alerts the person.

When the water level is between upper floater and lower floater indicator, then logic 1 is applied from upper indicator and logic 0 from lower indicator, to the input of XNOR gate. At this position no alarm will ring as the water level will is at required level. This is how the logic XNOR gate used as a Heat exchanger circuit.

AND-OR-Invert

AOI also known as AND-OR-Inverter. AND-OR-Invert (AOI) logic or say gates are two-level logic functions constructed from the combination of one or more AND gates followed by a NOR gate. If we construct AND, OR and NOT gate separately, Number of transistor in AOI gates are less.

You might be thinking why need individual logic gate, why can't we implement it using just 2 AND gate and 1 NOR gate. Yes, you are right. But now think from CMOS point of view. In CMOS, we can implement AND gate using 1 CMOS NAND gate and 1 Inverter. It means above 2 AND gate changes into 2 NAND and 2 Inverter.

Let's take an example and try to understand it.

For example: 2-2 AOI: ((A.B) + (C.D))'

Let's see how this function be implemented using logic gates (separately) Vs AOI cells.

Using Individual NAND, NOT and NOR Gate: First we have to change the function as per logic gates availability. Y== ((A.B) + (C.D))' == (((A.B)')' + ((C.D)')')'

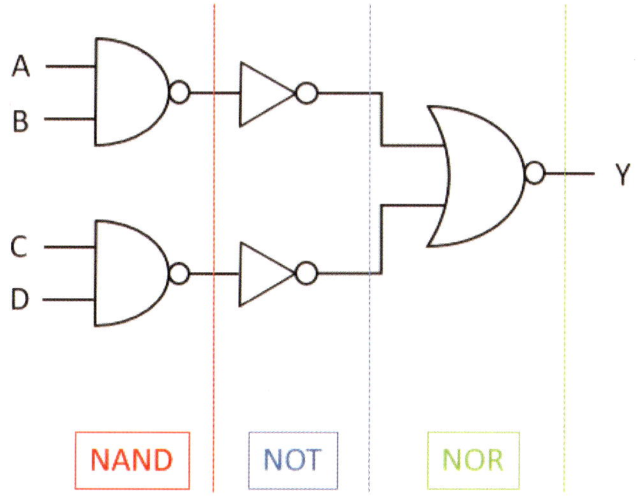

Function (((A.B)')' + ((C.D)')')' can be implemented as:

- NAND Gates:
  - 2 NAND gates: 1st for (A.B)' and 2nd for (C.D)' (Assume X=(A.B)' and Y=(C.D)').
  - 1 NAND gate uses 2 PMOS transistor and 2 NMOS transistor.
  - So, total Transistors in 2 2-input NAND gate are 8 Transistors.
- Inverter:
  - 2 Inverter: 1st (X)' and 2nd for (Y)'
  - 1 Inverter uses 1 PMOS and 1 NMOS
  - So, total Transistors in 1 Inverter are 2 Transistors.
- NOR Gates:
  - 1 NOR Gate: (X' + Y')'
  - 1 NOR gate uses 2 PMOS transistor and 2 NMOS transistor.
  - So, total Transistors in 1 2-input NOR gate are 4 Transistors.

Total Transistor in case of Individually implementing ((A.B) + (C.D))' = 14 Transistor.

CMOS Representation of above function is given below.

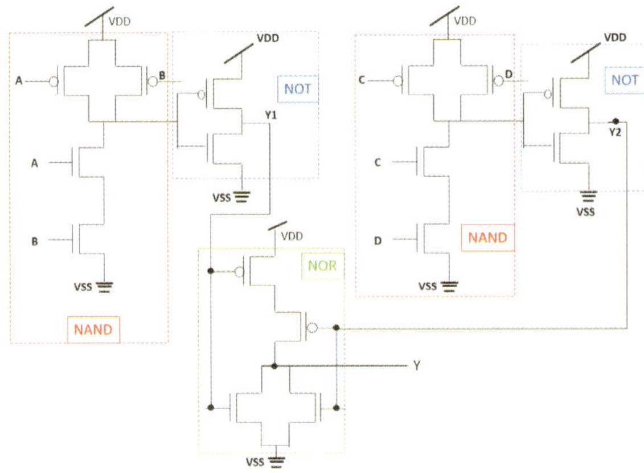

Using AOI Logic Gates

Implementation using AOI cells are very easy. In this case, we are not suppose to change the function. Y == ((A.B) + (C.D))'

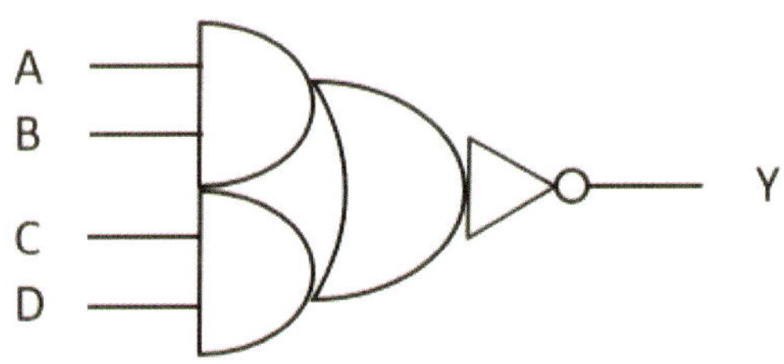

AOI (AND-OR-Inverter) Cell

Below diagram can help you to understand how cells can implement using CMOS.

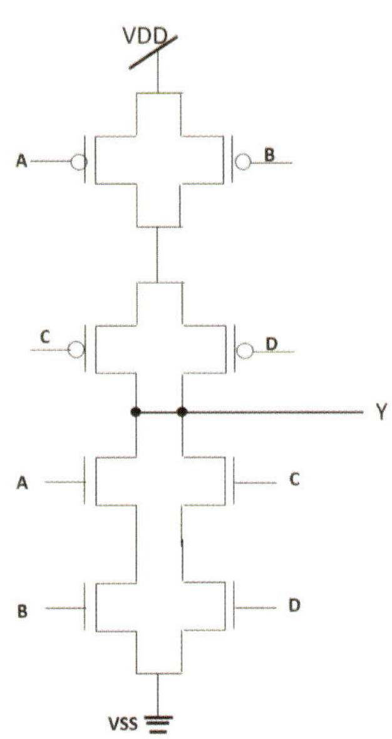

Logic Operations

AOI gates perform one or more AND operations followed by an OR operation and then an inversion. For example, a 2-2 AOI gate can be represented by the boolean equation and truth table:

$$F = \overline{(A \wedge B) \vee (C \wedge D)}$$

2-2 AOI				
INPUT A B C D				OUTPUT F
0	X	X	0	1
X	0	X	0	1
0	X	0	X	1
X	0	0	X	1
1	1	X	X	0
X	X	1	1	0

A 2-1 AOI gate can be represented by following the boolean equation and truth table:

$$F = \overline{A \vee (B \wedge C)}$$

2-1 AOI			
INPUT A B C			OUTPUT F
0	0	0	1
0	0	1	1
0	1	0	1
0	1	1	0
1	0	0	0
1	0	1	0
1	1	0	0
1	1	1	0

Larger AOI gates, such as 4-3 AOI or 3-3-3 AOI can also be used.

Electronic Implementation

AND-OR-INVERT(AOI) and OAI gates can be readily implemented in CMOS circuitry. AOI gates are particularly advantaged in that the total number of transistors (or gates) is less than if the AND, NOT, and OR functions were implemented separately. This results in increased speed, reduced power, smaller area, and potentially lower fabrication cost. For example, a 2-1 AOI gate can be constructed with 6 transistors in CMOS compared to 10 transistors using a 2-input NAND gate (4 transistors), an inverter (2 transistors), and a 2-input NOR gate (4 transistors).

In NMOS logic, the lower half of the CMOS circuit is used in combination with a load device or pull-up transistor (typically a depletion load or a dynamic load).

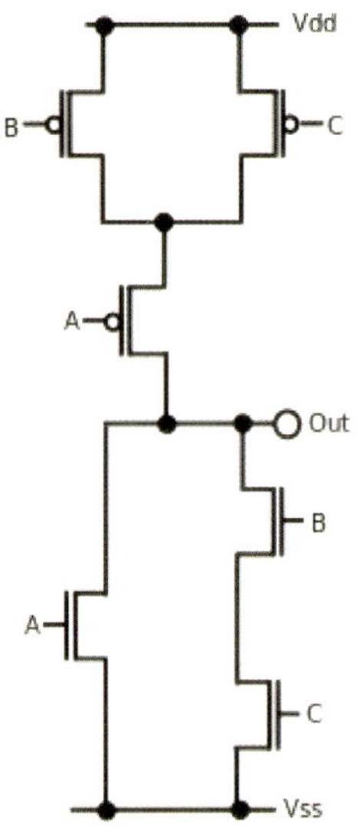

CMOS implementation of a 2-1 AOI gate

AOI gates are similarly efficient in transistor–transistor logic (TTL). The TTL 7400 line included a number of AOI gate parts, such as the 7451 dual 2-wide 2-input AND-OR-invert gate and the 7464 4-2-3-2-input AND-OR-invert gate.

Resistor-Transistor Logic

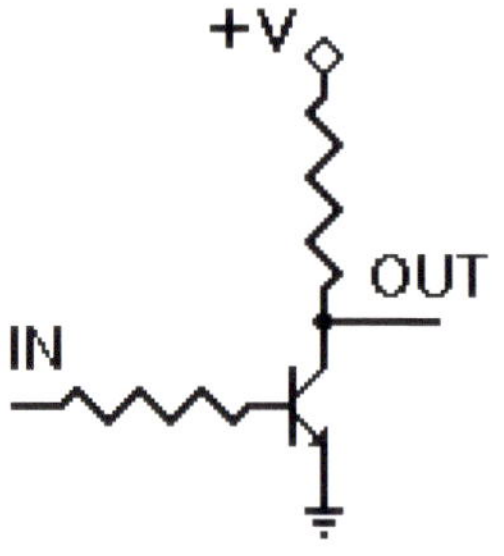

Consider the most basic transistor circuit, such as the one shown to the left. We will only be applying one of two voltages to the input I: 0 volts (logic 0) or +V volts (logic 1). The exact voltage used as +V depends on the circuit design parameters; in RTL integrated circuits, the usual voltage is +3.6v. We'll assume an ordinary NPN transistor here, with a reasonable dc current gain, an emitter-base forward voltage of 0.65 volt, and a collector-emitter saturation voltage no higher than 0.3 volt. In standard RTL ICs, the base resistor is 470Ω and the collector resistor is 640Ω.

When the input voltage is zero volts (actually, anything under 0.5 volt), there is no forward bias to the emitter-base junction, and the transistor does not conduct. Therefore no current flows through the collector resistor, and the output voltage is +V volts. Hence, a logic 0 input results in a logic 1 output.

When the input voltage is +V volts, the transistor's emitter-base junction will clearly be forward biased. For those who like the mathematics, we'll assume a similar output circuit connected to this input. Thus, we'll have a voltage of 3.6 - 0.65 = 2.95 volts applied across a series combination of a 640Ω output resistor and a 470Ω input resistor. This gives us a base current of:

$$2.95v / 1110\Omega = 0.0026576577 \text{ amperes} = 2.66 \text{ ma.}$$

RTL is a relatively old technology, and the transistors used in RTL ICs have a dc forward current gain of around 30. If we assume a current gain of 30, 2.66 ma base current will support a maximum of 79.8 ma collector current. However, if we drop all but 0.3 volts across the 640Ω collector resistor, it will carry 3.3/640Ω = 5.1 ma. Therefore this transistor is indeed fully saturated; it is turned on as hard as it can be.

With a logic 1 input, then, this circuit produces a logic 0 output. We have already seen that a logic 0 input will produce a logic 1 output. Hence, this is a basic inverter circuit.

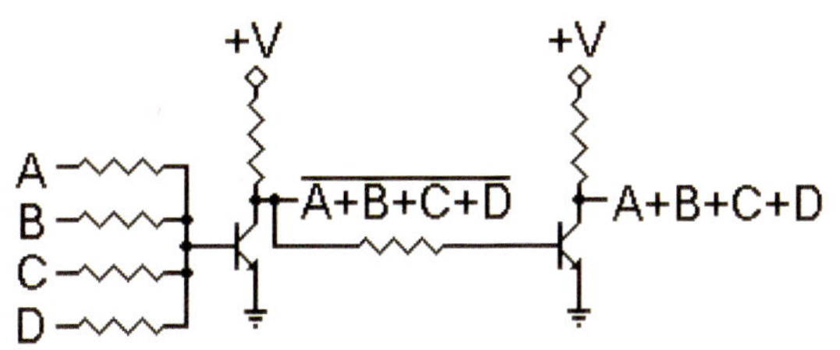

As we can see from the above calculations, the amount of current provided to the base of the transistor is far more than is necessary to drive the transistor into saturation. Therefore, we have the possibility of using one output to drive multiple inputs of other gates, and of having gates with multiple input resistors. Such a circuit is shown to the right.

In this circuit, we have four input resistors. Raising any one input to +3.6 volts will be sufficient to turn the transistor on, and applying additional logic 1 (+3.6 volt) inputs will not really have any appreciable effect on the output voltage. Remember that the forward bias voltage on the transistor's base will not exceed 0.65 volt, so the current through a grounded input resistor will not exceed 0.65v/470Ω = 1.383 ma. This does provide us with a practical limit on the number of allowable input resistors to a single transistor, but doesn't cause any serious problems within that limit.

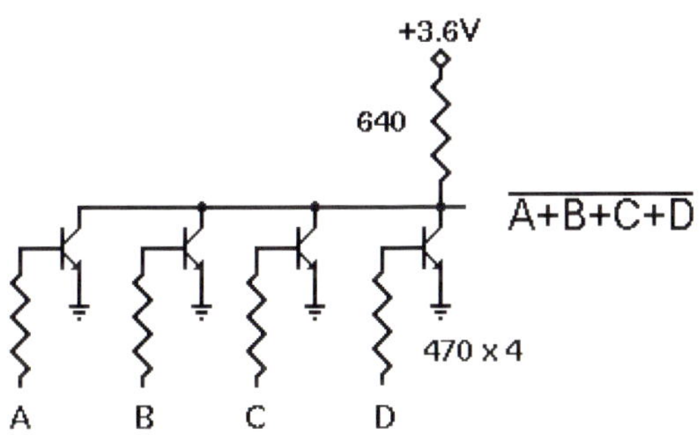

The RTL gate shown above will work, but has a problem due to possible signal interactions through the multiple input resistors. A better way to implement the NOR function is shown to the left.

Here, each transistor has only one input resistor, so there is no interaction between inputs. The NOR function is performed at the common collector connection of all transistors, which share a single collector load resistor.

This is in fact the pattern for all standard RTL ICs. The very commonly-used μL914 is a dual two-input NOR gate, where each gate is a two-transistor version of the circuit to the left. It is rated to draw 12 ma of current from the 3.6V power supply when both outputs are at logic 0. This corresponds quite well with the calculations we have already made.

Standard fan-out for RTL gates is rated at 16. However, the fan-in for a standard RTL gate input is 3. Thus, a gate can produce 16 units of drive current from the output, but requires 3 units to drive an input. There are low-power versions of these gates that increase the values of the base and collector resistors to 1.5K and 3.6K, respectively. Such gates demand less current, and typically have a fan-in of 1 and a fan-out of 2 or 3. They also have reduced frequency response, so they cannot operate as rapidly as the standard gates. To get greater output drive capabilities, *buffers* are used. These are typically inverters which have been designed with a fan-out of 80. They also have a fan-in requirement of 6, since they use pairs of input transistors to get increased drive.

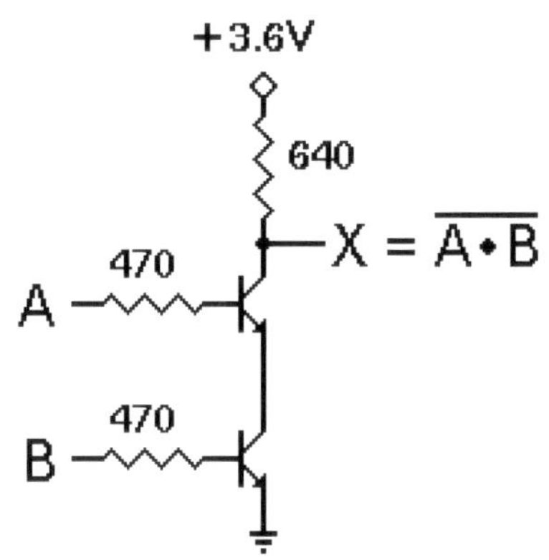

We can get a NAND function in either of two ways. We can simply invert the inputs to the NOR/OR gate, thus turning it into an AND/NAND gate, or we can use the circuit shown to the right.

In this circuit, each transistor has its own separate input resistor, so each is controlled by a different input signal. However, the only way the output can be pulled down to logic 0 is if both transistors are turned on by logic 1 inputs. If either input is a logic 0 that transistor cannot conduct, so there is no current through either one. The output is then a logic 1. This is the behavior of a NAND gate. Of course, an inverter can also be included to provide an AND output at the same time.

The problem with this NAND circuit stems from the fact that transistors are not ideal devices. Remember that 0.3 volt collector saturation voltage? Ideally it should be zero. Since it isn't, we need to look at what happens when we "stack" transistors this way. With two, the combined collector saturation voltage is 0.6 volt -- only slightly less than the 0.65 volt base voltage that will turn a transistor on.

If we stack three transistors for a 3-input NAND gate, the combined collector saturation voltage is 0.9 volt. This is too high; it will promote conduction in the next transistor no matter what. In addition, the load presented by the upper transistor to the gate that drives it will be different from the load presented by the lower transistor. This kind of unevenness can cause some odd problems to appear, especially as the frequency of operation increases. Because of these problems, this approach is not used in standard RTL ICs.

TTL Input & Output Voltage Levels

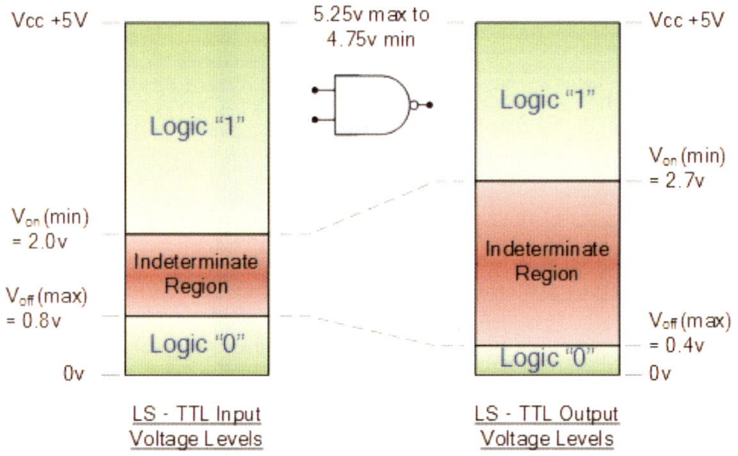

There are a large variety of logic gate types in both the bipolar 7400 and the CMOS 4000 families of digital logic gates such as 74Lxx, 74LSxx, 74ALSxx, 74HCxx, 74HCTxx, 74ACTxx etc, with each one having its own distinct advantages and disadvantages compared to the other. The exact switching voltage required to produce either a logic "0" or a logic "1" depends upon the specific logic group or family.

However, when using a standard +5 volt supply any TTL voltage input between 2.0v and 5v is considered to be a logic "1" or "HIGH" while any voltage input below 0.8v is recognised as a logic "0" or "LOW". The voltage region in between these two voltage levels either as an input or as an output is called the Indeterminate Region and operating within this region may cause the logic gate to produce a false output.

The CMOS 4000 logic family uses different levels of voltages compared to the TTL types as they are designed using field effect transistors, or FET's. In CMOS technology a logic "1" level operates between 3.0 and 18 volts and a logic "0" level is below 1.5 volts. Then the following table shows the difference between the logic levels of traditional TTL and CMOS logic gates.

TTL and CMOS Logic Levels

Device Type	Logic 0	Logic 1
TTL	0 to 0.8v	2.0 to 5v (V_{CC})
CMOS	0 to 1.5v	3.0 to 18v (V_{DD})

Then from the above observations, we can define the ideal TTL digital logic gate as one that has a "LOW" level logic "0" of 0 volts (ground) and a "HIGH" level logic "1" of +5 volts and this can be demonstrated as:

Ideal TTL Digital Logic Gate Voltage Levels

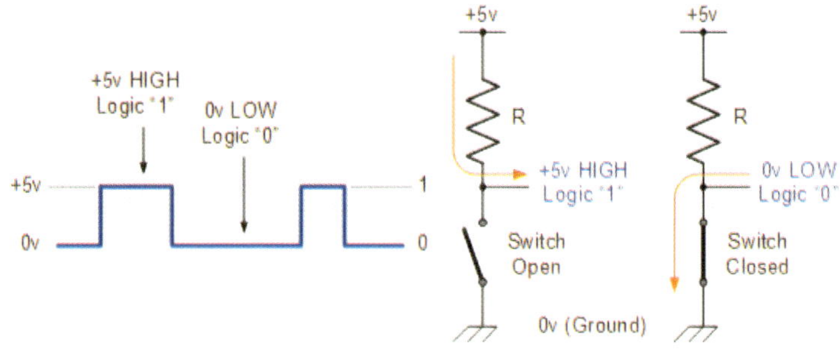

Where the opening or closing of the switch produces either a logic level "1" or a logic level "0" with the resistor R being known as a "pull-up" resistor.

Digital Logic Noise

However, between these defined HIGH and LOW values lies what is generally called a "no-man's land" (the blue area's above) and if we apply a signal voltage of a value within this no-man's land area we do not know whether the logic gate will respond to it as a level "0" or as a level "1", and the output will become unpredictable.

Noise is the name given to a random and unwanted voltage that is induced into electronic circuits by external interference, such as from nearby switches, power supply fluctuations or from wires and other conductors that pick-up stray electromagnetic radiation. Then in order for a logic gate not to be influence by noise in must have a certain amount of noise margin or noise immunity.

Digital Logic Gate Noise Immunity

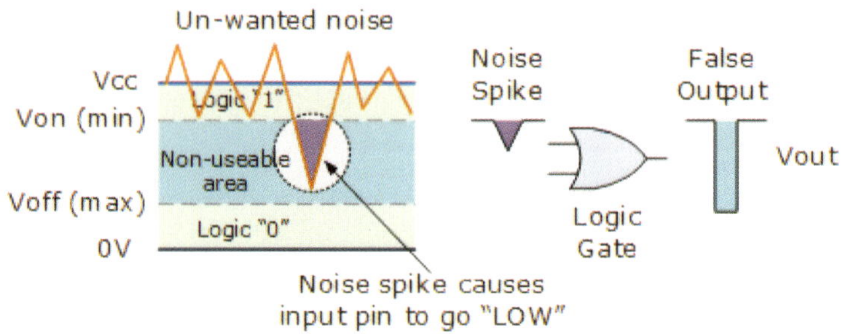

In the example above, the noise signal is superimposed onto the Vcc supply voltage and as long as it stays above the minimum level ($V_{ON(min)}$) the input an corresponding output of the logic gate are unaffected. But when the noise level becomes large enough and a noise spike causes the HIGH voltage level to drop below this minimum level, the logic gate may interpret this spike as a LOW level input and switch the output accordingly producing a false output switching. Then in order for the logic gate not to be affected by noise it must be able to tolerate a certain amount of unwanted noise on its input without changing the state of its output.

Advantages

The primary advantage of RTL technology was that it used a minimum number of transistors. In circuits using discrete components, before integrated circuits, transistors were the most expensive component to produce. Early IC logic production (such as Fairchild's in 1961) used the same approach briefly, but quickly transitioned to higher-performance circuits such as diode–transistor logic and then transistor–transistor logic (starting 1963 at Sylvania), since diodes and transistors were no more expensive than resistors in the IC.

Limitations

The disadvantage of RTL is its high power dissipation when the transistor is switched on, by current flowing in the collector and base resistors. This requires that more current be supplied to and heat be removed from RTL circuits. In contrast, TTL circuits with "totem-pole" output stage minimize both of these requirements.

Another limitation of RTL is its limited fan-in: 3 inputs being the limit for many circuit designs, before it completely loses usable noise immunity. It has a low noise margin. Lancaster says that integrated circuit RTL NOR gates (which have one transistor per input) may be constructed with "any reasonable number" of logic inputs, and gives an example of an 8-input NOR gate.

A standard integrated circuit RTL NOR gate can drive up to 3 other similar gates. Alternatively, it has enough output to drive up to 2 standard integrated circuit RTL "buffers", each of which can drive up to 25 other standard RTL NOR gates.

Speeding up RTL

Various companies applied the following speed-up methods to discrete RTL.

Transistor switching speed has increased steadily from the first transistorized computers through the present. The *GE Transistor Manual* recommends gaining speed by using higher-frequency transistors, or capacitors, or a diode from base to collector to prevent saturation.

Placing a capacitor in parallel with each input resistor decreases the time needed for a driving stage to forward-bias a driven stage's base-emitter junction. Engineers and technicians use "RCTL" (resistor capacitor transistor logic) to designate gates equipped with "speed-up capacitors". The Lincoln Laboratory TX-0 computer's circuits included some RCTL. However, methods involving capacitors were unsuitable for integrated circuits.

Using a high collector supply voltage and diode clamping decreased collector-base and wiring capacitance charging time. This arrangement required diode clamping the collector to the design logic level. This method was also applied to discrete DTL (diode–transistor logic).

Another method that was familiar in discrete-device logic circuits used a diode and a resistor, a germanium and a silicon diode, or three diodes in a negative feedback arrangement. These diode networks known as various Baker clamps reduced the voltage applied to the base as the collector approached saturation. Because the transistor went less deeply into saturation, the transistor accumulated fewer stored charge carriers. Therefore, less time was required to clear stored charge during transistor turn off. A low-voltage diode arranged to prevent saturation of the transistor was applied to integrated logic families by using Schottky diodes, as in Schottky TTL.

Diode-Transistor Logic

Resistor-Transistor Logic was an early form of logic used in the 1950's and early 1960's. RTL was made from discrete transistors and resistors and manufactured on printed circuit boards with several gates per board. These boards were plugged into board sockets with wiring on the socket pins determining the system function. RTL was a big improvement over vacuum tube technology previously used, requiring less than one quarter the space and one tenth the power dissipation. RTL was superceded in the 1960's by Diode Transistor Logic and then Transistor-Transistor Logic, DTL and TTL respectively. DTL was initially made with discrete transistors and resistors before being integrated onto silicon. One early form of DTL, used by IBM Corp in the 360 family of computers, was really a hybrid technology. Transistor and diode chips were glued to a ceramic substrate and aluminum resistor paste was deposited on the substrate to make resistors. Finally the ceramic base and components were hermetically sealed in an aluminum can. This family was used extensively in IBM products in the middle to late 1960's. While this family was not a true integrated circuit, it was very successful and was less expensive than true integrated circuits for several years. By the early 1970's integrated circuits became quite common and DTL gave way to TTL which was more appropriate to integrated circuit technology.

While DTL is no longer commercially used, we will discuss it because it is similar to and easier to understand than TTL, and because designers still find the configuration of value. First, however, we will discuss diode logic which is the front end of the DTL gate and performs the actual logic operation.

Diode Logic

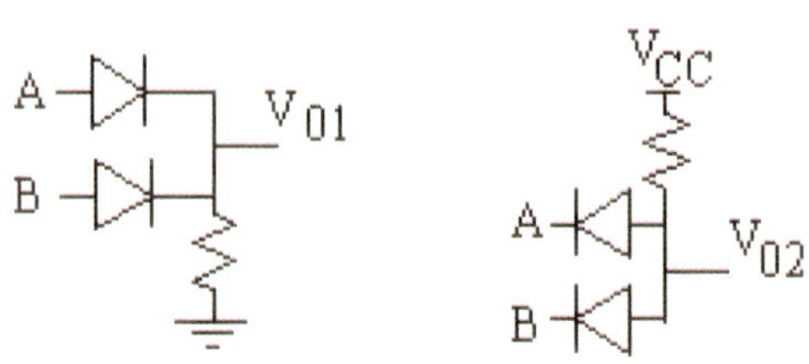

A	B	$V_{01}=A+B$
L	L	L
L	H	H
H	L	H
H	H	H

A	B	$V_{02}=A \cdot B$
L	L	L
L	H	L
H	L	L
H	H	H

Diode logic and truth tables

Two diode logic configurations are shown in above. The truth tables show that the first circuit performes a logical OR and the second circuit performs a logical AND. One problem with these circuits is that there is a voltage level shift through the circuit. If several circuits are cascaded as shown in figure, the output voltage for each stage is approximately one diode voltage drop further away from the rail. Logic "0" rises for the AND gate and logic "1" drops for the OR gate. Only a limited number of series cascaded circuits can be used before the logic levels must be restored to the rails. Even more serious is the voltage degradation for the cascaded AND-OR function.

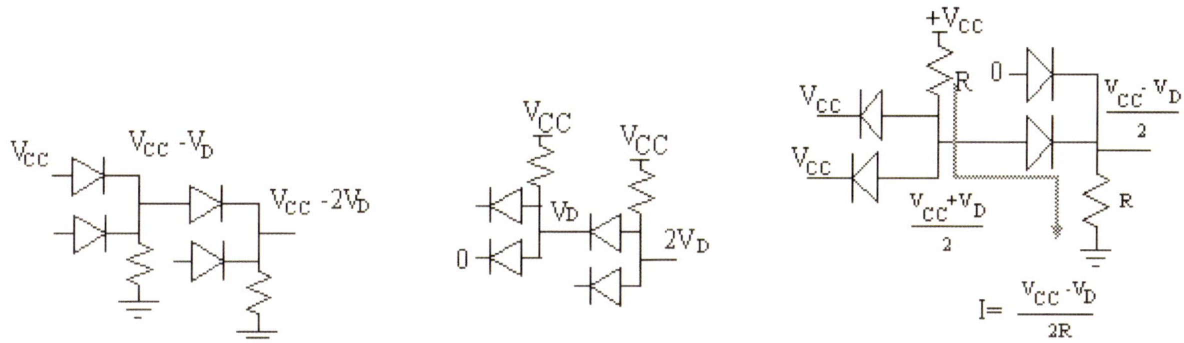

Fig. Cascaded diode logic showing level shifts.

The obvious way to restore the logic level voltages is to use a double transistor inverter. However, it does not take a genius to notice the a single inversion on the end of each gate will make a NAND or a NOR gate and the logic level is automatically restored for each gate. The NAND gate is used extensively. As you no doubt recall, any logic expression can be implemented using just NAND gates. Thus, other forms are not necessary.

Diode Transistor Logic Circuits

A typical DTL NAND gate is shown in. Observe the diode AND function on the front end and the transistor NOT at the output end. The extra resistors and diodes are used to maintain appropriate currents, to maintain proper functioning, and to guarantee certain noise margins. We will completely analyze this circuit, but will be particularly interested in the "terminal" characteristics of the gate. Our ultimate goal is to understand the operation of the gate so that the manufacturer's specifications can be understood.

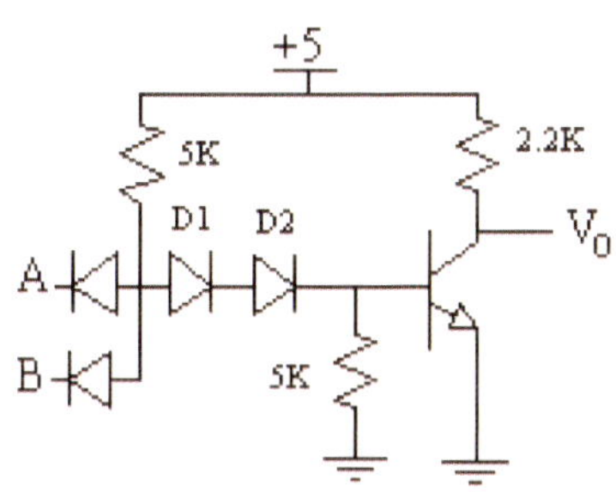

Fig. DTL Circuit

Speed Improvement

The DTL propagation delay is relatively large. When the transistor goes into saturation from all inputs being high, charge is stored in the base region. When it comes out of saturation (one input goes low) this charge has to be removed and will dominate the propagation time.

One way to speed up DTL is to add a small "speed-up" capacitor across R3. The capacitor helps to turn off the transistor by removing the stored base charge; the capacitor also helps to turn on the transistor by increasing the initial base drive.

Another way to speed up DTL is to avoid saturating the switching transistor. That can be done with a Baker clamp. The Baker clamp is named for Richard H. Baker, who described it in his 1956 technical report "Maximum Efficiency Switching Circuits."

A digital clock made only with discrete transistors, diodes and resistors, no integrated circuits. This clock uses 550 switching diodes and 196 transistors to divide 60 Hz power line frequency down to one pulse per second and provide a display of hours, minutes and seconds.

In 1964, James R. Biard filed a patent for the Schottky transistor. In his patent the Schottky diode prevented the transistor from saturating by minimizing the forward bias on the collector-base transistor junction, thus reducing the minority carrier injection to a negligible amount. The diode could also be integrated on the same die, it had a compact layout, it had no minority carrier charge storage, and it was faster than a conventional junction diode. His patent also showed how the Schottky transistor could be used in DTL circuits and improve the switching speed of other saturated logic designs, such as Schottky-TTL, at a low cost.

Transistor-Transistor Logic

Transistor-Transistor Logic, or TTL, refers to the technology for designing and fabricating digital integrated circuits that employ logic gates consisting primarily of bipolar transistors. It overcomes the main problem associated with DTL, i.e., lack of speed.

The input to a TTL circuit is always through the emitter(s) of the input transistor, which exhibits a low input resistance. The base of the input transistor, on the other hand, is connected to the Vcc line, which causes the input transistor to pass a current of about 1.6 mA when the input voltage to the emitter(s) is logic '0', i.e., near ground. Letting a TTL input 'float' (left unconnected) will usually make it go to logic '1', but such a state is vulnerable to stray signals, which is why it is good practice to connect TTL inputs to Vcc using 1 kohm pull-up resistors.

The most basic TTL circuit has a single output transistor configured as an inverter with its emitter grounded and its collector tied to Vcc with a pull-up resistor, and with the output taken from its collector. Most TTL circuits, however, use a totem pole output circuit, which replaces the pull-up resistor with a Vcc-side transistor sitting on top of the GND-side output transistor. The emitter of the Vcc-side transistor (whose collector is tied to Vcc) is connected to the collector of the GND-side transistor (whose emitter is grounded) by a diode. The output is taken from the collector of the GND-side transistor. Figure below shows a basic 2-input TTL NAND gate with a totem-pole output.

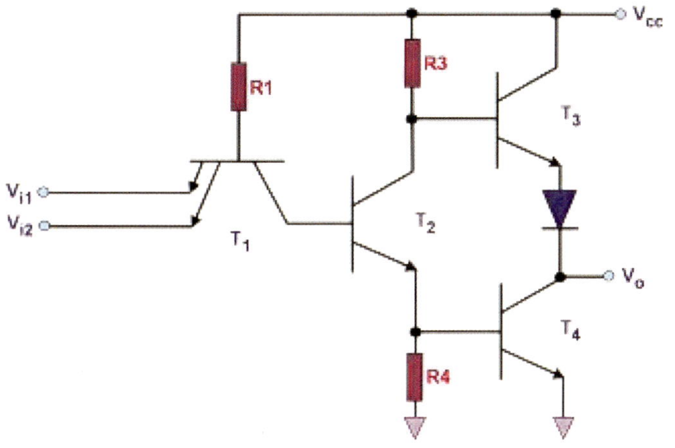

Figure: 2-input TTL NAND Gate with a Totem Pole Output Stage

In the TTL NAND gate of Figure, applying a logic '1' input voltage to both emitter inputs of T1 reverse-biases both base-emitter junctions, causing current to flow through R1 into the base of T2, which is driven into saturation. When T2 starts conducting, the stored base charge of T3 dissipates through the T2 collector, driving T3 into cut-off. On the other hand, current flows into the base of T4, causing it to saturate and pull down the output voltage Vo to logic '0', or near ground. Also, since T3 is in cut-off, no current will flow from Vcc to the output, keeping it at logic '0'. Note that T2 always provides complementary inputs to the bases of T3 and T4, such that T3 and T4 always operate in opposite regions, except during momentary transition between regions.

On the other hand, applying a logic '0' input voltage to at least one emitter input of T1 will forward-bias the corresponding base-emitter junction, causing current to flow out of that emitter. This causes the stored base charge of T2 to discharge through T1, driving T2 into-cut-off. Now that T2 is in cut-off, current from Vcc will be diverted to the base of T3 through R3, causing T3 to saturate. On the other hand, the base of T4 will be deprived of current, causing T to go into cut-off. With T4 in cut-off and T3 in saturation, the output Vo is pulled up to logic '1', or closer to Vcc.

Outputs of different TTL gates that employ the totem-pole configuration must not be connected together since differences in their output logic will cause large currents to flow from the logic '1' output to the logic '0' output, destroying both output stages. The output of a typical TTL gate under normal operation can sink currents of up to 16 mA.

The noise margin of a logic gate for logic level '0', Δ0, is defined as the difference between the maximum input voltage that it will recognize as a '0' (Vil) and the maximum voltage that may be applied to it as a '0' (Vol of the gate driving it). For logic level '1', the noise margin Δ1 is the difference between the minimum input voltage that may be applied to it as a '1' (Voh of the gate driving it) and the minimum input voltage that it will recognize as a '1' (Vih). Mathematically, Δ0 = Vil-Vol and Δ1 = Voh-Vih. Any noise that causes a noise margin to be overcome will result in a ‹0› being erroneously read as a ‹1› or vice versa. In other words, noise margin is a measure of the immunity of a gate from reading an input logic level incorrectly. For TTL, Vil = 0.8V and

Vol = 0.4V, so $\Delta 0$ = 0.4V, and Voh = 2.4V and Vih = 2.0 V, so $\Delta 1$ = 0.4V. These noise margins are not as good as the noise margins exhibited by DTL.

As mentioned earlier, TTL has a much higher speed than DTL. This is due to the fact that when the output transistor is turned off, there is a path for the stored charge in its base to dissipate through, allowing it to reach cut-off faster than a DTL output transistor. At the same time, the equivalent capacitance of the output is charged from Vcc through T3 and the output diode, allowing the output voltage to rise more quickly to logic '1' than in a DTL output wherein the output capacitance is charged through a resistor.

The commercial names of digital IC's that employ TTL start with '74', e.g., 7400, 74244, etc. Most TTL devices nowadays, however, are named '74LSXXX', with the 'LS' standing for 'low power Schottky'. Low power schottky TTL devices employ a Schottly diode, which is used to limit the voltage between the collector and the base of a transistor, making it possible to design TTL gates that use significantly less power to operate while allowing higher switching speeds.

Packaging

Like most integrated circuits of the period 1963–1990, commercial TTL devices are usually packaged in dual in-line packages (DIPs), usually with 14 to 24 pins, for through-hole or socket mounting. The DIPs were usually made of epoxy plastic (PDIP) for commercial-grade parts or of ceramic (CDIP) for military-grade parts.

Beam-lead chip dies without packages were made for assembly into larger arrays as hybrid integrated circuits. Parts for military and aerospace applications were packaged in flatpacks, a form of surface-mount package, with leads suitable for welding or soldering to printed circuit boards. Today, many TTL-compatible devices are available in surface-mount packages, which are available in a wider array of types than through-hole packages.

TTL is particularly well suited to bipolar integrated circuits because additional inputs to a gate merely required additional emitters on a shared base region of the input transistor. If individually packaged transistors were used, the cost of all the transistors would discourage one from using such an input structure. But in an integrated circuit, the additional emitters for extra gate inputs add only a small area.

At least one computer manufacturer, IBM, built its own flip chip integrated circuits with TTL; these chips were mounted on ceramic multi-chip modules.

Comparison with other Logic Families

TTL devices consume substantially more power than equivalent CMOS devices at rest, but power consumption does not increase with clock speed as rapidly as for CMOS devices. Compared to contemporary ECL circuits, TTL uses less power and has easier design rules but is substantially slower. Designers can combine ECL and TTL devices in the same system to achieve best overall performance and economy, but level-shifting devices are required between the two logic families. TTL is less sensitive to damage from electrostatic discharge than early CMOS devices.

Due to the output structure of TTL devices, the output impedance is asymmetrical between the high and low state, making them unsuitable for driving transmission lines. This drawback is usually overcome by buffering the outputs with special line-driver devices where signals need to be sent through cables. ECL, by virtue of its symmetric low-impedance output structure, does not have this drawback.

The TTL "totem-pole" output structure often has a momentary overlap when both the upper and lower transistors are conducting, resulting in a substantial pulse of current drawn from the power supply. These pulses can couple in unexpected ways between multiple integrated circuit packages, resulting in reduced noise margin and lower performance. TTL systems usually have a decoupling capacitor for every one or two IC packages, so that a current pulse from one TTL chip does not momentarily reduce the supply voltage to another.

Several manufacturers now supply CMOS logic equivalents with TTL-compatible input and output levels, usually bearing part numbers similar to the equivalent TTL component and with the same pinouts. For example, the 74HCT00 series provides many drop-in replacements for bipolar 7400 series parts, but uses CMOS technology.

Sub-types

Successive generations of technology produced compatible parts with improved power consumption or switching speed, or both. Although vendors uniformly marketed these various product lines as TTL with Schottky diodes, some of the underlying circuits, such as used in the LS family, could rather be considered DTL.

Variations of and successors to the basic TTL family, which has a typical gate propagation delay of 10ns and a power dissipation of 10 mW per gate, for a power–delay product (PDP) or switching energy of about 100 pJ, include:

- Low-power TTL (L), which traded switching speed (33ns) for a reduction in power consumption (1 mW) (now essentially replaced by CMOS logic)
- High-speed TTL (H), with faster switching than standard TTL (6ns) but significantly higher power dissipation (22 mW)
- Schottky TTL (S), introduced in 1969, which used Schottky diode clamps at gate inputs to prevent charge storage and improve switching time. These gates operated more quickly (3ns) but had higher power dissipation (19 mW)
- Low-power Schottky TTL (LS) – used the higher resistance values of low-power TTL and the Schottky diodes to provide a good combination of speed (9.5ns) and reduced power consumption (2 mW), and PDP of about 20 pJ. Probably the most common type of TTL, these were used as glue logic in microcomputers, essentially replacing the former H, L, and S sub-families.
- Fast (F) and Advanced-Schottky (AS) variants of LS from Fairchild and TI, respectively, circa 1985, with "Miller-killer" circuits to speed up the low-to-high transition. These families achieved PDPs of 10 pJ and 4 pJ, respectively, the lowest of all the TTL families.
- Low-voltage TTL (LVTTL) for 3.3-volt power supplies and memory interfacing.

Most manufacturers offer commercial and extended temperature ranges: for example Texas Instruments 7400 series parts are rated from 0 to 70 °C, and 5400 series devices over the military-specification temperature range of −55 to +125 °C.

Special quality levels and high-reliability parts are available for military and aerospace applications.

Radiation-hardened devices (for example from the SNJ54 series) are offered for space applications.

Applications

Before the advent of VLSI devices, TTL integrated circuits were a standard method of construction for the processors of mini-computer and mainframe processors; such as the DEC VAX and Data General Eclipse, and for equipment such as machine tool numerical controls, printers and video display terminals. As microprocessors became more functional, TTL devices became important for "glue logic" applications, such as fast bus drivers on a motherboard, which tie together the function blocks realized in VLSI elements.

Direct-coupled Transistor Logic

DIRECT-COUPLED transistor logic circuitry (DCTL), first announced in 1955, has provided many desirable features for digital computer use. Among the principal characteristics of DCTL are extremely simple circuitry, a minimum number of auxiliary components, very low dissipation and power-supply requirements, and the possibility of extraordinary compactness. Various special-purpose machines using DCTL have been built, and at least one class of general-purpose computer has been designed and is in operation.

Circuit Description

DCTL circuitry takes advantage of the fact that certain types of transistors have useful current gains and impedance ratios with the collector junction slightly forward biased (in saturation), whereas the usual mode of operation calls for a reverse bias on the collector. DCTL operation is unique to transistors; a vacuum-tube equivalent would require useful operation with the control grid more positive than the plate.

Given the possibility of a transistor running with zero or slightly forward collector bias, the basic inverter circuit shown in figure becomes possible. This figure shows three cascaded commonemitter inverters, using p-n-p transistors. If the first stage is conducting (in saturation) , its col1ector-to-emitter voltage is sufficiently low that the second stage is held off (nonconducting). With the second transistor not conducting, a sufficiently large current is drawn from the base of the third stage, through the secondstage load resistor, to put the third stage in saturation.

In order to appreciate the possibility of this modus operandi, consider the transistor characteristic curves shown in Fig. These curves are representative of most "wafer" transistors, including alloy-junction. surface-barrier, and similar types, which have negligible series resistance in the emitter and co11ector leads. On the left of Fig. is shown a set of col1ector characteristics for the commonemitter connection. These plots of collector current versus collector-to-emitter voltage, with base current the fixed input parameter, show that the transistor maintains good current gain and output impedance with collector voltages of 0.15 volt or less; such curves are representative of both silicon and germanium wafer transistors. The common-emitter input characteristics, shown on the right, indicate that the threshold of significant conduction occurs only with forward biases (negative for p-n-p transistors) greater than 0.1 volt for germanium or 0.6 volt for silicon. Thus it is obvious that the collector-to-emitter voltage of a conducting transistor can be maintained at a lower value than the threshold required for conduction at the input of a succeeding transistor.

Given a DCTL inverter, it is simple to make a flip-flop by connecting two inverters together, as shown in Fig. The mechanism by which the gates affect the flip-flop is as follows: When a gate conducts, its collector-to-emitter voltage becomes low, thus turning off the flip-flop transistor whose base is connected to that gate, in turn causing the other flip-flop transistor (the one whose collector is tied to the gate in conduction) to conduct. Quite obviously, more than one gating circuit could be connected in parallel to either side of the flip-flop, so that the activation of anyone or more of the gates on one side would set the flip-flop.

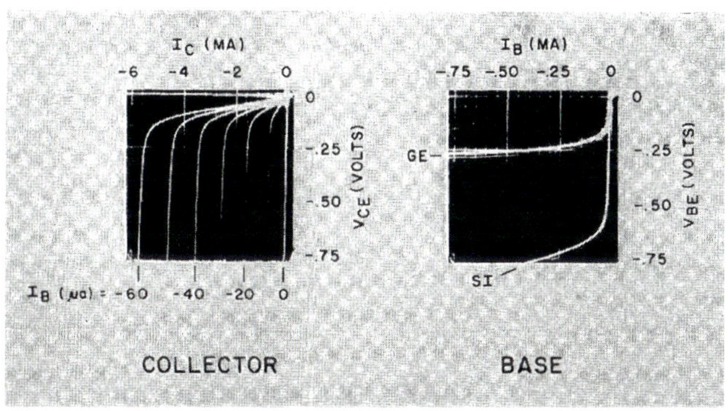

Fig. Transistor common-emitter characteristics

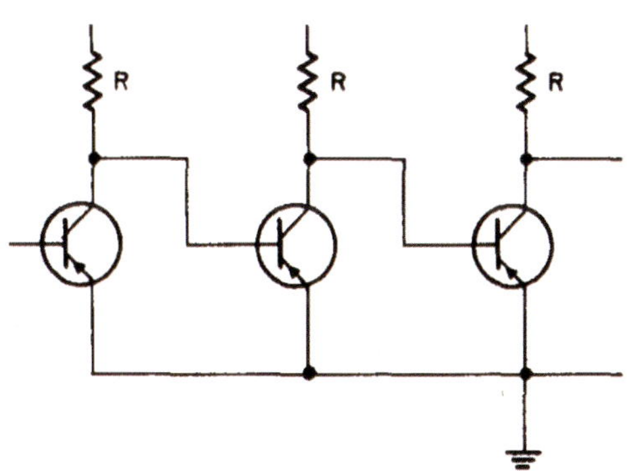

DCTL inverters

Gating functions can be achieved with transistors either connected in parallel or stacked in series. Examples of parallel and series gates are shown in figures above. and. If the transistors are normally in the nonconducting state, the parallel gate could be called an "or" gate, since the output circuit conducts if one or more of the inputs are activated; similarly, the series gate is an "and" gate, since its output conducts only if all the inputs are simultaneously activated. An example of the simultaneous use of both gates to form a half-adder is shown in Fig. The complementary inputs are normally derived from the two sides of flip-flops storing the appropriate input characters. It should be noted that an output signal in the sum or carry channel is the complement of the usually desired signal, in that a "one" corresponds to the low-impedance (conducting), or low-voltage condition. Therefore, it may be necessary in some cases to use an inverter to obtain the appropriate sense for the output signal.

It can be seen that, in general, fewer resistors than transistors are employed in these circuits, and that there is relatively little need for capacitors. One circuit in which a capacitor is required is the oneshot multivibrator, shown in Fig. This circuit is used to generate delays for timing signals, and is triggered by the gating transistor shown in the figure, as suggested by the waveforms sketched there.

Typical practice shows that an over-all -system uses roughly 400 resistors per 1,000 transistors and less than 10 capacitors per 1,000 transistors.

Diode Logic

Diode Logic makes use of the fact that the electronic device known as a *diode* will conduct an electrical current in one direction, but not in the other. In this manner, the diode acts as an electronic switch.

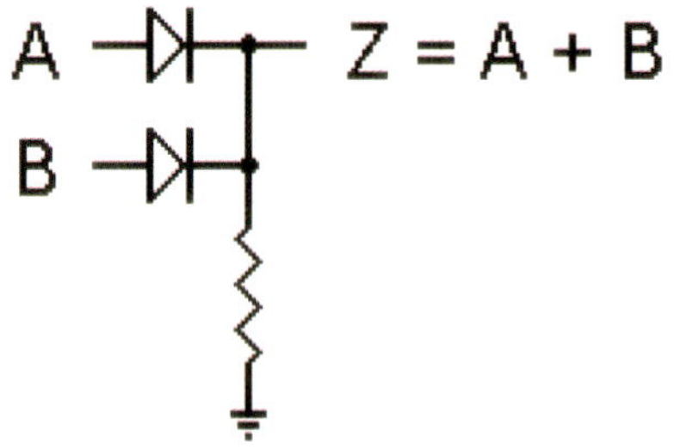

To the left a basic Diode Logic OR gate. We'll assume that a logic 1 is represented by +5 volts, and a logic 0 is represented by ground, or zero volts. In this figure, if both inputs are left unconnected or are both at logic 0, output Z will also be held at zero volts by the resistor, and will thus be a logic 0 as well. However, if either input is raised to +5 volts, its diode will become forward biased and will therefore conduct. This in turn will force the output up to logic 1. If both inputs are logic 1, the output will still be logic 1. Hence, this gate correctly performs a logical OR function.

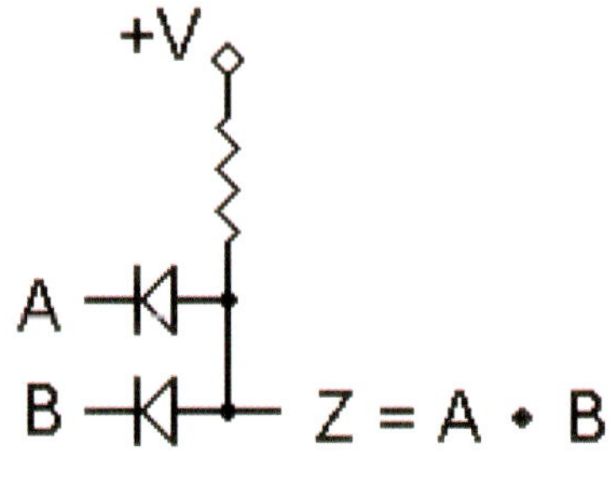

To the right is the equivalent AND gate. We use the same logic levels, but the diodes are reversed and the resistor is set to pull the output voltage up to a logic 1 state. For this example, +V = +5 volts, although other voltages can just as easily be used. Now, if both inputs are unconnected or if they are both at logic 1, output Z will be at logic 1. If either input is grounded (logic 0), that diode will conduct and will pull the output down to logic 0 as well. Both inputs must be logic 1 in order for the output to be logic 1, so this circuit performs the logical AND function.

In both of these gates, we have made the assumption that the diodes do not introduce any errors or losses into the circuit. This is not really the case; a silicon diode will experience a forward voltage drop of about 0.65v to 0.7v while conducting. But we can get around this very nicely by specifying that any voltage above +3.5 volts shall be logic 1, and any voltage below +1.5 volts shall be logic 0. It is illegal in this system for an output voltage to be between +1.5 and +3.5 volts; this is the undefined voltage region.

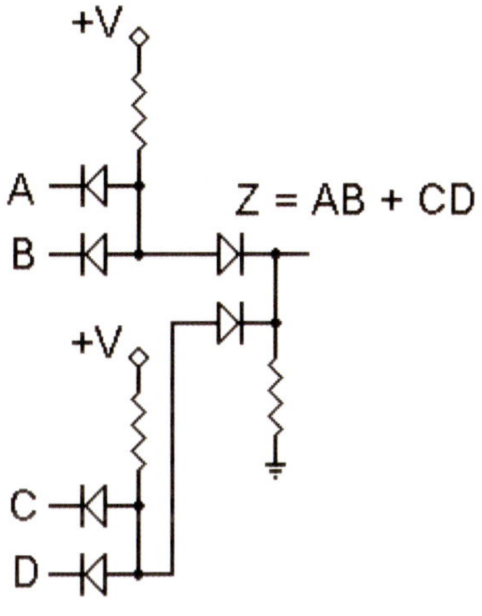

Individual gates like the two above can be used to advantage in specific circumstances. However, when DL gates are cascaded, as shown to the left, some additional problems occur. Here, we have two AND gates, whose outputs are connected to the inputs of an OR gate. Very simple and apparently reasonable.

If we pull the inputs down to logic 0, sure enough the output will be held at logic 0. However, if both inputs of either AND gate are at +5 volts, what will the output voltage be? That diode in the OR gate will immediately be forward biased, and current will flow through the AND gate resistor, through the diode, and through the OR gate resistor.

If we assume that all resistors are of equal value (typically, they are), they will act as a voltage divider and equally share the +5 volt supply voltage. The OR gate diode will insert its small loss into the system, and the output voltage will be about 2.1 to 2.2 volts. If both AND gates have logic 1 inputs, the output voltage can rise to about 2.8 to 2.9 volts. Clearly, this is in the "forbidden zone," which is not supposed to be permitted.

If we go one step further and connect the outputs of two or more of these structures to another AND gate, we will have lost all control over the output voltage; there will always be a reverse-biased diode somewhere blocking the input signals and preventing the circuit from operating correctly. This is why Diode Logic is used only for single gates, and only in specific circumstances.

Negative Logic

The assignment of 1 and 0 to the positive and negative signal levels respectively is an option of the logic designer using the AND or OR circuits. With this assignment it assumes that the logic is positive. It is just as likely that the assignment might be the reversed where 1 is the negative voltage and 0 is the positive voltage. This would be negative logic. Switching between positive and negative logic is commonly used to achieve a more efficient logic design.

In Boolean algebra it is recognized that a positive logic OR is a negative logic AND. Similarly a positive logic AND is a negative logic OR.

This relationship can easily be recognized by reading the above description of their operation. In the OR it stated, "Only if all inputs, A and B and C are 0 will the output be 0." In negative logic each node at the lower voltage would become a logic 1, making the statement, "Only if all inputs, A AND B AND C are 1 will the output be 1." That is the definition of an AND function.

Similarly for the AND it was stated, "If input A or B or C is 0 the output will be 0." In negative logic each node at the lower voltage would become a logic 1, making the statement, "If input A OR B OR C is 1 the output will be 1." That is the definition of an OR function.

The logical function of any arrangement of diodes can only be established if the representation of logic states by voltage levels is known.

Diode Logic with Real Diodes

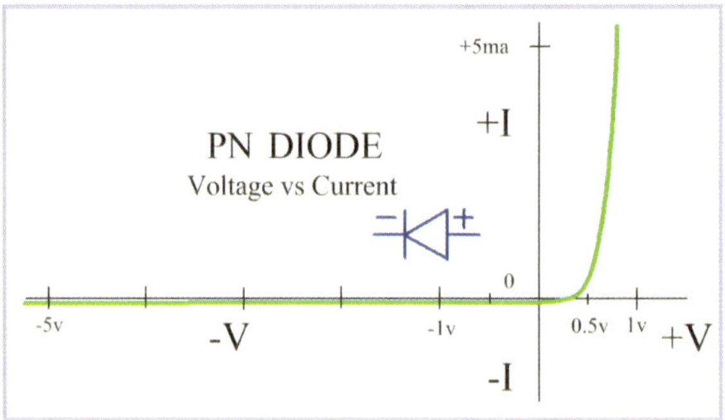

Diode approximation of voltage vs. current

The above descriptions assumed an ideal diode with zero resistance in the forward direction and infinite resistance in the reverse direction. The circuit designer must concern himself with real diodes. The articles p-n diode and a less detailed article p-n junction describe the physics of the PN diode. After all the discussion of electrons, holes, majority and minority carriers etc. each come down to an equation that most directly relates to the circuit designer. The real PN diode actually has a voltage current characteristic similar to the curve on the right. A more specific definition can be found in the Shockley diode equation. The designer of a reliable diode logic circuit is usually limited to what the diode specification provides which is often less than the equation suggests. Typically the specification will primarily provide a maximum forward voltage drop at one or more forward currents and a reverse leakage current. It will also provide a maximum reverse voltage limited by zener or avalanche breakdown. Typical worst case specifications are shown below for both germanium and silicon PN diodes.

Germanium diode:

Max forward voltage at 10 ma = 1 volt @ 0 to 85 °C

Max reverse leakage current at 15 volts = 100 microamps @ 85 °C

Silicon diode:

Max forward voltage at 10 ma = 1 volt @ 0 to 125 °C

Max reverse leakage current at 15 volts = 1 microamps @ 85 °C

Effects of component manufacturing variations and temperature are usually included in these specifications.

More realistically the germanium forward voltage might be 0.25 to 0.4 volts but this is often not specified. The silicon leakage current might be much lower possibly 1 to 100 nanoamps.

PN diodes also have transient behaviors that might be of concern with the design. The capacitance of a PN diode between anode and cathode is inversely proportional to the reverse voltage, growing as it approaches zero volts and into forward bias. There is also a recovery concern where the current will not decrease immediately when it is switched from forward bias to reverse bias. In the case of the diode OR if two or more of the inputs are at the 1 level and one switches to 0 it will cause a glitch or increase in current in the diodes that remain at 1. This can cause a short term dip in the output voltage. In practice if the diode logic gate drives a transistor inverter, as it usually does, and the diode and transistor are of similar construction the transistor will have a similar base collector capacitance that is amplified by the transistor gain so that it will be too slow to pass the glitch. Only when the diode is of a much slower construction will it become any concern at all. In one unusual design small selenium diode discs were used with germanium transistors. The recovery time of the very slow selenium diodes caused a glitch on the inverter output. It was fixed by placing a selenium diode across the base emitter junction of the transistor making it "think" it was a selenium transistor (if there could ever be one).

Properties

Nonrestoring Logic

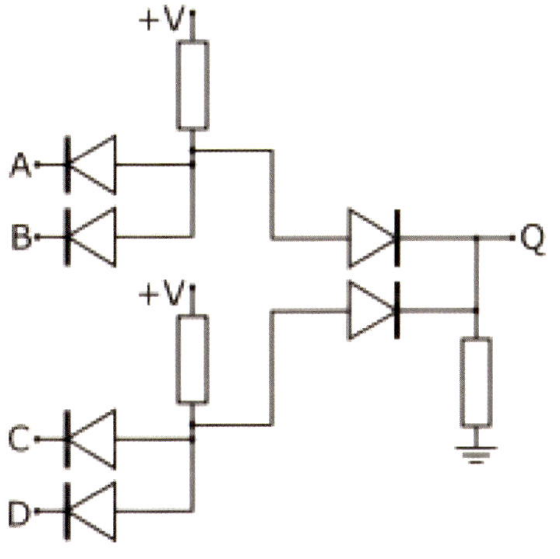

In cascaded AND-OR diode gates, the high voltage level is decreased more than two times.

Digital logic implemented by active elements is characterized by signal restoration. *True* and *false* or *1* and *0* are represented by two specific voltage levels. If the inputs to a digital logic gate is close to their respective levels, the output will be closer or exactly equal to its desired level. Active logic gates may be integrated in large numbers because each gate tends to remove noise at its input. Diode logic gates are implemented by passive elements; so, they have two restoration problems.

Forward voltage drop: The first restoration problem of diode logic is that there is a voltage drop V_F about 0.6 V across the forward-biased diode. This voltage is added to or subtracted from the input of every gate so that it accumulates when identical diode gates are cascaded. In an OR gate, V_F decreases the high voltage level (the logical *1*) while in an AND gate, it increases the low voltage level (the logical *0*). The feasible number of logic stages thus depends on the difference between the high and low voltages.

Source resistance: Another problem of diode logic is the internal resistance of the input voltage sources. Together with the gate resistor, it constitutes a voltage divider that worsens the voltage levels. In an OR gate, the source resistance decreases the high voltage level (the logical *1*) while in an AND gate, it increases the low voltage level (the logical *0*). In the cascaded AND-OR diode gates in the picture on the right, the AND high output voltages are decreased because of the internal voltage drops across the AND pull-up resistances.

Applications

Diode logic gates are used to build diode–transistor logic (DTL) gates as integrated circuits.

The outputs of conventional ICs (with complementary output drive stages) are never directly connected together since they act as voltage sources. However, diodes can be used to combine two or more digital (high/low) outputs from an IC such as a counter. This wired logic connection can be a useful way of producing simple logic functions without using additional logic gates.

Most circuit families are designed to have compatible inputs and outputs depending on these signal levels to achieve reliable performance. Adding diode logic will degrade the signal level and result in poor noise rejection and possible failure.

Programmable Logic Device

A PLD, or programmable logic device, is an electronic component that is used in order to build digital circuits that are reprogrammable. A programmable logic device does not have a defined function once manufactured, unlike a logic gate and has to be programmed before it can be used.

Because of the advantage of re-programmability, they have replaced special purpose logic devices like Logic gates, flip-flops, counters and multiplexers in many semicustom applications. It reduces design time and thus reduces time for the product to reach the market. It consists of arrays of AND and OR gates, which can be programmed to realize required logic function.

Device programmer blows fuses on the PLD to control each gate operation. Inexpensive software tools are used for quick development, simulation and testing, therefore design cost is comparatively low. Another important advantage is that customer can modify their design, based on changes in requirement.

The most commonly used Programmable Logic Devices are listed here

Programmable Read Only Memory

Programmable read only memory is a memory chip, on which data can be written only once. Once a data is written onto a PROM, it remains there forever. Therefore, it is also called as a one-time programmable memory. Memory chip is delivered blank and the programmer transfers the data on to it. A blank PROM consists of many fuses which can be selectively burned out during the first programming. They are used in applications like, computer bios where reprogramming is not required.

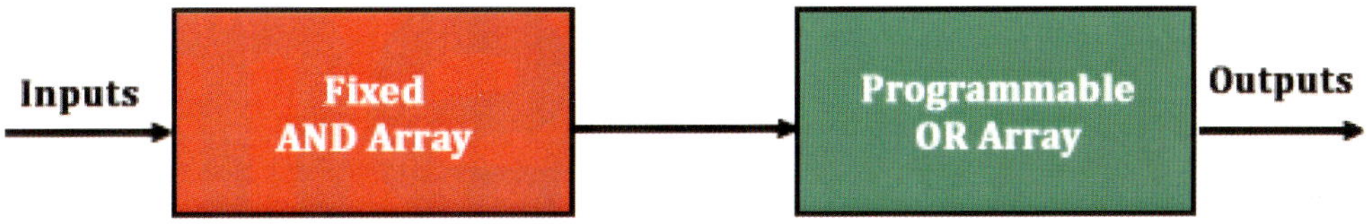

The above figure shows the block diagram of a PROM. It consists of a fixed AND gate array followed by a Programmable OR gate array. AND gate array is used as the address decoder which selects the corresponding address location based on the input address provided to it. Data is stored in the OR gate array. It contains programmable fuses, which can be burned off depending on the data values that are to be stored.

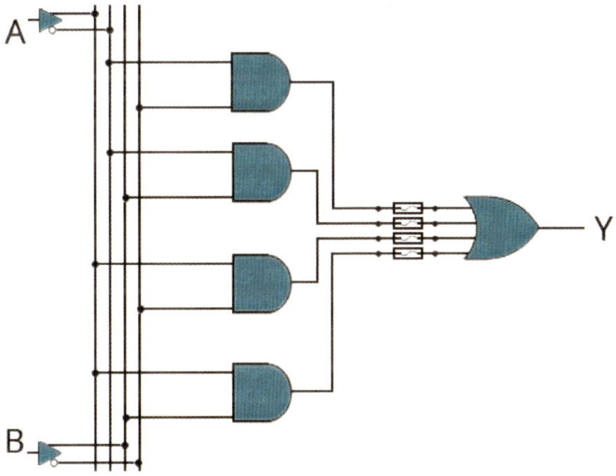

Internal structure of each block in the PROM is shown in the above figure. Here, A and B are the address inputs and Y is the data output. AND arrays are fixed to select each row (address location) for corresponding inputs. As shown in the figure, data in each memory location is determined by the fuses in the OR array. If the fuse is not burned off, charge send through the row is received at the output, indicating a logic one and when the fuse is burned off, the signal cannot reach the data output and is detected as logic zero. Thus, in PROM binary data is stored using fuses.

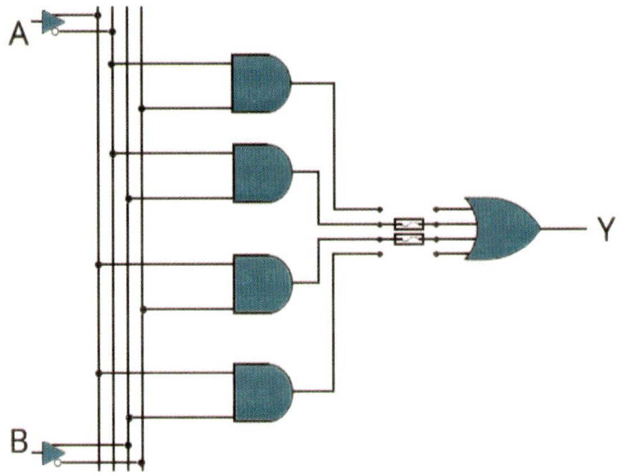

For example in this PROM, single bit data can be stored in each memory location (fuses). Since there are four address locations to store data, two bit address inputs are required to select these locations. Working of a PROM can be better understood if it is explained with an example. Consider address input to be "00". As shown in the above figure, the inverted values of both address inputs are given to the first AND gate. That is, both inputs of first AND is now at logic one. This means that, first address location is selected. Now, data in the first address location is determined by the presence or absence of fuses in the OR array. Like in above figure, if fuse at this location is burned off, then, output data is '0'. Similarly in this example, logic one is stored in second and third location and logic zero in the fourth location.

Programmable Read Only Memory as Logic Device

Programmable Read Only Memory can also be used as a logic device. A logic device performs an operation on one or more inputs, and produces a single output. Output is constant for respective input combinations. Therefore, if the device behavior is stored, the PROM can work like a logic device.

For this, address pins of the PROM are used as the inputs of the logic device and data out is same as PROM output data. Outputs of respective inputs are stored in the corresponding memory locations. For example an XOR gate can be implemented by storing its output values in the respective address location.

Address	Data
00	0
01	1
10	1
11	0

We know that, output of a two input XOR gate is logic one, if exactly one of its input is at logic one state. The table shows a two bit PROM with four address locations ('00', '01', '10', '11') in which data '0', '1', '1', and '0' are stored. Therefore, if input is '00' (address input for PROM) the data ('0') stored in the memory location '00' is fetched and outputted. Similarly, we get logic one at the output for the input combinations '01' and '10' and logic zero for '11'. Thus device works like a two input XOR gate. Any other two input logic device can be implemented using this PROM by changing the data stored in the memory. Now refer the figure of the example that was discussed earlier.

If there are m address bits, then 2^m locations can be addressed with that. This 2^m address locations can store 2^(2^m) different values, therefore 2^(2^m) logic function can be implemented using an m bit PROM. In our example, we used a simple PROM with two address bits. That means, 2 ^ 2 data can be stored in it. Each data combination is a particular logic function.

Programmable Array Logic (PAL)

PAL or Programmable Array Logic is used to implement logic functions in digital circuits. Structure of a PAL can be divided into two parts, AND and OR array. In this, AND array is programmable that means connections to the inputs of each AND gate is through fuses. Therefore, when a particular input is not required to implement, using a specific logic function, it can be burned off. AND array is followed by a fixed OR array. It is used to sum off outputs from all AND gates. Input connections are fixed in the OR gate array, therefore, no changes can be made in the topic of PAL device.

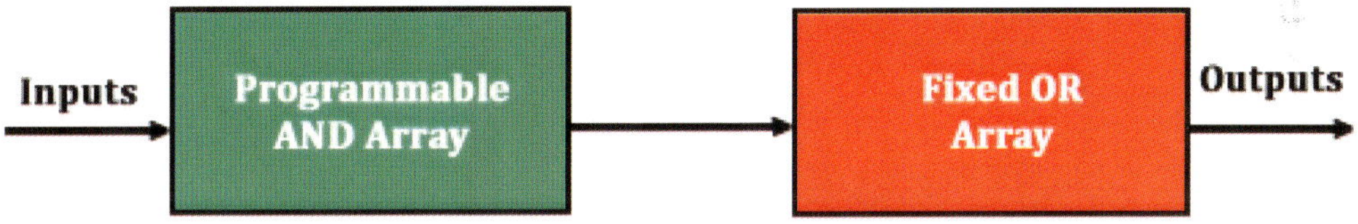

Figure: Block Diagram of Programmable Array Logic

Internal structure of a two input PAL logic.

Figure. shows the simplest Programmable Array Logic device with only two inputs. Inverted and non-inverted inputs (four inputs) are connected to the AND through fuses. Output from the AND is then directly connected to the OR gate. Any two input logic function can be realized using this device by burning off the unwanted fuses at the input of each AND gates.

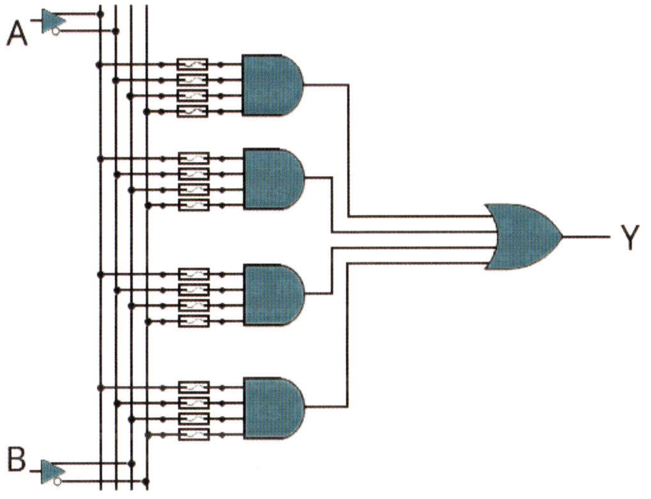

Fig: Programmable Array Logic

For example consider an XOR gate. If the inputs are labeled as 'A' and 'B' then the output can be given by

$$Y = AB' + A'B$$

To implement this, we only need inputs A and B' in the first (top) AND gate and A' and B inputs in the second (bottom) AND gate. Rest two inputs in each AND gate is not needed, therefore, respective fuses are burned off. Two input XOR gates using Programmable Array Logic is shown below.

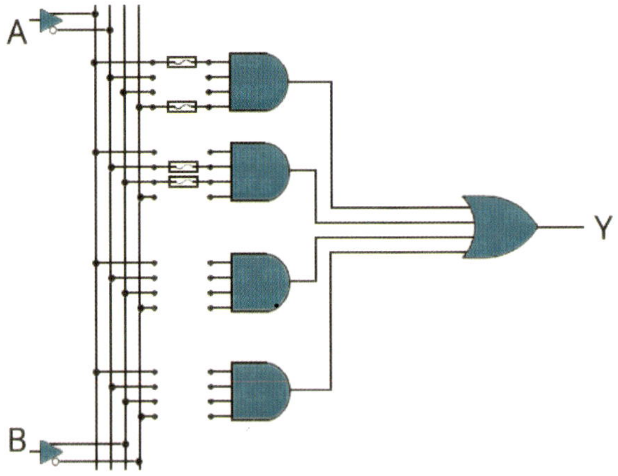

Fig: XOR gate Implementation using Programmable Array Logic

Programmable Logic Array (PLA)

Programmable Logic Array or PLA is used to implement logic functions in digital circuits. The structure has programmable AND-matrix, programmable OR-matrix, input and output buffers. Block diagram of a PLA device is as shown below.

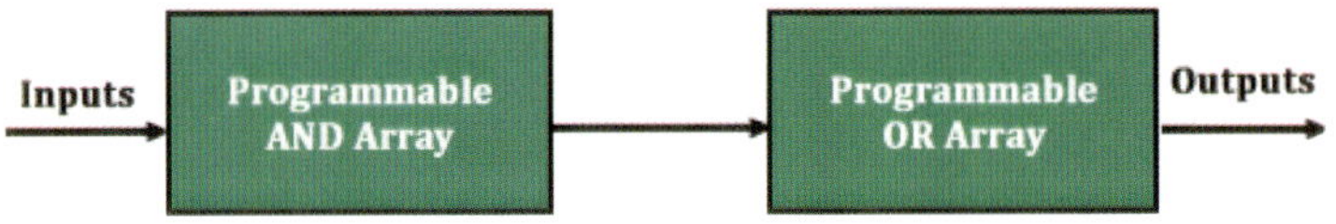

Both inverted and original values of each PLA inputs are provided by input buffers. Input to both AND and OR gates are through fuse and therefore they can be burned off depending on our requirements. Structure of a PLA with all fuses (before programming) is shown below.

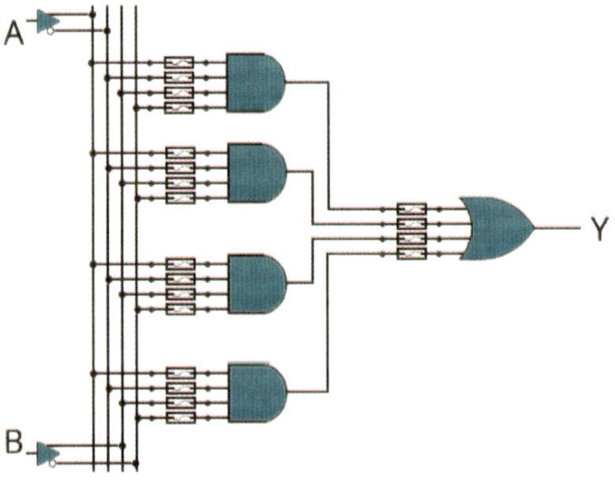

Two input PLA structure can be used to realize any two input logic gates. For that, fuses which are not required to realize that particular logic function are burned off. For example, XOR gate realized using a programmable Logic device is shown below.

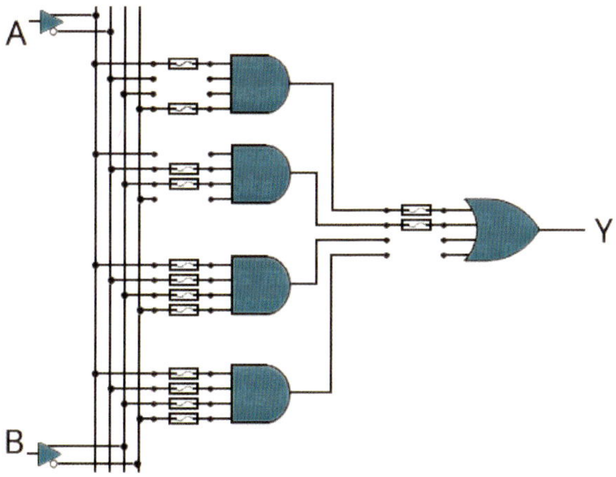

Characteristic equation of a XOR gate contains of two min terms. Given by

$$Y = AB' + A'B$$

Each AND is used to generate a particular min term and required min terms can be selected using the fuses in the input of OR gate.

Generic Array Logic (GAL)

A GAL or Generic Array Logic device consists of a re-programmable *PAL* matrix and a programmable output-cell. GAL is an improved form of PAL which uses electrically erasable CMOS cells instead of fuses. Therefore, AND matrix of GAL can be re-programmed several times unlike one time programmable PAL devices. AND matrix is followed by fixed OR matrix (inside output cell), used to sum off all min terms from the AND output. Block diagram of a GAL device is shown here.

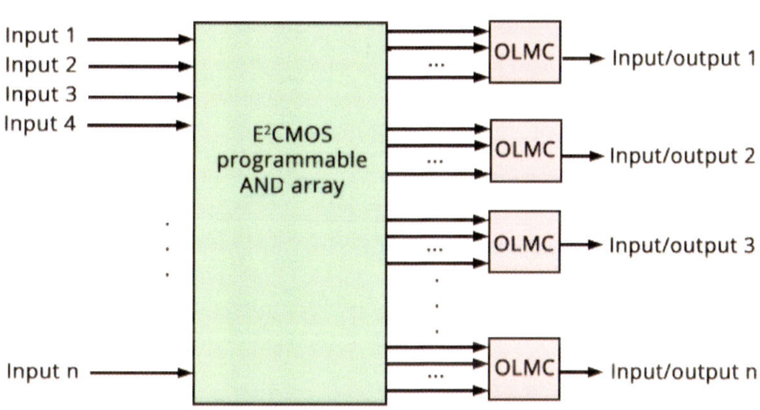

Block of Generic Array Logic

Output Logic Macrocell

Another added feature of GAL is that it also has reprogrammable output logic called OLMC (Output Logic Macrocell). Internal sturcture of an output cell is shown below.

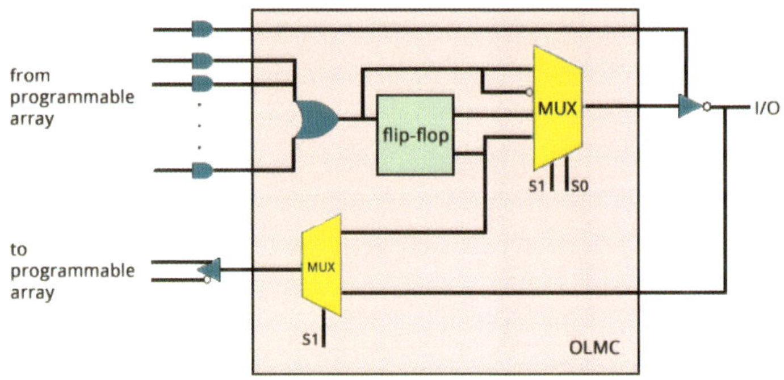

As shown in the figure three main components of an output cell are:

(a) N-input OR,

(b) D-flip-flop,

(c) Multiplexers.

Like in PAL, OR gates are used to sum off min terms from the output of the AND gates. An OLMC cell consists of a D-flip-flop, which is used to implement sequential circuits. Multiplexers in the OLMC cells are used to select the routing of the input signals to the external output or to the feedback output. It is also used to select from the sequential and non-sequential output taken from the input and output of the D-flip-flop depending on the requirement.

Complex Programmable Logic Device (CPLD)

CPLD is defined as the network of PLDs that are connected together through a switching matrix. General block diagram of a CPLD is shown here. The global interconnection matrix, as shown in the figure, is reconfigurable and so we can change the connections between the Functional Blocks depending on our requirement.

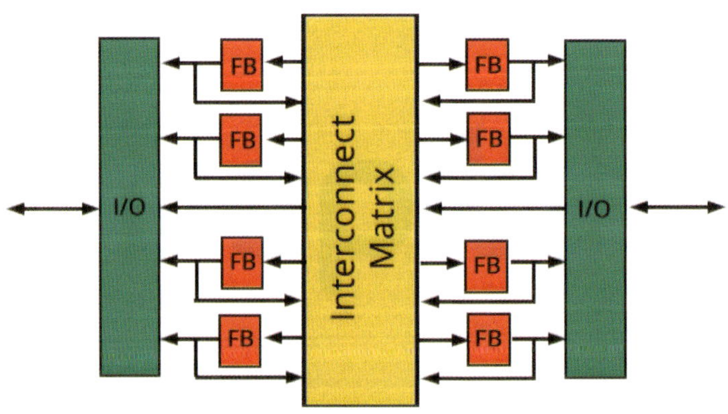

Blocks of Complex Programmable Logic Device

Each Functional Block (FB) in the CPLD contains a re-programmable AND/OR array along with a bank of macro-cells. Therefore, multiple types of logic functions, both combinational and sequential circuits, can be implemented using CPLD. As shown in the figure, it is connected to the external world through the I/O blocks. The entire device contains thousands to tens of thousands of logic gates. Therefore, more complex designs, other than PLD devices, can be implemented using CPLD.

Field-programmable Gate Array (FPGA)

FPGA or Field Programmable Gate Array is now used in the mainstream of modern IC verification. Circuit design is done in Hardware Description Language, which is then synthesized to bit streams before burning into FPGA core. It can be used to realize simple digital logic gates to complex mathematical equations.

Logic blocks in FPGA architecture are arranged in two dimensional arrays. Hierarchy of reconfigurable interconnects is then programmed to implement complex circuits. Desired logic function is implemented using this logic block, which is then connected together using programmable switch boxes. Figure shows, architecture of FPGA containing array of Logic blocks, interconnects, Switch blocks and I/O blocks.

Complex designs are first divided into small functions. Logic blocks are used to implement these sub functions and connections are made using programmable interconnects. The figure shows the architecture of programmed FPGA. Required sub functions are implemented using each logic blocks, which is then programmed and interconnected using switch boxes.

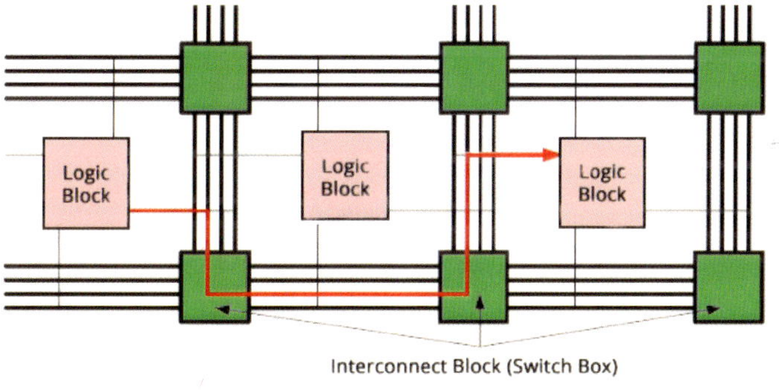

Logic Block

Logic Blocks in the FPGA are used to implement sub functions. Any type of logic function (both combinational and sequential) circuits, can be implemented using a logic block. Therefore, logic blocks are commonly referred to as configurable logic blocks (CLBs). A basic Logic block contains

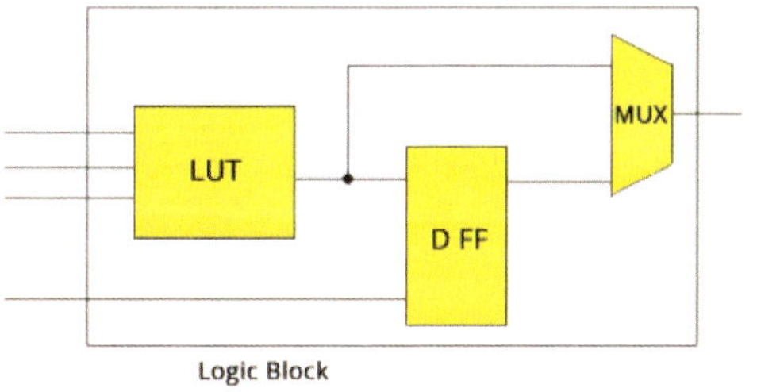

Logic Block

- Lookup table (LUT): to implement the combinational logic functions.
- Register (D flip flop): to store the output from the Lookup Table.
- Multiplexer: to select the output from the LUT.

A simple block diagram of a logic block consists of a lookup table, register and a multiplexer as shown in the figure. SRAM is used to implement lookup table. Therefore, desired logic function can be implemented by varying the data stored in the SRAM. Output from the lookup table is given as inputs to both multiplexer and D flip flop. D flip flop is used to delay the output. And depending on the application, the multiplexer selects LUT output or delayed output. Therefore, by using select input of the multiplexer we can implement both combinational and sequential circuits using logic blocks. Many such logic blocks are configured and finally interconnected using the switch box to build the desired complex circuits.

Compared to other logic devices FPGA has very high logic density. Which means, a single FPGA chip contains ten thousand to eight million gates. Therefore, more complex logic circuits can be implemented using FPGA. FPGA undergo concurrent processing which is faster and more efficient than other pipeline architectures.

After manufacturing, customer configures desired circuit in to FPGA. The main advantage of FPGA is its ability to reprogram. Therefore, it is mostly preferred during the design phase where continuous changes in the requirements can occur. Whereas, Custom ICs are expensive, not programmable and takes long time to design. Disadvantages of FPGA are that, they are slow and they draw more power. The configuration of FPGA is stored in the RAM (volatile), so once they lose power its configuration is lost. Therefore in practical applications, configurations are externally stored in non-volatile flash memories and from there data is automatically restored after retaining the power.

Applications for Programmable Logic Devices

There are many applications that use programmable logic devices including:

- Networking
- Aerospace and defence
- Automotive
- Consumer electronics
- Computing
- Distributed monetary systems
- Communications
- Audio
- Computing
- Avionics

References

- Tinder, Richard F. (2000). Engineering digital design: Revised Second Edition. pp. 317–319. ISBN 0-12-691295-5. Retrieved 2008-07-04
- Rymaszewski, E. J.; Walsh, J. L.; Leehan, G. W. (1981), "Semiconductor Logic Technology in IBM", IBM Journal of Research and Development, 25 (5): 603–616, doi:10.1147/rd.255.0603
- Basic-logic-gates-with-truth-tables: elprocus.com, Retrieved 19 May 2018
- IBM (1960). Customer Engineering Manual of Instruction: Transistor Component Circuits (PDF). IBM. Form 223-688 (5M-11R-156). Retrieved 2012-04-24
- Digital-logic-not-gate: electronicshub.org, Retrieved 29 June 2018
- Nylan, Michael (2001). The Five "Confucian" Classics. Yale University Press. pp. 204–206. ISBN 978-0-300-08185-5. Retrieved 8 June 2010
- Universal-gates-nand-gate: electronicshub.org, Retrieved 06 July 2018
- Pittler, M. S.; Powers, D. M.; Schnabel, D. L. (1982), "System development and technology aspects of the IBM 3081 Processor Complex" (PDF), IBM Journal of Research and Development, 26 (1): 2–11, doi:10.1147/rd.261.0002, p. 5
- Aoi-and-or-inverter-cell: vlsi-expert.com, Retrieved 26 June 2018
- Ayers, J. UConn EE 215 notes for lecture 4. Harvard University faculty web page. Archive of web page from University of Connecticut. n.d. Retrieved 17 September 2008
- Hans Kleine Büning; Theodor Lettmann (1999). Propositional logic: deduction and algorithms. Cambridge University Press. p. 2. ISBN 978-0-521-63017-7

3

Number Systems

In order to completely understand the use of number systems in digital electronics, it is vital to understand the fundamentals of binary code, Boolean algebra, bitwise operation and binary-coded decimal. The following chapter discusses in detail about the different number systems and in particular, about the binary number system and its use in digital electronic.

Any symbol can be represented as a number and the set of symbols and digits to represent information in a code system can be expressed as a set of mathematical numbers. Every number in a code system or encoding scheme refers to a unique symbol or digit and can be expressed in binary form i.e. in the form of bits or bytes.

There are many number systems in use, commonly decimal, octal and hexadecimal. The numbers in any number system can be converted to binary system and so represented in binary form, and so in digital electronic form.

Most number systems follow a common pattern for writing down the value of a number.

A fixed number of values can be written with a single numerical character, then a new column is used to count how many times the highest value in the counting system has been reached. The number of numerical values the system uses is called the base of the system. For example, the decimal system has 10 numerical characters and so has a base of 10:

0 1 2 3 4 5 6 7 8 9

For writing numbers greater than 9 a second column is added to the left, and this column has 10 times the value of the column immediately to its right.

Because number systems commonly used in digital electronics have different base values to the decimal system, they look less familiar, but work in essentially the same way.

Decimal, (Base 10)

Decimal has ten values 0 to 9. If larger values than 9 are needed, extra columns are added to the left. Each column value is ten times the value of the column to its right. For example the decimal value twenty-two is written 22 (2 tens + 2 ones).

Binary, (Base 2)

Binary has only two values 0 and 1. If larger values than 1 are needed, extra columns are added to the left. Each column value is now twice the value of the column to its right. For example the decimal value three is written 11 in binary (1 two + 1 one).

Octal, (Base 8)

Octal has eight values 0 to 7. If larger values than 7 are needed, extra columns are added to the left. Each column value is now 8 times the value of the column to its right. For example the decimal value twenty-seven is written 33 in octal (3 eights + 3 ones).

Hexadecimal, (Base 16)

Hexadecimal has sixteen values 0 to 15, but to keep all these values in a single column, the 16 values (0 to 15) are written as 0 to F, using the letters A to F to represent numbers 10 to 15, so avoiding the use of a second column. Again, if higher values than 15 (F in hexadecimal) are needed, extra columns to the left are used. Each column value is sixteen times that of the column to its right. For example the decimal value sixty-eight is written as 44 in hexadecimal (4 sixteens + 4 ones).

Some column values of different number systems				
Decimal	1000	100	10	1
Binary	8	4	2	1
Octal	512	64	8	1
Hexadecimal	4096	256	16	1

Each of these different number systems works in the same way, it is just that each system has a different base, and the column values in each system increase by multiples of the base number as columns are added to the left.

Because this topic describes several different number systems, it is important to know which system is being described. Therefore if there is some doubt which system a number is in, the base of the system, written as a subscript immediately after the value, is used to identify the number system.

For example:

10_{10} represents the decimal value ten. (1 ten + 0 units)

10_{2} represents the binary value two. (1 two + 0 units)

10_{8} represents the octal value eight. (1 eight + 0 units)

10_{16} represents the hexadecimal value sixteen. (1 sixteen + 0 units)

System Radix

The base of a system, more properly called the RADIX, is the number of different values that can be expressed using a single digit. Therefore the decimal system has a radix of 10, the octal system has a radix of 8, hexadecimal is radix 16, and binary radix 2.

Decimal (Radix 10)	Binary (Radix 2)	Octal (Radix 8)	Hexadecimal (Radix 16)
0	0	0	0
1	1	1	1
2		2	2
3		3	3
4		4	4
5		5	5
6		6	6
7		7	7
8			8
9			9
			A
			B
			C
			D
			E
			F

The range of number values in different number systems is shown in table above, Notice that because the hexadecimal system must express 16 values using only one column, it uses the letters A B C D E & F to represent the numbers 10 to 15.

Radix Point

When writing a number, the digits used give its value, but the number is 'scaled' by its RADIX POINT.

For example, 456.2_{10} is ten times bigger than 45.62_{10} although the digits are the same.

Notice also that when using multiple number systems, the term 'RADIX point' instead of 'DECIMAL point' is used. When using decimal numbers, a decimal point is used, but if a different system is used, it would be wrong to call the point a decimal point, it would need to be called "Binary point" or "Octal point" etc. The simplest way around this is to refer to the point in any system (which will of course have its value labelled with its radix) as the RADIX POINT.

Exponents

A decimal number such as 456.2_{10} can be considered as the sum of the values of its individual digits, where each digit has a value dependent on its position within the number (the value of the column):

$$= 456.2_{10}$$

Col 2	Col 1	Col 0	Col -1
4 hundreds	+ 5 tens	+ 6 units	+ 2 tenths
(4×10^2)	$+ (5 \times 10^1)$	$+ (6 \times 10^0)$	$+ (2 \times 10^{-1})$
400	+ 50	+ 6	+ 0.2

Each digit in the number is multiplied by the system radix raised to a power depending on its position relative to the radix point. This is called the EXPONENT. The digit immediately to the left of the radix point has the exponent 0 applied to its radix, and for each place to the left, the exponent increases by one. The first place to the right of the radix point has the exponent -1 and so on, positive exponents to the left of the radix point and negative exponents to the right.

This method of writing numbers is widely used in electronics with decimal numbers, but can be used with any number system. Only the radix is different.

Hexadecimal exponents $98.2_{16} = (9 \times 16^1) + (8 \times 16^0) + (2 \times 16^{-1})$

Octal exponents $56.2_8 = (5 \times 8^1) + (6 \times 8^0) + (2 \times 8^{-1})$

Binary Exponents $10.1_2 = (1 \times 2^1) + (0 \times 2^0) + (1 \times 2^{-1})$

When using your calculator for the above examples you may find that it does not like radix points in anything other than decimal mode. This is common with many electronic calculators.

Floating Point Notation

If electronic calculators cannot use radix points other than in decimal, this could be a problem. Fortunately for every problem there is a solution. The radix exponent can also be used to eliminate the radix point, without altering the value of the number. In the example below, see how the value remains the same while the radix point moves. It is all done by changing the radix exponent.

$102.6_{10} = 102.6 \times 10^0 = 10.26 \times 10^1 = 1.026 \times 10^2 = .1026 \times 10^3$

The radix point is moved one place to the left by increasing the exponent by one.

It is also possible to move the radix point to the right by decreasing the exponent. In this way the radix point can be positioned wherever it is required, in any number system, simply by changing the exponent. This is called FLOATING POINT NOTATION and it is how calculators handle decimal points in calculations.

Normalised Form

By putting the radix point at the front of the number, and keeping it there by changing the exponent, calculations become easier to do electronically, in any radix.

Electronic Storage of Numbers

A number written (or stored) in this way, with the radix point at the left of the most significant digit is said to be in NORMALISED FORM. For example $.11011_2 \times 2^3$ is the normalised form of the binary number 110.11_2. Because numbers in electronic systems are stored as binary digits, and a binary digit can only be 1 or 0, it is not possible to store the radix point within the number. Therefore the number is stored in its normalised form and the exponent is stored separately. The exponent is then reused to restore the radix point to its correct position when the number is displayed.

In electronics systems a single binary digit is called a bit (short for Binary DigIT), but as using a single digit would seriously limit the maths that could be performed, binary bits are normally used in groups.

4 bits = 1 nibble

8 bits = 1 byte

Multiple bytes, such as 16 bits, 32 bits, 64 bits are usually called 'words', e.g. a 32 bit word. The length of the word depends on how many bits can be physically handled or stored by the system at one time.

4 Bit Binary Representation

Decimal	MSB	4 Bit Binary		LSB
	$2^3 = 8$	$2^2 = 4$	$2^1 = 2$	$2^0 = 1$
0	0	0	0	0
1	0	0	0	1
2	0	0	1	0
3	0	0	1	1
4	0	1	0	0
5	0	1	0	1
6	0	1	1	0
7	0	1	1	1
8	1	0	0	0
9	1	0	0	1
10	1	0	1	0
11	1	0	1	1
12	1	1	0	0
13	1	1	0	1
14	1	1	1	0
15	1	1	1	1

When a number is stored in an electronic system, it is stored in a memory location having a fixed number of binary bits. Some of these memory locations are used for general storage whilst others, having some special function, are called registers. Wherever a number is stored, it will be held in some form of binary, and must always have a set number of bits. Therefore a decimal number such as 13, which can be expressed in four binary bits as 1101_2 becomes 00001101_2 when stored in an eight-bit register. This is achieved by adding four NON SIGNIFICANT ZEROS to the left of the most significant '1' digit.

Using this system, a binary register that is n bits wide can hold 2^n values.

Therefore an 8 bit register can hold 2^8 values = 256 values (0 to 255)

A 4 bit register can hold 2^4 values = 16 values (0 to 15)

Binary Number System

Binary numbers are the flow of information in the form of zeros and ones used by digital computers and systems.

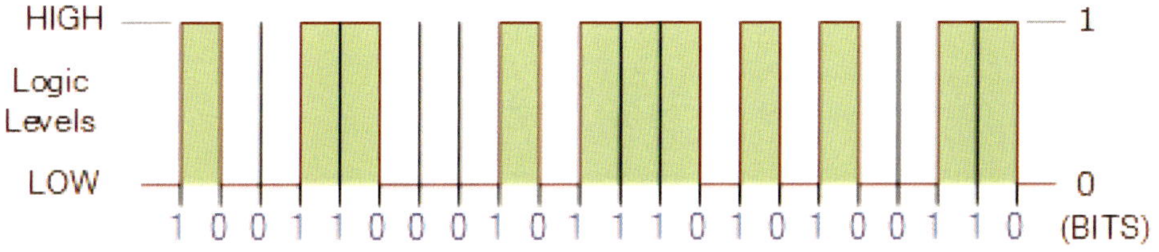

Unlike a linear, or analogue circuits, such as AC amplifiers, which process signals that are constantly changing from one value to another, for example amplitude or frequency, digital circuits process signals that contain just two voltage levels or states, labelled, Logic "0" and Logic "1".

Generally, a logic "1" represents a higher voltage, such as 5 volts, which is commonly referred to as a HIGH value, while a logic "0" represents a low voltage, such as 0 volts or ground, and is commonly referred to as a LOW value. These two discrete voltage levels representing the digital values of "1's" (one's) and "0's" (zero's) are commonly called: Binary digits, and in digital and computational circuits and applications they are normally referred to as binary BITS.

Binary Bits of Zeros and Ones

Because there are only two valid Boolean values for representing either a logic "1" or a logic "0", makes the system of using Binary Numbers ideal for use in digital or electronic circuits and systems.

The *binary number system* is a Base-2 numbering system which follows the same set of rules in mathematics as the commonly used decimal or base-10 number system. So instead of powers of ten, (10^n) for example: 1, 10, 100, 1000 etc, binary numbers use powers of two, (2^n) effectively doubling the value of each successive bit as it goes, for example: 1, 2, 4, 8, 16, 32 etc.

The voltages used to represent a digital circuit can be of any value, but generally in digital and computer systems they are kept well below 10 volts. In digital systems theses voltages are called "logic levels" and ideally one voltage level represents a "HIGH" state, while another different and lower voltage level represents a "LOW" state. A binary number system uses both of these two states.

Digital waveforms or signals consist of discrete or distinctive voltage levels that are changing back and forth between these two "HIGH" and "LOW" states. But what makes a signal or voltage "Digital" and how can we represent these "HIGH" and "LOW" voltage levels.

Digital Voltage Output

In this digital circuit example, the potentiometer wiper has been replaced by a single rotary switch which is connected in turn to each junction of the series resistor chain, forming a basic potential divider network. As the switch is rotated from one position (or node) to the next the output voltage, V_{out} changes quickly in discrete and distinctive voltage levels representing multiples of 1.0 volts on each switching action or step, as shown in the output graph.

So for example, the output voltage will be 2 volts, 3 volts, 5 volts, etc. but NOT 2.5V, 3.1V or 4.6V. Finer output voltage levels could easily be produced by using a multi-positional switch and increasing the number of resistive elements within the potential divider network, therefore increasing the number of discrete switching steps.

Digital Voltage Output Representation

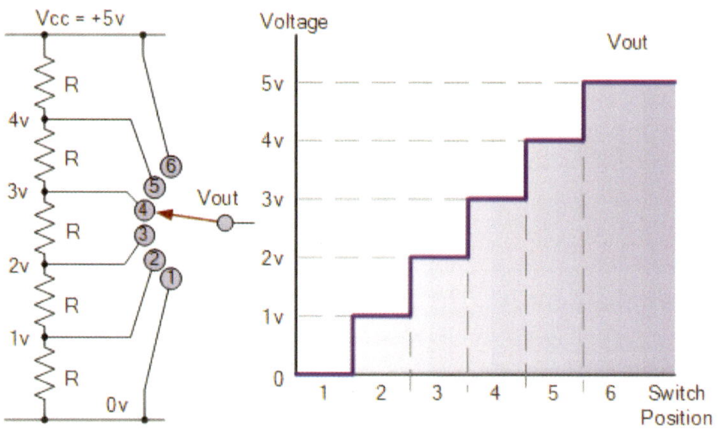

Digital Voltage Output

Then we can see that the major difference between an analogue signal or quantity and a digital quantity is that an "Analogue" quantity is continuously changing over time while a "Digital" quantity has discrete (step by step) values. "LOW" to "HIGH" or "HIGH" to "LOW".

A good example of this would be a light dimmer in your house that varies the lights intensity (brightness) up or down as it is rotated between fully-ON (maximum brightness) and fully-OFF, producing an analogue output that varies continuously. While on the other hand, with a standard wall mounted light switch, the light is either "ON" (HIGH) or it is "OFF" (LOW) when the switch is operated. The result is that there is no in between producing a form of ON-OFF digital output.

Some circuits combine both analogue and digital signals such as an analogue to digital converter (ADC) or a digital to analogue converter (DAC). Either way, the digital input or output signal represents a binary number value equivalent of an analogue signal.

Digital Value Representation

First State	Second State
Logic "0"	Logic "1"
LOW	HIGH
FALSE	TRUE
Low Level Voltage Output	High Level Voltage Output
0V or Ground	+5 Volts
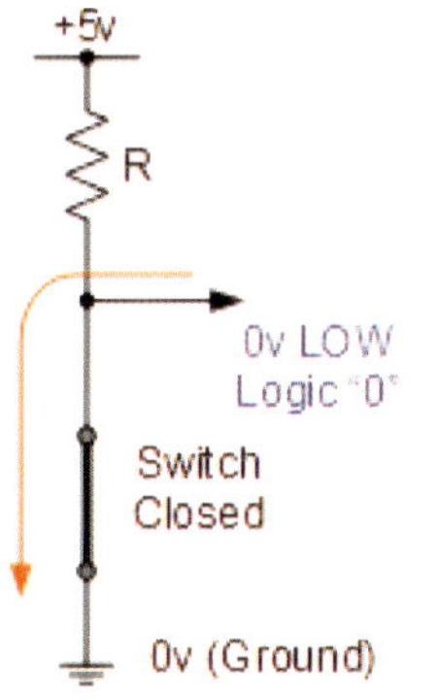	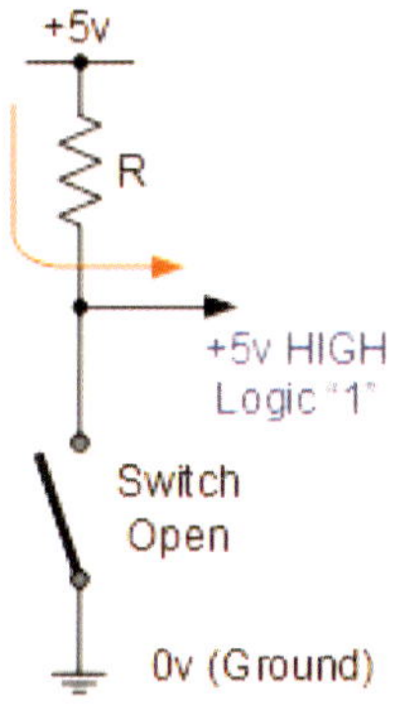

Generally the switching from one voltage level, ">0" to "1" or "1" to "0" is made as quickly as possible to prevent miss switching of the logic circuit. In standard TTL (transistor-transistor-logic) IC's there is a pre-defined range of input and output voltage limits for defining what exactly is a logic "1" value and what is a logic "0" value as shown below.

Representation

Any number can be represented by a sequence of bits (binary digits), which in turn may be represented by any mechanism capable of being in two mutually exclusive states. Any of the following rows of symbols can be interpreted as the binary numeric value of 667:

1	0	1	0	0	1	1	0	1	1
|	—	|	—	—	|	|	—	|	|
☒	☐	☒	☐	☐	☒	☒	☐	☒	☒
y	n	y	n	n	y	y	n	y	y

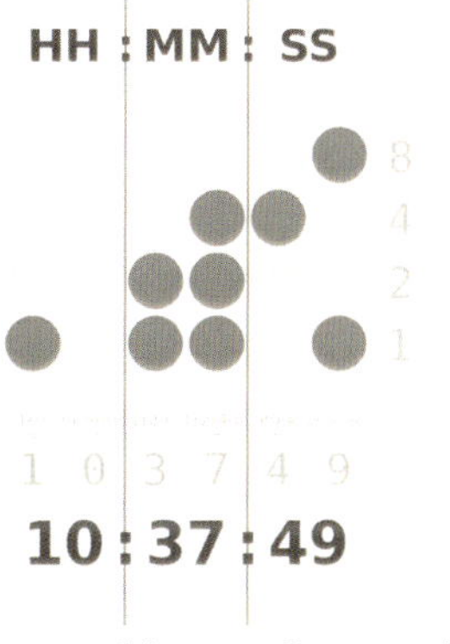

A binary clock might use LEDs to express binary values. In this clock, each column of LEDs shows a binary-coded decimal numeral of the traditional sexagesimal time.

The numeric value represented in each case is dependent upon the value assigned to each symbol. In a computer, the numeric values may be represented by two different voltages; on a magnetic disk, magnetic polarities may be used. A "positive", "yes", or "on" state is not necessarily equivalent to the numerical value of one; it depends on the architecture in use.

In keeping with customary representation of numerals using Arabic numerals, binary numbers are commonly written using the symbols 0 and 1. When written, binary numerals are often subscripted, prefixed or suffixed in order to indicate their base, or radix. The following notations are equivalent:

- 100101 binary (explicit statement of format)
- 100101b (a suffix indicating binary format; also known as Intel convention)
- 100101B (a suffix indicating binary format)
- bin 100101 (a prefix indicating binary format)
- 100101_2 (a subscript indicating base-2 (binary) notation)
- %100101 (a prefix indicating binary format; also known as Motorola convention)
- 0b100101 (a prefix indicating binary format, common in programming languages)
- 6b100101 (a prefix indicating number of bits in binary format, common in programming languages)

When spoken, binary numerals are usually read digit-by-digit, in order to distinguish them from decimal numerals. For example, the binary numeral 100 is pronounced *one zero zero*, rather than *one hundred*, to make its binary nature explicit, and for purposes of correctness. Since the binary numeral 100 represents the value four, it would be confusing to refer to the numeral as *one hundred* (a word that represents a completely different value, or amount). Alternatively, the binary numeral 100 can be read out as "four" (the correct *value*), but this does not make its binary nature explicit.

Counting in Binary

Decimal pattern	Binary number
0	0
1	1
2	10
3	11
4	100
5	101
6	110
7	111
8	1000

9	1001
10	1010
11	1011
12	1100
13	1101
14	1110
15	1111

Counting in binary is similar to counting in any other number system. Beginning with a single digit, counting proceeds through each symbol, in increasing order. Before examining binary counting, it is useful to briefly discuss the more familiar decimal counting system as a frame of reference.

Decimal Counting

Decimal counting uses the ten symbols *0* through *9*. Counting begins with the incremental substitution of the least significant digit (rightmost digit) which is often called the *first digit*. When the available symbols for this position are exhausted, the least significant digit is reset to *0*, and the next digit of higher significance (one position to the left) is incremented (*overflow*), and incremental substitution of the low-order digit resumes. This method of reset and overflow is repeated for each digit of significance. Counting progresses as follows:

> 000, 001, 002, 007, 008, 009, (rightmost digit is reset to zero, and the digit to its left is incremented)
>
> 010, 011, 012
>
> 090, 091, 092, 097, 098, 099, (rightmost two digits are reset to zeroes, and next digit is incremented)
>
> 100, 101, 102

Binary Counting

2^4	2^3	2^2	2^1	2^0	
16	8	4	2	1	
0	0	0	0	0	00

This counter shows how to count in binary from numbers zero through thirty-one.

Binary counting follows the same procedure, except that only the two symbols *0* and *1* are available. Thus, after a digit reaches 1 in binary, an increment resets it to 0 but also causes an increment of the next digit to the left:

> 0000,
>
> 0001, (rightmost digit starts over, and next digit is incremented)
>
> 0010, 0011, (rightmost two digits start over, and next digit is incremented)

0100, 0101, 0110, 0111, (rightmost three digits start over, and the next digit is incremented)

1000, 1001, 1010, 1011, 1100, 1101, 1110, 1111

In the binary system, each digit represents an increasing power of 2, with the rightmost digit representing 2^0, the next representing 2^1, then 2^2, and so on. The equivalent decimal representation of a binary number is sum of the powers of 2 which each digit represents. For example, the binary number 100101 is converted to decimal form as follows:

$$100101_2 = [(1) \times 2^5] + [(0) \times 2^4] + [(0) \times 2^3] + [(1) \times 2^2] + [(0) \times 2^1] + [(1) \times 2^0]$$

$$100101_2 = [1 \times 32] + [0 \times 16] + [0 \times 8] + [1 \times 4] + [0 \times 2] + [1 \times 1]$$

$$100101_2 = 37_{10}$$

Fractions

Fractions in binary only terminate if the denominator has 2 as the only prime factor. As a result, 1/10 does not have a finite binary representation, and this causes 10 × 0.1 not to be precisely equal to 1 in floating point arithmetic. As an example, to interpret the binary expression for 1/3 = .010101..., this means: $1/3 = 0 \times 2^{-1} + 1 \times 2^{-2} + 0 \times 2^{-3} + 1 \times 2^{-4} + ... = 0.3125 + ...$ An exact value cannot be found with a sum of a finite number of inverse powers of two, the zeros and ones in the binary representation of 1/3 alternate forever.

Fraction	Decimal	Binary	Fractional approximation
1/1	1 or 0.999...	1 or 0.111...	1/2 + 1/4 + 1/8...
1/2	0.5 or 0.4999...	0.1 or 0.0111...	1/4 + 1/8 + 1/16 . . .
1/3	0.333...	0.010101...	1/4 + 1/16 + 1/64 . . .
1/4	0.25 or 0.24999...	0.01 or 0.00111...	1/8 + 1/16 + 1/32 . . .
1/5	0.2 or 0.1999...	0.00110011...	1/8 + 1/16 + 1/128 . . .
1/6	0.1666...	0.0010101...	1/8 + 1/32 + 1/128 . . .
1/7	0.142857142857...	0.001001...	1/8 + 1/64 + 1/512 . . .
1/8	0.125 or 0.124999...	0.001 or 0.000111...	1/16 + 1/32 + 1/64 . . .
1/9	0.111...	0.000111000111...	1/16 + 1/32 + 1/64 . . .
1/10	0.1 or 0.0999...	0.000110011...	1/16 + 1/32 + 1/256 . . .
1/11	0.090909...	0.00010111010001011101...	1/16 + 1/64 + 1/128 . . .
1/12	0.08333...	0.00010101...	1/16 + 1/64 + 1/256 . . .
1/13	0.076923076923...	0.000100111011000100111011...	1/16 + 1/128 + 1/256 . . .
1/14	0.0714285714285...	0.0001001001...	1/16 + 1/128 + 1/1024 . . .
1/15	0.0666...	0.00010001...	1/16 + 1/256 . . .
1/16	0.0625 or 0.0624999...	0.0001 or 0.0000111...	1/32 + 1/64 + 1/128 . . .

Binary Arithmetic

Arithmetic in binary is much like arithmetic in other numeral systems. Addition, subtraction, multiplication, and division can be performed on binary numerals.

Addition

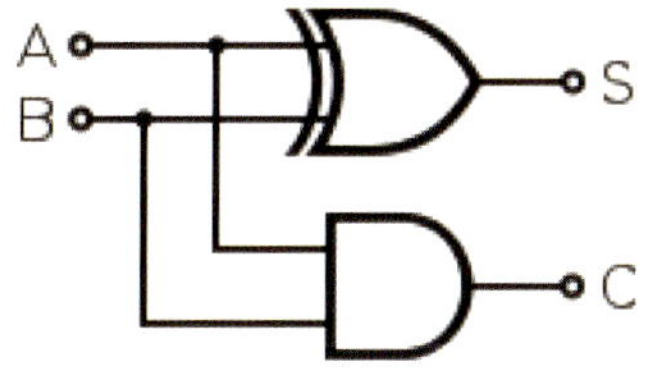

The circuit diagram for a binary half adder, which adds two bits together, producing sum and carry bits.

The simplest arithmetic operation in binary is addition. Adding two single-digit binary numbers is relatively simple, using a form of carrying:

$0 + 0 \rightarrow 0$

$0 + 1 \rightarrow 1$

$1 + 0 \rightarrow 1$

$1 + 1 \rightarrow 0$, carry 1 (since $1 + 1 = 2 = 0 + (1 \times 2^1)$)

Adding two "1" digits produces a digit "0", while 1 will have to be added to the next column. This is similar to what happens in decimal when certain single-digit numbers are added together; if the result equals or exceeds the value of the radix (10), the digit to the left is incremented:

$5 + 5 \rightarrow 0$, carry 1 (since $5 + 5 = 10 = 0 + (1 \times 10^1)$)

$7 + 9 \rightarrow 6$, carry 1 (since $7 + 9 = 16 = 6 + (1 \times 10^1)$)

This is known as *carrying*. When the result of an addition exceeds the value of a digit, the procedure is to "carry" the excess amount divided by the radix (that is, 10/10) to the left, adding it to the next positional value. This is correct since the next position has a weight that is higher by a factor equal to the radix. Carrying works the same way in binary:

```
  1 1 1 1 1     (carried digits)
   0 1 1 0 1
+  1 0 1 1 1
-------------
= 1 0 0 1 0 0 = 36
```

In this example, two numerals are being added together: 01101_2 (13_{10}) and 10111_2 (23_{10}). The top row shows the carry bits used. Starting in the rightmost column, $1 + 1 = 10_2$. The 1 is carried to the left, and the 0 is written at the bottom of the rightmost column. The second column from the right is added: $1 + 0 + 1 = 10_2$ again; the 1 is carried, and 0 is written at the bottom. The third column: $1 + 1 + 1 = 11_2$. This time, a 1 is carried, and a 1 is written in the bottom row. Proceeding like this gives the final answer 100100_2 (36 decimal).

When computers must add two numbers, the rule that: x xor y = (x + y) mod 2 for any two bits x and y allows for very fast calculation, as well.

Long Carry Method

A simplification for many binary addition problems is the Long Carry Method or Brookhouse Method of Binary Addition. This method is generally useful in any binary addition where one of the numbers contains a long "string" of ones. It is based on the simple premise that under the binary system, when given a "string" of digits composed entirely of n ones (*where:* n is any integer length), adding 1 will result in the number 1 followed by a string of n zeros. That concept follows, logically, just as in the decimal system, where adding 1 to a string of n 9s will result in the number 1 followed by a string of n 0s:

Binary		Decimal
11111	likewise	99999
+ 1		+ 1
------		-------
100000		100000

Such long strings are quite common in the binary system. From that one finds that large binary numbers can be added using two simple steps, without excessive carry operations. In the following example, two numerals are being added together: 1110111110_2 (958_{10}) and 1010110011_2 (691_{10}), using the traditional carry method on the left, and the long carry method on the right:

Traditional Carry Method vs. Long Carry Method

Traditional Carry Method	Long Carry Method	
111 11111 (carried digits)	1← 1←	carry the 1 until it is one digit past the "string" below
111 0111110	~~111~~0~~11111~~0	cross out the "string",
+ 1010110011	+ 10~~1~~01100~~1~~1	and cross out the digit that was added to it
-----------	-----------	
= 11001110001	11001110001	

The top row shows the carry bits used. Instead of the standard carry from one column to the next, the lowest-ordered "1" with a "1" in the corresponding place value beneath it may be added and a "1" may be carried to one digit past the end of the series. The "used" numbers must be crossed off, since they are already added. Other long strings may likewise be cancelled using the same technique. Then, simply add together any remaining digits normally. Proceeding in this manner gives the final answer of 11001110001_2 (1649_{10}). In our simple example using small numbers, the traditional carry method required eight carry operations, yet the long carry method required only two, representing a substantial reduction of effort.

Addition Table

	0	1
0	0	1
1	1	10

The binary addition table is similar, but not the same, as the truth table of the logical disjunction operation $\vee$. The difference is that $\vee 1 = 1$, while $1 + 1 = 10$.

Subtraction

Subtraction works in much the same way:

0 − 0 → 0

0 − 1 → 1, borrow 1

1 − 0 → 1

1 − 1 → 0

Subtracting a "1" digit from a "0" digit produces the digit "1", while 1 will have to be subtracted from the next column. This is known as *borrowing*. The principle is the same as for carrying. When the result of a subtraction is less than 0, the least possible value of a digit, the procedure is to "borrow" the deficit divided by the radix (that is, 10/10) from the left, subtracting it from the next positional value.

```
     *   * * *  (starred columns are borrowed from)
    1 1 0 1 1 1 0
  −     1 0 1 1 1
----------------
  = 1 0 1 0 1 1 1
```

```
     *          (starred columns are borrowed from)
    1 0 1 1 1 1 1
  -   1 0 1 0 1 1
----------------
  = 0 1 1 0 1 0 0
```

Subtracting a positive number is equivalent to *adding* a negative number of equal absolute value. Computers use signed number representations to handle negative numbers—most commonly the two's complement notation. Such representations eliminate the need for a separate "subtract" operation. Using two's complement notation subtraction can be summarized by the following formula:

A − B = A + not B + 1

Multiplication

Multiplication in binary is similar to its decimal counterpart. Two numbers A and B can be multiplied by partial products: for each digit in B, the product of that digit in A is calculated and written on a new line, shifted leftward so that its rightmost digit lines up with the digit in B that was used. The sum of all these partial products gives the final result.

Since there are only two digits in binary, there are only two possible outcomes of each partial multiplication:

- If the digit in B is 0, the partial product is also 0

- If the digit in B is 1, the partial product is equal to A

For example, the binary numbers 1011 and 1010 are multiplied as follows:

```
      1011  (A)
     ×1010  (B)
     ---------
      0000    ← Corresponds to the rightmost 'zero' in B
  +  1011     ← Corresponds to the next 'one' in B
 +  0000
 +1011
 ---------------
 =1101110
```

Binary numbers can also be multiplied with bits after a binary point:

```
         101.101         A (5.625 in decimal)
        ×110.01          B (6.25  in decimal)
       -------------------
             1.01101  ← Corresponds to a 'one' in B
 +          00.0000   ← Corresponds to a 'zero' in B
 +        000.000
 +       1011.01
 +     10110.1
 ---------------------------
 =   100011.00101 (35.15625 in decimal)
```

Multiplication Table

	0	1
0	0	0
1	0	1

The binary multiplication table is the same as the truth table of the logical conjunction operation .

Division

Long division in binary is again similar to its decimal counterpart.

In the example below, the divisor is 101_2, or 5 decimal, while the dividend is 11011_2, or 27 decimal. The procedure is the same as that of decimal long division; here, the divisor 101_2 goes into the first three digits 110_2 of the dividend one time, so a "1" is written on the top line. This result is multiplied by the divisor, and subtracted from the first three digits of the dividend; the next digit (a "1") is included to obtain a new three-digit sequence:

```
              1
         ___________
101 )11011
      -101
       -----
       001
```

The procedure is then repeated with the new sequence, continuing until the digits in the dividend have been exhausted:

```
             101
         ___________
101 )11011
     -101
      ------
         111
    -   101
        ------
          10
```

Thus, the quotient of 11011_2 divided by 101_2 is 101_2, as shown on the top line, while the remainder, shown on the bottom line, is 10_2. In decimal, 27 divided by 5 is 5, with a remainder of 2.

Square Root

The process of taking a binary square root digit by digit is the same as for a decimal square root, and is explained here. An example is:

```
          1001
         ---------
       √1010001
          1
        ---------
  101     01
           0
         --------
 1001     100
             0
           --------
10001     10001
          10001
          -------
                0
```

Conversion to and from other Numeral Systems

Decimal

To convert from a base-10 integer to its base-2 (binary) equivalent, the number is divided by two. The remainder is the least-significant bit. The quotient is again divided by two; its remainder becomes the next least significant bit. This process repeats until a quotient of one is reached. The sequence of remainders (including the final quotient of one) forms the binary value, as each remainder must be either zero or one when dividing by two. For example, $(357)_{10}$ is expressed as $(101100101)_2$

Conversion from base-2 to base-10 simply inverts the preceding algorithm. The bits o.f the binary number are used one by one, starting with the most significant (leftmost) bit. Beginning with the value 0, the prior value is doubled, and the next bit is then added to produce the next value. This can be organized in a multi-column table. For example, to convert 10010101101_2 to decimal:

Prior value	× 2 +	Next bit	Next value
0	× 2 +	1	= 1
1	× 2 +	0	= 2
2	× 2 +	0	= 4
4	× 2 +	1	= 9
9	× 2 +	0	= 18
18	× 2 +	1	= 37
37	× 2 +	0	= 74
74	× 2 +	1	= 149
149	× 2 +	1	= 299
299	× 2 +	0	= 598
598	× 2 +	1	= 1197

The result is 1197_{10}. Note that the first Prior Value of 0 is simply an initial decimal value. This method is an application of the Horner scheme.

Binary	1	0	0	1	0	1	0	1	1	0	1	
Decimal	1×2^{10} +	0×2^9 +	0×2^8 +	1×2^7 +	0×2^6 +	1×2^5 +	0×2^4 +	1×2^3 +	1×2^2 +	0×2^1 +	1×2^0 =	1197

The fractional parts of a number are converted with similar methods. They are again based on the equivalence of shifting with doubling or halving.

In a fractional binary number such as 0.11010110101_2, the first digit is $\frac{1}{2}$, the second $(\frac{1}{2})^2 = \frac{1}{4}$, etc. So if there is a 1 in the first place after the decimal, then the number is at least $\frac{1}{2}$, and vice versa. Double that number is at least 1. This suggests the algorithm: Repeatedly double the number to be converted, record if the result is at least 1, and then throw away the integer part.

For example, $(\frac{1}{3})_{10}$, in binary, is:

Converting	Result
$(\frac{1}{3})$	0.
$\frac{1}{3}\times 2=\frac{2}{3}<1$	0.0
$\frac{2}{3}\times 2=1\frac{1}{3}\geq 1$	0.01
$\frac{1}{3}\times 2=\frac{2}{3}<1$	0.010
$\frac{2}{3}\times 2=1\frac{1}{3}\geq 1$	0.0101

Thus the repeating decimal fraction $0.\overline{3}$... is equivalent to the repeating binary fraction 0.01... .

Or for example, 0.1_{10}, in binary, is:

Converting	Result
0.1	0.
0.1 × 2 = 0.2 < 1	0.0
0.2 × 2 = 0.4 < 1	0.00
0.4 × 2 = 0.8 < 1	0.000
0.8 × 2 = 1.6 ≥ 1	0.0001
0.6 × 2 = 1.2 ≥ 1	0.00011
0.2 × 2 = 0.4 < 1	0.000110
0.4 × 2 = 0.8 < 1	0.0001100
0.8 × 2 = 1.6 ≥ 1	0.00011001
0.6 × 2 = 1.2 ≥ 1	0.000110011
0.2 × 2 = 0.4 < 1	0.0001100110

This is also a repeating binary fraction $0.0\overline{0011}$... . It may come as a surprise that terminating decimal fractions can have repeating expansions in binary. It is for this reason that many are surprised to discover that 0.1 + ... + 0.1, (10 additions) differs from 1 in floating point arithmetic. In fact, the only binary fractions with terminating expansions are of the form of an integer divided by a power of 2, which 1/10 is not.

The final conversion is from binary to decimal fractions. The only difficulty arises with repeating fractions, but otherwise the method is to shift the fraction to an integer, convert it as above, and then divide by the appropriate power of two in the decimal base. For example:

$$
\begin{aligned}
x &= 1100.1\overline{01110}\ldots \\
x \times 2^{6} &= 1100101110.\overline{01110}\ldots \\
x \times 2 &= 11001.\overline{01110}\ldots \\
x \times (2^{6} - 2) &= 1100010101 \\
x &= 1100010101 / 111110 \\
x &= (789/62)_{10}
\end{aligned}
$$

Another way of converting from binary to decimal, often quicker for a person familiar with hexadecimal, is to do so indirectly—first converting (x in binary) into (x in hexadecimal) and then converting (x in hexadecimal) into (x in decimal).

For very large numbers, these simple methods are inefficient because they perform a large number of multiplications or divisions where one operand is very large. A simple divide-and-conquer algorithm is more effective asymptotically: given a binary number, it is divided by 10^k, where k is chosen so that the quotient roughly equals the remainder; then each of these pieces is converted to decimal and the two are concatenated. Given a decimal number, it can be split into two pieces of about the same size, each of which is converted to binary, whereupon the first converted piece is multiplied by 10^k and added to the second converted piece, where k is the number of decimal digits in the second, least-significant piece before conversion.

Hexadecimal

0_{hex}	=	0_{dec}	=	0_{oct}	0	0	0	0
1_{hex}	=	1_{dec}	=	1_{oct}	0	0	0	1
2_{hex}	=	2_{dec}	=	2_{oct}	0	0	1	0
3_{hex}	=	3_{dec}	=	3_{oct}	0	0	1	1
4_{hex}	=	4_{dec}	=	4_{oct}	0	1	0	0
5_{hex}	=	5_{dec}	=	5_{oct}	0	1	0	1
6_{hex}	=	6_{dec}	=	6_{oct}	0	1	1	0
7_{hex}	=	7_{dec}	=	7_{oct}	0	1	1	1
8_{hex}	=	8_{dec}	=	10_{oct}	1	0	0	0
9_{hex}	=	9_{dec}	=	11_{oct}	1	0	0	1
A_{hex}	=	10_{dec}	=	12_{oct}	1	0	1	0
B_{hex}	=	11_{dec}	=	13_{oct}	1	0	1	1
C_{hex}	=	12_{dec}	=	14_{oct}	1	1	0	0
D_{hex}	=	13_{dec}	=	15_{oct}	1	1	0	1
E_{hex}	=	14_{dec}	=	16_{oct}	1	1	1	0
F_{hex}	=	15_{dec}	=	17_{oct}	1	1	1	1

Binary may be converted to and from hexadecimal more easily. This is because the radix of the hexadecimal system (16) is a power of the radix of the binary system (2). More specifically, 16 = 2^4, so it takes four digits of binary to represent one digit of hexadecimal, as shown in the adjacent table.

To convert a hexadecimal number into its binary equivalent, simply substitute the corresponding binary digits:

$3A_{16} = 0011\ 1010_2$

$E7_{16} = 1110\ 0111_2$

To convert a binary number into its hexadecimal equivalent, divide it into groups of four bits. If the number of bits isn't a multiple of four, simply insert extra 0 bits at the left (called padding). For example:

1010010_2 = 0101 0010 grouped with padding = 52_{16}

11011101_2 = 1101 1101 grouped = DD_{16}

To convert a hexadecimal number into its decimal equivalent, multiply the decimal equivalent of each hexadecimal digit by the corresponding power of 16 and add the resulting values:

$C0E7_{16} = (12 \times 16^3) + (0 \times 16^2) + (14 \times 16^1) + (7 \times 16^0) = (12 \times 4096) + (0 \times 256) + (14 \times 16) + (7 \times 1) = 49{,}383_{10}$

Octal

Binary is also easily converted to the octal numeral system, since octal uses a radix of 8, which is a power of two (namely, 2^3, so it takes exactly three binary digits to represent an octal digit). The correspondence between octal and binary numerals is the same as for the first eight digits of hexadecimal in the table above. Binary 000 is equivalent to the octal digit 0, binary 111 is equivalent to octal 7, and so forth.

Octal	Binary
0	000
1	001
2	010
3	011
4	100
5	101
6	110
7	111

Converting from octal to binary proceeds in the same fashion as it does for hexadecimal:

$65_8 = 110\ 101_2$

$17_8 = 001\ 111_2$

And from binary to octal:

$101100_2 = 101\ 100_2$ grouped = 54_8

$10011_2 = 010\ 011_2$ grouped with padding = 23_8

And from octal to decimal:

$$65_8 = (6 \times 8^1) + (5 \times 8^0) = (6 \times 8) + (5 \times 1) = 53_{10}$$

$$127_8 = (1 \times 8^2) + (2 \times 8^1) + (7 \times 8^0) = (1 \times 64) + (2 \times 8) + (7 \times 1) = 87_{10}$$

Representing Real Numbers

Non-integers can be represented by using negative powers, which are set off from the other digits by means of a radix point (called a decimal point in the decimal system). For example, the binary number 11.01_2 thus means:

1×2^1	$(1 \times 2 = 2)$	plus
1×2^0	$(1 \times 1 = 1)$	plus
0×2^{-1}	$(0 \times \frac{1}{2} = 0)$	plus
1×2^{-2}	$(1 \times \frac{1}{4} = 0.25)$	

For a total of 3.25 decimal.

All dyadic rational numbers $\frac{p}{2^a}$ have a *terminating* binary numeral—the binary representation has a finite number of terms after the radix point. Other rational numbers have binary representation, but instead of terminating, they *recur*, with a finite sequence of digits repeating indefinitely. For instance

$$\frac{1_{10}}{3_{10}} = \frac{1_2}{11_2} = 0.0101010101\ldots_2$$

$$\frac{12_{10}}{17_{10}} = \frac{1100_2}{10001_2} = 0.10110100\ 10110100\ 10110100\ldots_2$$

The phenomenon that the binary representation of any rational is either terminating or recurring also occurs in other radix-based numeral systems. for instance, the explanation in decimal. Another similarity is the existence of alternative representations for any terminating representation, relying on the fact that 0.111111... is the sum of the geometric series $2^{-1} + 2^{-2} + 2^{-3} + \ldots$ which is 1.

Binary numerals which neither terminate nor recur represent irrational numbers. For instance,

- 0.10100100010000100000100... does have a pattern, but it is not a fixed-length recurring pattern, so the number is irrational
- 1.0110101000001001111001100110011111110... is the binary representation of $\sqrt{2}$, the square root of 2, another irrational. It has no discernible pattern.

Binary Code

Binary code is the simplest form of computer code or programming data. It is represented entirely by a binary system of digits consisting of a string of consecutive zeros and ones. Binary code is often associated with machine code in that binary sets can be combined to form raw code, which is interpreted by a computer or other piece of hardware.

Importance of Binary Code

In computers and other computerized devices (like calculators, printers, coffee makers, and microwaves), bits are usually transmitted electronically. But this electronic information is fleeting. For it to exist for any length of time—and without a power supply—it must be stored physically within the device's hardware. That means that every piece of binary code in a computer must be converted into a physical object or state. Binary code, as it turns out, is easy to convert from electronic information (e.g., 0s and 1s) to physical information, because only two kinds of physical objects or states are needed.

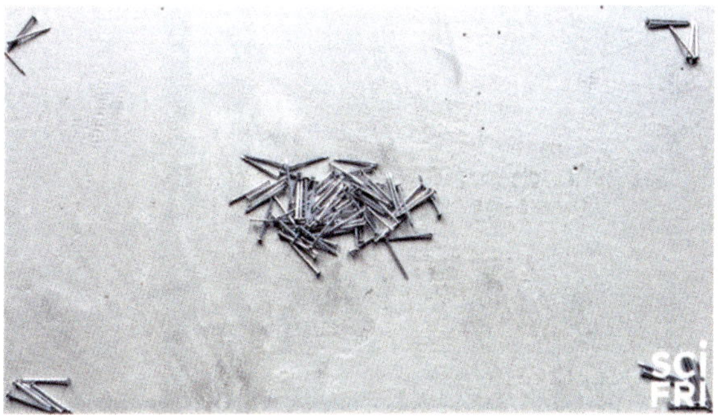

The conversion from electrical information to physical information storage is analogous to someone speaking the binary code for "dog" in 0s and 1s while you write them down on a piece of paper. The spoken 0s and 1s can't be heard indefinitely after they are spoken, but by writing them physically on a piece of paper, you can refer back to them over and over again. In the case of a computer, that binary code might be stored in high and low voltages, in magnetized or demagnetized segments of a metallic disk, or, in super old computers, in punched and unpunched holes in cardstock.

Regardless of the medium, binary code has been the gold standard of physical information storage in computational devices from calculators to super computers.

Importance of Bit Number

Arranging and reading bits in ordered groups is what makes binary exceptionally powerful for storing and transmitting huge amounts of information. To understand why, it helps to consider the alternative: what if only one bit was used at a time? Well, you'd only be able to share two types of information—one type represented by 0 and the other by 1. Forget encoding the entire alphabet or punctuation signs—you just get two kinds of information.

But when you group bits by two, you get four kinds of information:

00, 01, 10, 11

By increasing from two-bit groups to three-bit groups, you double the amount of information you can encode:

000, 001, 010, 011, 100, 101, 110, 111

While eight different kinds of information are still not enough for representing a whole alphabet, perhaps you can see where the pattern is headed.

Using any binary code representation you'd like, try to figure out how many possible combinations of bits you can make out using bits grouped by four. Then try again using bits grouped by five. How many possible combinations do you think you can get using six bits at a time, or 64? By grouping single bits together in larger and larger groups, computers can use binary code to find, organize, send, and store more and more kinds of information.

As computer technology has advanced, computer engineers have needed ways of sending and storing greater amounts of information at a time. As a result, the bit-length used by computers has been growing steadily over the course of computer history. If you have a new iPhone, it is using a 64-bit microprocessor, which means that it stores and accesses information in groups of 64 binary digits—which means that it's capable of storing 2^{64}, or more than 18,000,000,000,000,000,000 unique 64-bit combinations of binary integers.

This idea of coding information with more bits at a time to improve the power and efficiency of computers has driven computer engineering from the beginning, and still does. Though this excerpt from The Soul of a New Machine was first published in 1981, the basic principle of encoding information in binary code with increasing complexity is still representative of the progression of computational power today:

> "Inside certain crucial parts of a typical modern computer, the bits – the electrical symbols – are handled in packets. Like phone numbers the packets are of a standard size. IBM's machines have traditionally handled information in packages 32 bits long. Data General's NOVA and most minicomputers after it, including the Eclipses, deal with packages only 16 bits long. The distinction is inconsequential in theory, since any computer is hypothetically capable of doing what any other computer may do. But the ease and speed with which different computers can be made to perform the same piece of work vary widely, and in general a machine that handles symbols in chunks of 32 bits runs faster, and for some purposes – usually large ones – it is easier to program than a machine that handles only 16 bits at a time."

Advantages of Binary Code

Following is the list of advantages that binary code offers.

- Binary codes are suitable for the computer applications.
- Binary codes are suitable for the digital communications.
- Binary codes make the analysis and designing of digital circuits if we use the binary codes.
- Since only 0 & 1 are being used, implementation becomes easy.

Weighted Codes

Weighted binary codes are those binary codes which obey the positional weight principle. Each position of the number represents a specific weight. Several systems of the codes are used to express the decimal digits 0 through 9. In these codes each decimal digit is represented by a group of four bits.

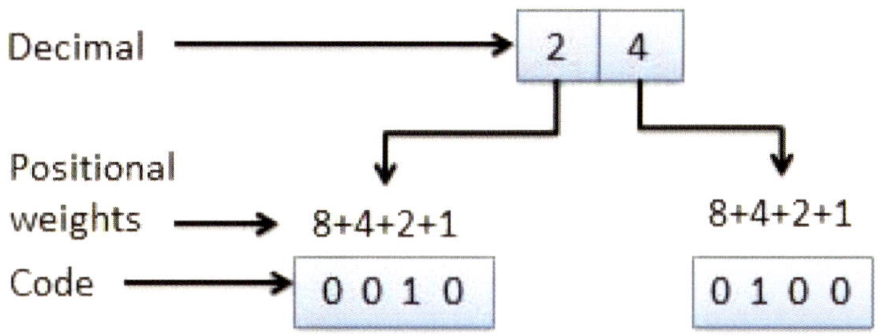

Non-Weighted Codes

In this type of binary codes, the positional weights are not assigned. The examples of non-weighted codes are Excess-3 code and Gray code.

Excess-3 code

The Excess-3 code is also called as XS-3 code. It is non-weighted code used to express decimal numbers. The Excess-3 code words are derived from the 8421 BCD code words adding $(0011)_2$ or $(3)_{10}$ to each code word in 8421. The excess-3 codes are obtained as follows –

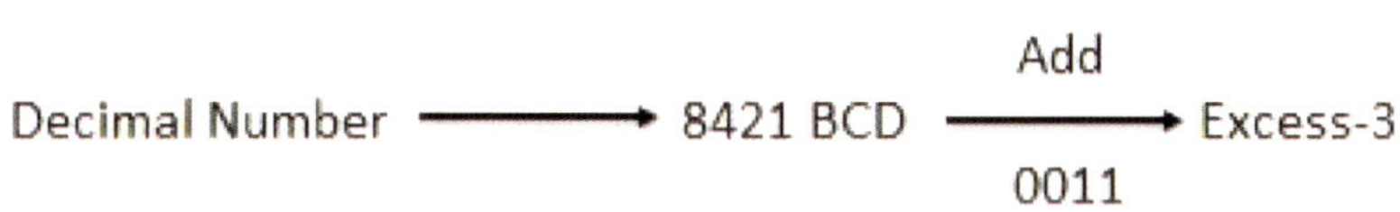

Example

Decimal	BCD 8 4 2 1	Excess-3 BCD + 0011
0	0 0 0 0	0 0 1 1
1	0 0 0 1	0 1 0 0
2	0 0 1 0	0 1 0 1
3	0 0 1 1	0 1 1 0
4	0 1 0 0	0 1 1 1
5	0 1 0 1	1 0 0 0
6	0 1 1 0	1 0 0 1
7	0 1 1 1	1 0 1 0
8	1 0 0 0	1 0 1 1
9	1 0 0 1	1 1 0 0

Gray Code

It is the non-weighted code and it is not arithmetic codes. That means there are no specific weights assigned to the bit position. It has a very special feature that, only one bit will change each time the decimal number is incremented as shown in fig. As only one bit changes at a time, the gray code is called as a unit distance code. The gray code is a cyclic code. Gray code cannot be used for arithmetic operation.

Decimal	BCD	Gray
0	0 0 0 0	0 0 0 0
1	0 0 0 1	0 0 0 1
2	0 0 1 0	0 0 1 1
3	0 0 1 1	0 0 1 0
4	0 1 0 0	0 1 1 0
5	0 1 0 1	0 1 1 1
6	0 1 1 0	0 1 0 1
7	0 1 1 1	0 1 0 0
8	1 0 0 0	1 1 0 0
9	1 0 0 1	1 1 0 1

Application of Gray Code

- Gray code is popularly used in the shaft position encoders.
- A shaft position encoder produces a code word which represents the angular position of the shaft.

Alphanumeric Codes

A binary digit or bit can represent only two symbols as it has only two states '0' or '1'. But this is not enough for communication between two computers because there we need many more symbols for communication. These symbols are required to represent 26 alphabets with capital and small letters, numbers from 0 to 9, punctuation marks and other symbols.

The alphanumeric codes are the codes that represent numbers and alphabetic characters. Mostly such codes also represent other characters such as symbol and various instructions necessary for conveying information. An alphanumeric code should at least represent 10 digits and 26 letters of alphabet i.e. total 36 items. The following three alphanumeric codes are very commonly used for the data representation.

- American Standard Code for Information Interchange (ASCII).
- Extended Binary Coded Decimal Interchange Code (EBCDIC).
- Five bit Baudot Code.

ASCII code is a 7-bit code whereas EBCDIC is an 8-bit code. ASCII code is more commonly used worldwide while EBCDIC is used primarily in large IBM computers.

Binary-coded Decimal

Binary Coded Decimal or BCD as it is more commonly called, is another process for converting decimal numbers into their binary equivalents.

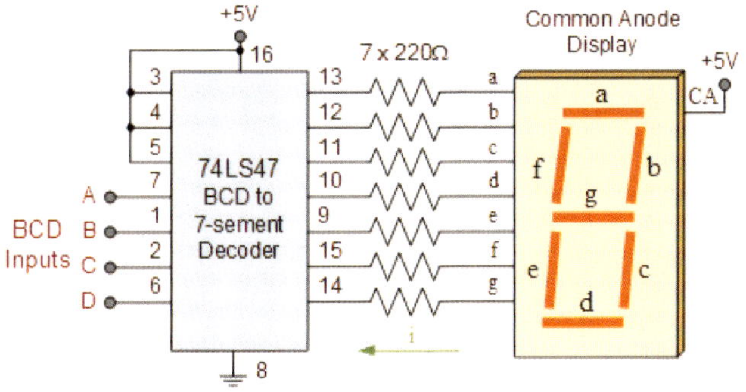

As we have seen in this Binary Numbers section of topic, there are many different binary codes used in digital and electronic circuits, each with its own specific use.

As we naturally live in a decimal (base-10) world we need some way of converting these decimal numbers into a binary (base-2) environment that computers and digital electronic devices understand, and binary coded decimal code allows us to do that.

We have seen previously that an n-bit binary code is a group of "n" bits that assume up to 2^n distinct combinations of 1's and 0's. The advantage of the Binary Coded Decimal system is that each decimal digit is represented by a group of 4 binary digits or bits in much the same way as Hexadecimal. So for the 10 decimal digits (0-to-9) we need a 4-bit binary code.

But do not get confused, binary coded decimal is not the same as hexadecimal. Whereas a 4-bit hexadecimal number is valid up to F_{16} representing binary 1111_2, (decimal 15), binary coded decimal numbers stop at 9 binary 1001_2. This means that although 16 numbers (2^4) can be represented using four binary digits, in the BCD numbering system the six binary code combinations of: 1010 (decimal 10), 1011 (decimal 11), 1100 (decimal 12), 1101 (decimal 13), 1110 (decimal 14), and 1111 (decimal 15) are classed as forbidden numbers and can not be used.

The main advantage of binary coded decimal is that it allows easy conversion between decimal (base-10) and binary (base-2) form. However, the disadvantage is that BCD code is wasteful as the states between 1010(decimal 10), and 1111 (decimal 15) are not used. Nevertheless, binary coded decimal has many important applications especially using digital displays.

In the BCD numbering system, a decimal number is separated into four bits for each decimal digit within the number. Each decimal digit is represented by its weighted binary value performing a direct translation of the number. So a 4-bit group represents each displayed decimal digit from 0000 for a zero to 1001 for a nine.

So for example, 357_{10} (Three Hundred and Fifty Seven) in decimal would be presented in Binary Coded Decimal as:

$$357_{10} = 0011\ 0101\ 0111\ (BCD)$$

Then we can see that BCD uses weighted codification, because the binary bit of each 4-bit group represents a given weight of the final value. In other words, the BCD is a weighted code and the weights used in binary coded decimal code are 8, 4, 2, 1, commonly called the 8421 code as it forms the 4-bit binary representation of the relevant decimal digit.

Binary Coded Decimal Representation of a Decimal Number

Binary Power	2^3	2^2	2^1	2^0
Binary Weight:	8	4	2	1

The decimal weight of each decimal digit to the left increases by a factor of 10. In the BCD number system, the binary weight of each digit increases by a factor of 2 as shown. Then the first digit has a weight of 1 (2^0), the second digit has a weight of 2 (2^1), the third a weight of 4 (2^2), the fourth a weight of 8 (2^3).

Then the relationship between decimal (denary) numbers and weighted binary coded decimal digits is given below.

Truth Table for Binary Coded Decimal

Decimal Number	BCD 8421 Code
0	0000 0000
1	0000 0001
2	0000 0010
3	0000 0011
4	0000 0100
5	0000 0101
6	0000 0110
7	0000 0111
8	0000 1000
9	0000 1001
10 (1+0)	0001 0000
11 (1+1)	0001 0001
12 (1+2)	0001 0010
...	...
20 (2+0)	0010 0000
21 (2+1)	0010 0001
22 (2+2)	0010 0010
etc., continuing upwards in groups of four	

Then we can see that 8421 BCD code is nothing more than the weights of each binary digit, with each decimal (denary) number expressed as its four-bit pure binary equivalent.

Decimal-to-BCD Conversion

As we have seen above, the conversion of decimal to binary coded decimal is very similar to the conversion of hexadecimal to binary. Firstly, separate the decimal number into its weighted digits and then write down the equivalent 4-bit 8421 BCD code representing each decimal digit as shown.

Binary Coded Decimal Example

Using the above table, convert the following decimal (denary) numbers: 85_{10}, 572_{10} and 8579_{10} into their 8421 BCD equivalents.

8510 = 1000 0101 (BCD)

57210 = 0101 0111 0010 (BCD)

857910 = 1000 0101 0111 1001 (BCD)

Note that the resulting binary number after the conversion will be a true binary translation of decimal digits. This is because the binary code translates as a true binary count.

BCD-to-Decimal Conversion

The conversion from binary coded decimal to decimal is the exact opposite of the above. Simply divide the binary number into groups of four digits, starting with the least significant digit and then write the decimal digit represented by each 4-bit group. Add additional zero's at the end if required to produce a complete 4-bit grouping. So for example, 110101_2 would become: $0011\ 0101_2$ or 35_{10} in decimal.

Binary Coded Decimal Example

Convert the following binary numbers: 1001_2, 1010_2, 1000111_2 and 10100111000.101_2 into their decimal equivalents.

$1001_2 = 1001_{BCD} = 9_{10}$

1010_2 = this will produce an error as it is decimal 10_{10} and not a valid BCD number

$1000111_2 = 0100\ 0111_{BCD} = 47_{10}$

$10100111000.101_2 = 0101\ 0011\ 0001.1010_{BCD} = 538.625_{10}$

The conversion of BCD-to-decimal or decimal-to-BCD is a relatively straight forward task but we need to remember that BCD numbers are decimal numbers and not binary numbers, even though they are represented using bits. The BCD representation of a decimal number is important to understand, because microprocessor based systems used by most people needs to be in the decimal system.

However, while BCD is easy to code and decode, it is not an efficient way to store numbers. In the standard 8421 BCD encoding of decimal numbers, the number of individual data bits needed to represent a given decimal number will always be greater than the number of bits required for an equivalent binary encoding.

For example, in binary a three digit decimal number from 0-to-999 requires only 10-bits (1111100111_2), whereas in binary coded decimal, the same number requires a minimum of 12-bits ($0011\ 1110\ 0111_{BCD}$) for the same representation.

Also, performing arithmetic tasks using binary coded decimal numbers can be a bit awkward since each digit can not exceed 9. The addition of two decimal digits in BCD, will create a possible carry bit of 1 which needs to be added to the next group of 4-bits.

If the binary sum with the added carry bit is equal to or less than 9 (1001), the corresponding BCD digit is correct. But when the binary sum is greater than 9 the result is an invalid BCD digit. Therefore it is better to convert BCD numbers into pure binary, perform the required addition, and then convert the back to BCD before displaying the results.

Nevertheless, the use of a BCD coding system in both microelectronics and computer systems is particularly useful in situations where the binary coded decimal is intended to be displayed on one or more 7-segment LED or LCD displays and there are many popular integrated circuits available that are configured to give a BCD output or outputs.

One common IC is the 74LS90 asynchronous counter/divider that contains independent divide-by-2 and divide-by-5 counters that can be used together to produce a divide-by-10 decade counter with BCD outputs. Another is the 74LS390 which is a dual version of the basic 74LS90, and can also be configured to produce a BCD output.

But the most commonly used BCD encoded IC's are the 74LS47 and the 74LS48 BCD to 7-segment decoder/driver, which converts a 4-bit BCD code of a counter, etc. and converts it into the required display code to drive the individual segments of a 7-segment LED display. While both IC's are functionally the same, the 74LS47 has active-low outputs for driving common-anode displays, while the 74LS48 has active-high outputs for driving common-cathode displays.

Binary Coded Decimal Decoder IC

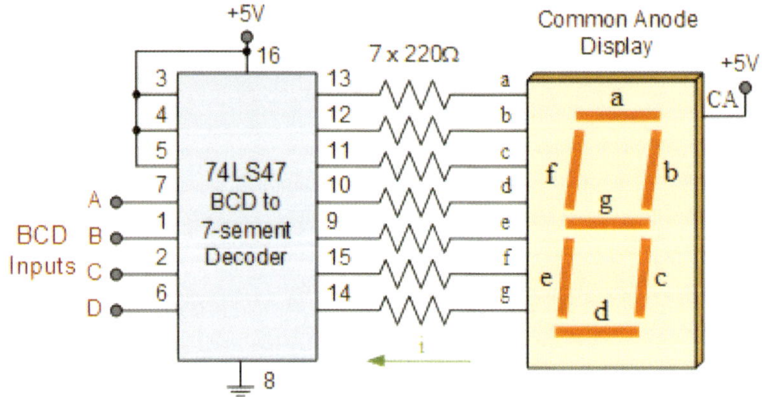

Binary Coded Decimal

We have seen here that Binary Coded Decimal or BCD is simply the 4-bit binary code representation of a decimal digit with each decimal digit replaced in the integer and fractional parts with its binary equivalent. BCD Code uses four bits to represent the 10 decimal digits of 0 to 9.

So for example, if we wanted to display decimal numbers in the range of 0-to-9, (one digit) we would need 4 data bits (a nibble), decimal numbers in the range of 0-to-99, (two digits) we would need 8 bits (one byte), decimal numbers in the range of 0-to-999, (three digits) we would need 12 bits, and so on. The use of a single byte (8-bits) to store or display two BCD digits, allowing a byte to hold a BCD number in the range of 00 – 99, is known as packed BCD.

Standard binary coded decimal code is commonly known as a weighted 8421 BCD code, with 8, 4, 2 and 1 representing the weights of the different bits starting from the most significant bit (MSB) and proceeding towards the least significant bit (LSB). The weights of the individual positions of the bits of a BCD code are: $2^3 = 8$, $2^2= 4$, $2^1= 2$, $2^0 = 1$.

The main advantage of the Binary Coded Decimal system is that it is a fast and efficient system to convert the decimal numbers into binary numbers as compared to the pure binary system. But the BCD code is wasteful as many of the 4-bit states (10-to-16) are not used but decimal displays have important applications.

Bitwise Operation

A bitwise operator is an operator used to perform bitwise operations on bit patterns or binary numerals that involve the manipulation of individual bits.

Bitwise operators are used in:

- Communication stacks where the individual bits in the header attached to the data signify important information
- Embedded software for controlling different functions in the chip and indicating the status of hardware by manipulating the individual bits of hardware registers of embedded microcontrollers
- Low-level programming for applications such as device drivers, cryptographic software, video decoding software, memory allocators, compression software and graphics
- Maintaining large sets of integers efficiently in search and optimization problems
- Bitwise operations performed on bit flags, which can enable an instance of enumeration type to store any combination of values defined in an enumerator list

Bitwise operators are used to change individual bits in an operand. A single byte of computer memory-when viewed as 8 bits-can signify the true/false status of 8 flags because each bit can be used as a boolean variable that can hold one of two values: true or false. A *flag* variable is typically used to indicate the status of something. For instance, computer files can be marked as read-only. So you might have a $fReadOnly variable whose job would be to hold the read-only status of a file. This variable is called a flag variable because when $fReadOnly has a true value, it's equivalent to a football referee throwing a flag. The variable says, "Whoa! Don't modify this file."

When you have more than one flag variable, it might be more efficient to use a single variable to indicate the value of more than one flag.

Bitwise Operators

In the explanations below, any indication of a bit's position is counted from the right (least significant) side, advancing left. For example, the binary value 0001 (decimal 1) has zeroes at every position but the first one.

NOT

The bitwise NOT, or complement, is a unary operation that performs logical negation on each bit, forming the ones' complement of the given binary value. Bits that are 0 become 1, and those that are 1 become 0. For example:

NOT 0111 (decimal 7)

= 1000 (decimal 8)

NOT 10101011

= 01010100

The bitwise complement is equal to the two's complement of the value minus one. If two's complement arithmetic is used, then NOT x = -x − 1.

For unsigned integers, the bitwise complement of a number is the "mirror reflection" of the number across the half-way point of the unsigned integer's range. For example, for 8-bit unsigned integers, NOT x = 255 - x, which can be visualized on a graph as a downward line that effectively "flips" an increasing range from 0 to 255, to a decreasing range from 255 to 0. A simple but illustrative example use is to invert a grayscale image where each pixel is stored as an unsigned integer.

AND

A bitwise AND takes two equal-length binary representations and performs the logical AND operation on each pair of the corresponding bits, by multiplying them. Thus, if both bits in the compared position are 1, the bit in the resulting binary representation is 1 ($1 \times 1 = 1$); otherwise, the result is 0 ($1 \times 0 = 0$ and $0 \times 0 = 0$). For example:

0101 (decimal 5)

AND 0011 (decimal 3)

= 0001 (decimal 1)

The operation may be used to determine whether a particular bit is *set* (1) or *clear* (0). For example, given a bit pattern 0011 (decimal 3), to determine whether the second bit is set we use a bitwise AND with a bit pattern containing 1 only in the second bit:

0011 (decimal 3)

AND 0010 (decimal 2)

= 0010 (decimal 2)

Because the result 0010 is non-zero, we know the second bit in the original pattern was set. This is often called *bit masking*. (By analogy, the use of masking tape covers, or *masks*, portions that should not be altered or portions that are not of interest. In this case, the 0 values mask the bits that are not of interest).

The bitwise AND may be used to clear selected bits (or flags) of a register in which each bit represents an individual Boolean state. This technique is an efficient way to store a number of Boolean values using as little memory as possible.

For example, 0110 (decimal 6) can be considered a set of four flags, where the first and fourth flags are clear (0), and the second and third flags are set (1). The second bit may be cleared by using a bitwise AND with the pattern that has a zero only in the second bit:

0110 (decimal 6)

AND 1101 (decimal 13)

= 0100 (decimal 4)

Because of this property, it becomes easy to check the parity of a binary number by checking the value of the lowest valued bit. Using the example above:

0110 (decimal 6)

AND 0001 (decimal 1)

= 0000 (decimal 0)

Because 6 AND 1 is zero, 6 is divisible by two and therefore even.

OR

A bitwise OR takes two bit patterns of equal length and performs the logical inclusive OR operation on each pair of corresponding bits. The result in each position is 0 if both bits are 0, while otherwise the result is 1. For example:

0101 (decimal 5)

OR 0011 (decimal 3)

= 0111 (decimal 7)

The bitwise OR may be used to set to 1 the selected bits of the register described above. For example, the fourth bit of 0010 (decimal 2) may be set by performing a bitwise OR with the pattern with only the fourth bit set:

0010 (decimal 2)

OR 1000 (decimal 8)

= 1010 (decimal 10)

XOR

A bitwise XOR takes two bit patterns of equal length and performs the logical exclusive OR operation on each pair of corresponding bits. The result in each position is 1 if only the first bit is 1 *or* only the second bit is 1, but will be 0 if both are 0 or both are 1. In this we perform the comparison of two bits, being 1 if the two bits are different, and 0 if they are the same. For example:

0101 (decimal 5)

XOR 0011 (decimal 3)

= 0110 (decimal 6)

The bitwise XOR may be used to invert selected bits in a register (also called toggle or flip). Any bit may be toggled by XORing it with 1. For example, given the bit pattern 0010 (decimal 2) the second and fourth bits may be toggled by a bitwise XOR with a bit pattern containing 1 in the second and fourth positions:

0010 (decimal 2)

XOR 1010 (decimal 10)

= 1000 (decimal 8)

This technique may be used to manipulate bit patterns representing sets of Boolean states.

Assembly language programmers and optimizing compilers sometimes use XOR as a short-cut to setting the value of a register to zero. Performing XOR on a value against itself always yields zero, and on many architectures this operation requires fewer clock cycles and memory than loading a zero value and saving it to the register.

Mathematical Equivalents

Assuming $x \geq y$, for the non-negative integers, the bitwise operations can be written as follows:

$$\text{NOT}\, x = \sum_{n=0}^{\lfloor \log_2(x) \rfloor} 2^n \left[\left(\left\lfloor \frac{x}{2^n} \right\rfloor \bmod 2 + 1 \right) \bmod 2 \right] = 2^{\lfloor \log_2(x) \rfloor + 1} - 1 - x$$

$$x\, \text{AND}\, y = \sum_{n=0}^{\lfloor \log_2(x) \rfloor} 2^n \left(\left\lfloor \frac{x}{2^n} \right\rfloor \bmod 2 \right) \left(\left\lfloor \frac{y}{2^n} \right\rfloor \bmod 2 \right)$$

$$x\, \text{OR}\, y = \sum_{n=0}^{\lfloor \log_2(x) \rfloor} 2^n \left[\left[\left(\left\lfloor \frac{x}{2^n} \right\rfloor \bmod 2 \right) + \left(\left\lfloor \frac{y}{2^n} \right\rfloor \bmod 2 \right) + \left(\left\lfloor \frac{x}{2^n} \right\rfloor \bmod 2 \right) \left(\left\lfloor \frac{y}{2^n} \right\rfloor \bmod 2 \right) \right] \bmod 2 \right]$$

$$x\, \text{XOR}\, y = \sum_{n=0}^{\lfloor \log_2(x) \rfloor} 2^n \left[\left[\left(\left\lfloor \frac{x}{2^n} \right\rfloor \bmod 2 \right) + \left(\left\lfloor \frac{y}{2^n} \right\rfloor \bmod 2 \right) \right] \bmod 2 \right] \sum_{n=0}^{\lfloor \log_2(x) \rfloor} 2^n \left[\left(\left\lfloor \frac{x}{2^n} \right\rfloor + \left\lfloor \frac{y}{2^n} \right\rfloor \right) \bmod 2 \right]$$

Bit Shifts

The bit shifts are sometimes considered bitwise operations, because they treat a value as a series of bits rather than as a numerical quantity. In these operations the digits are moved, or *shifted*, to the left or right. Registers in a computer processor have a fixed width, so some bits will be "shifted out" of the register at one end, while the same number of bits are "shifted in" from the other end; the differences between bit shift operators lie in how they determine the values of the shifted-in bits.

Arithmetic Shift

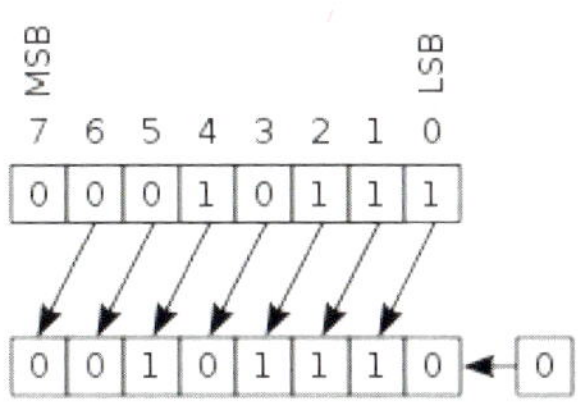

Left arithmetic shift

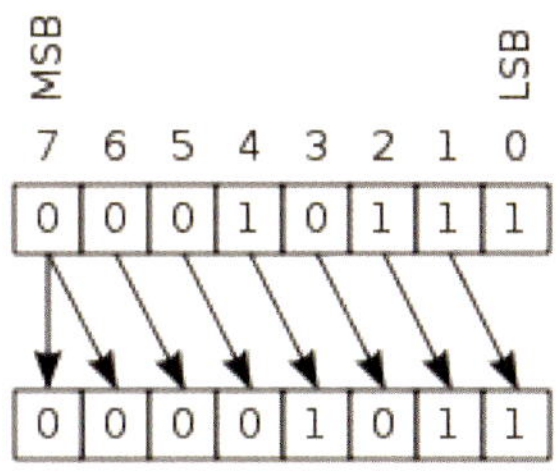

Right arithmetic shift

In an *arithmetic shift*, the bits that are shifted out of either end are discarded. In a left arithmetic shift, zeros are shifted in on the right; in a right arithmetic shift, the sign bit (the MSB in two's complement) is shifted in on the left, thus preserving the sign of the operand.

This example uses an 8-bit register:

00010111 (decimal +23) LEFT-SHIFT

= 00101110 (decimal +46)

10010111 (decimal −105) RIGHT-SHIFT

= 11001011 (decimal −53)

In the first case, the leftmost digit was shifted past the end of the register, and a new 0 was shifted into the rightmost position. In the second case, the rightmost 1 was shifted out (perhaps into the carry flag), and a new 1 was copied into the leftmost position, preserving the sign of the number. Multiple shifts are sometimes shortened to a single shift by some number of digits. For example:

00010111 (decimal +23) LEFT-SHIFT-BY-TWO

= 01011100 (decimal +92)

A left arithmetic shift by n is equivalent to multiplying by 2^n (provided the value does not overflow), while a right arithmetic shift by n of a two's complement value is equivalent to dividing by 2^n and rounding toward negative infinity. If the binary number is treated as ones' complement, then the same right-shift operation results in division by 2^n and rounding toward zero.

Logical Shift

In a *logical shift*, zeros are shifted in to replace the discarded bits. Therefore, the logical and arithmetic left-shifts are exactly the same.

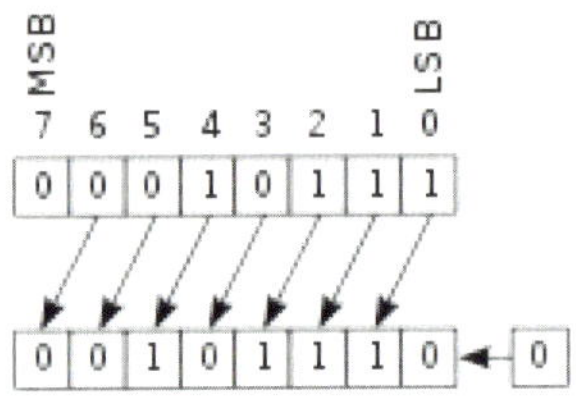

Left logical shift

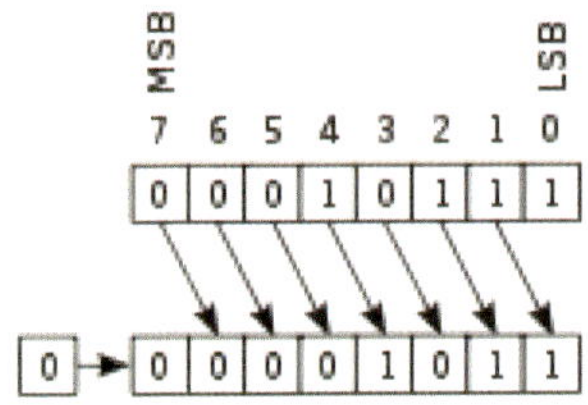

Right logical shift

However, as the logical right-shift inserts value 0 bits into the most significant bit, instead of copying the sign bit, it is ideal for unsigned binary numbers, while the arithmetic right-shift is ideal for signed two's complement binary numbers.

Circular Shift

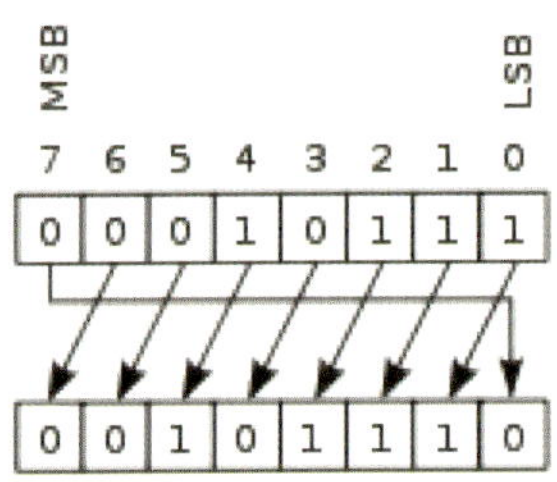

Left circular shift or rotate

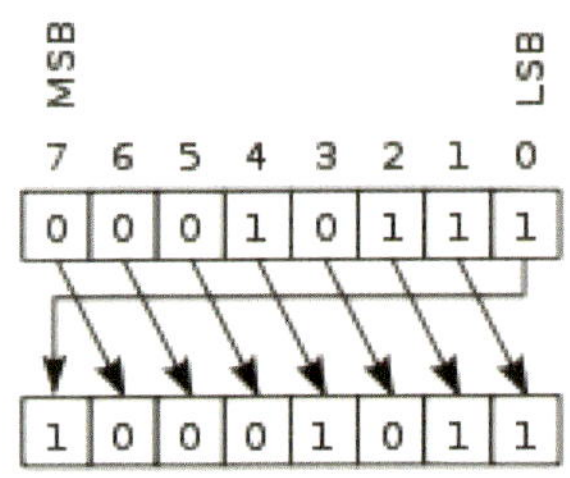

Right circular shift or rotate

Another form of shift is the *circular shift* or *bit rotation*. In this operation, the bits are "rotated" as if the left and right ends of the register were joined. The value that is shifted in on the right during a left-shift is whatever value was shifted out on the left, and vice versa. This operation is useful if it is necessary to retain all the existing bits, and is frequently used in digital cryptography.

Rotate through Carry

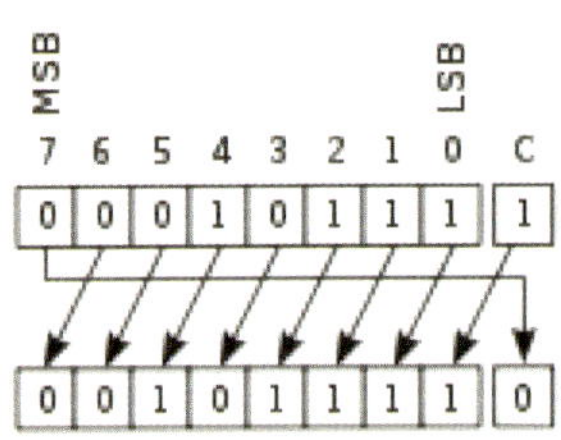

Left rotate through carry

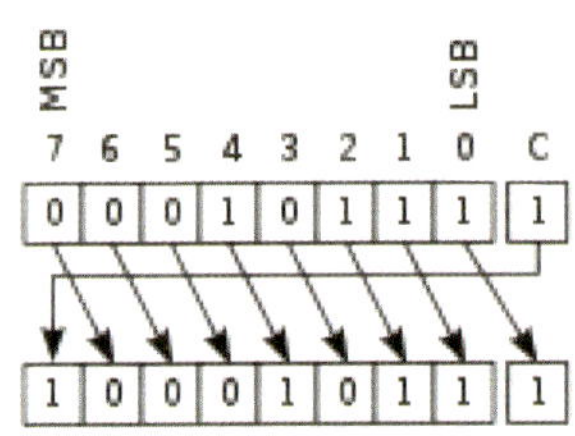

Right rotate through carry

Rotate through carry is similar to the *rotate no carry* operation, but the two ends of the register are separated by the carry flag. The bit that is shifted in (on either end) is the old value of the carry flag, and the bit that is shifted out (on the other end) becomes the new value of the carry flag.

A single *rotate through carry* can simulate a logical or arithmetic shift of one position by setting up the carry flag beforehand. For example, if the carry flag contains 0, then x RIGHT-ROTATE-THROUGH-CARRY-BY-ONE is a logical right-shift, and if the carry flag contains a copy of the sign bit, then x RIGHT-ROTATE-THROUGH-CARRY-BY-ONE is an arithmetic right-shift. For this reason, some microcontrollers such as low end PICs just have *rotate* and *rotate through carry*, and don't bother with arithmetic or logical shift instructions.

Rotate through carry is especially useful when performing shifts on numbers larger than the processor's native word size, because if a large number is stored in two registers, the bit that is shifted off the end of the first register must come in at the other end of the second. With rotate-through-carry, that bit is "saved" in the carry flag during the first shift, ready to shift in during the second shift without any extra preparation.

Shifts in C, C++, C#, Go, Java, JavaScript, Pascal, Perl, PHP, Python and Ruby

In C-inspired languages, the left and right shift operators are "<<" and ">>", respectively. The number of places to shift is given as the second argument to the shift operators. For example,

```
x = y << 2;
```

assigns x the result of shifting y to the left by two bits, which is equivalent to a multiplication by four.

Shifts can result in implementation-defined behavior or undefined behavior, so care must be taken when using them. The result of shifting by a bit count greater than or equal to the word's size is undefined behavior in C and C++. Right-shifting a negative value is implementation-defined and not recommended by good coding practice; the result of left-shifting a signed value is undefined if the result cannot be represented in the result type. In C#, the right-shift is an arithmetic shift when the first operand is an int or long. If the first operand is of type uint or ulong, the right-shift is a logical shift.

Circular Shifts in C-family Languages

The C-family of languages lack a rotate operator, but one can be synthesized from the shift operators. Care must be taken to ensure the statement is well formed to avoid undefined behavior and timing attacks in software with security requirements. For example, a naive implementation that left rotates a 32-bit unsigned value x by n positions is simply:

```
unsigned int x = ..., n = ...;

unsigned int y = (x << n) | (x >> (32 - n));
```

However, a shift by 0 bits results in undefined behavior in the right hand expression (x >> (32 - n)) because 32 - 0 is 32, and 32 is outside the range [0 - 31] inclusive. A second try might result in:

```
unsigned int x = ..., n = ...;

unsigned int y = n ? (x << n) | (x >> (32 - n)) : x;
```

where the shift amount is tested to ensure it does not introduce undefined behavior. However, the branch adds an additional code path and presents an opportunity for timing analysis and attack, which is often not acceptable in high integrity software. In addition, the code compiles to multiple machine instructions, which is often less efficient than the processor's native instruction.

To avoid the undefined behavior and branches under GCC and Clang, the following should be used. The pattern is recognized by many compilers, and the compiler will emit a single rotate instruction:

```
unsigned int x = ..., n = ...;
unsigned int y = (x << n) | (x >> (-n & 31));
```

There are also compiler-specific intrinsics implementing circular shifts, like _rotl8, _rotl16, _rotr8, _rotr16 in Microsoft Visual C++. Clang provides some rotate intrinsics for Microsoft compatibility that suffers the problems above. GCC does not offer rotate intrinsics. Intel also provides x86 Intrinsics.

Shifts in Java

In Java, all integer types are signed, so the "<<" and ">>" operators perform arithmetic shifts. Java adds the operator ">>>" to perform logical right shifts, but since the logical and arithmetic left-shift operations are identical for signed integer, there is no "<<<" operator in Java.

More details of Java shift operators:

- The operators << (left shift), >> (signed right shift), and >>> (unsigned right shift) are called the *shift operators*.
- The type of the shift expression is the promoted type of the left-hand operand. For example, aByte >>> 2 is equivalent to ((int) aByte) >>> 2.
- If the promoted type of the left-hand operand is int, only the five lowest-order bits of the right-hand operand are used as the shift distance. It is as if the right-hand operand were subjected to a bitwise logical AND operator & with the mask value 0x1f (0b11111). The shift distance actually used is therefore always in the range 0 to 31, inclusive.
- If the promoted type of the left-hand operand is long, then only the six lowest-order bits of the right-hand operand are used as the shift distance. It is as if the right-hand operand were subjected to a bitwise logical AND operator & with the mask value 0x3f (0b111111). The shift distance actually used is therefore always in the range 0 to 63, inclusive.
- The value of n >>> s is *n* right-shifted *s* bit positions with zero-extension.
- In bit and shift operations, the type *byte* is implicitly converted to *int*. If the byte value is negative, the highest bit is one, then ones are used to fill up the extra bytes in the int. So byte b1 = -5; int i = b1 | 0x0200; will give i == -5 as result.

Shifts in JavaScript

JavaScript uses bitwise operations to evaluate each of two or more units place to 1 or 0.

Shifts in Pascal

In Pascal, as well as in all its dialects (such as Object Pascal and Standard Pascal), the left and right shift operators are "shl" and "shr", respectively. The number of places to shift is given as the second argument. For example, the following assigns x the result of shifting y to the left by two bits:

```
x := y shl 2;
```

Applications

Bitwise operations are necessary particularly in lower-level programming such as device drivers, low-level graphics, communications protocol packet assembly, and decoding.

Although machines often have efficient built-in instructions for performing arithmetic and logical operations, all these operations can be performed by combining the bitwise operators and zero-testing in various ways. For example, here is a pseudocode implementation of ancient Egyptian multiplication showing how to multiply two arbitrary integers a and b (a greater than b) using only bitshifts and addition:

```
c ← 0
while b ≠ 0
    if (b and 1) ≠ 0
        c ← c + a
    left shift a by 1
    right shift b by 1
return c
```

Another example is a pseudocode implementation of addition, showing how to calculate a sum of two integers a and b using bitwise operators and zero-testing:

```
while a ≠ 0
    c ← b and a
    b ← b xor a
    left shift c by 1
    a ← c
return b
```

Boolean Algebra

A set of rules or Laws of Boolean Algebra expressions have been invented to help reduce the number of logic gates needed to perform a particular logic operation resulting in a list of functions or theorems known commonly as the Laws of Boolean Algebra.

Boolean Algebra is the mathematics we use to analyse digital gates and circuits. We can use these "Laws of Boolean" to both reduce and simplify a complex Boolean expression in an attempt to reduce the number of logic gates required. Boolean Algebra is therefore a system of mathematics based on logic that has its own set of rules or laws which are used to define and reduce Boolean expressions.

The variables used in Boolean Algebra only have one of two possible values, a logic "0" and a logic "1" but an expression can have an infinite number of variables all labelled individually to represent inputs to the expression, For example, variables A, B, C etc, giving us a logical expression of A + B = C, but each variable can ONLY be a 0 or a 1.

Examples of these individual laws of Boolean, rules and theorems for Boolean Algebra are given in the following table.

Truth Tables for the Laws of Boolean

Boolean Expression	Description	Equivalent Switching Circuit	Boolean Algebra Law or Rule
A + 1 = 1	A in parallel with closed = "CLOSED"	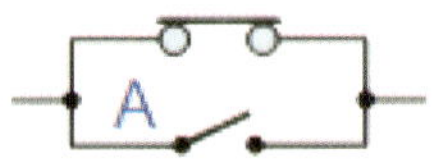	Annulment
A + 0 = A	A in parallel with open = "A"		Identity
A . 1 = A	A in series with closed = "A"		Identity
A . 0 = 0	A in series with open = "OPEN"		Annulment
A + A = A	A in parallel with A = "A"	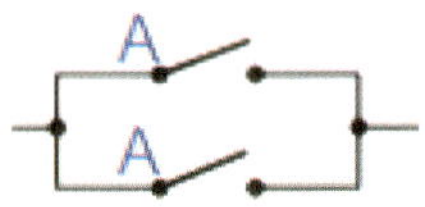	Idempotent
A . A = A	A in series with A = "A"		Idempotent
NOT A = A	NOT NOT A (double negative) = "A"		Double Negation
A + A = 1	A in parallel with NOT A = "CLOSED"	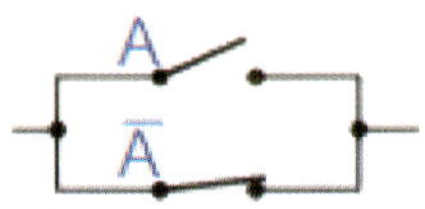	Complement

A . A = 0	A in series with NOT A = "OPEN"		Complement
A+B = B+A	A in parallel with B = B in parallel with A	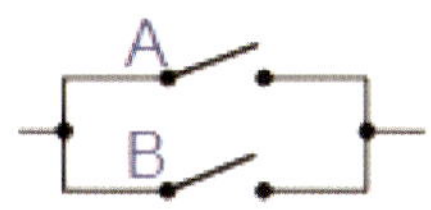	Commutative
A.B = B.A	A in series with B = B in series with A		Commutative
A+B = A.B	invert and replace OR with AND		de Morgan's Theorem
A.B = A+B	invert and replace AND with OR		de Morgan's Theorem

The basic Laws of Boolean Algebra that relate to the Commutative Law allowing a change in position for addition and multiplication, the Associative Law allowing the removal of brackets for addition and multiplication, as well as the Distributive Law allowing the factoring of an expression, are the same as in ordinary algebra.

Each of the Boolean Laws above are given with just a single or two variables, but the number of variables defined by a single law is not limited to this as there can be an infinite number of variables as inputs too the expression. These Boolean laws detailed above can be used to prove any given Boolean expression as well as for simplifying complicated digital circuits.

Operations

Basic Operations

The basic operations of Boolean algebra are as follows:

- AND (conjunction), denoted $x \wedge y$ (sometimes x AND y or Kxy), satisfies $x \wedge y = 1$ if $x = y = 1$ and $x \wedge y = 0$ otherwise.
- OR (disjunction), denoted $x \vee y$ (sometimes x OR y or Axy), satisfies $x \vee y = 0$ if $x = y = 0$ and $x \vee y = 1$ otherwise.
- NOT (negation), denoted $\neg x$ (sometimes NOT x, Nx or $!x$), satisfies $\neg x = 0$ if $x = 1$ and $\neg x = 1$ if $x = 0$.

Alternatively the values of $x \wedge y$, $x \vee y$, and $\neg x$ can be expressed by tabulating their values with truth tables as follows.

x	y	$x \wedge y$	$x \vee y$	x	$\neg x$
0	0	0	0	0	1
1	0	0	1	1	0
0	1	0	1		
1	1	1	1		

If the truth values 0 and 1 are interpreted as integers, these operations may be expressed with the ordinary operations of arithmetic (where $x + y$ uses addition and xy uses multiplication), or by the minimum/maximum functions:

$$x \wedge y = xy = \min(x, y)$$
$$x \vee y = x + y - xy = \max(x, y)$$
$$\neg x = 1 - x$$

One may consider that only the negation and one of the two other operations are basic, because of the following identities that allow to define the conjunction in terms of the negation and the disjunction, and vice versa:

$$x \wedge y = \neg(\neg x \vee \neg y)$$
$$x \vee y = \neg(\neg x \wedge \neg y)$$

Secondary operations

The three Boolean operations described above are referred to as basic, meaning that they can be taken as a basis for other Boolean operations that can be built up from them by composition, the manner in which operations are combined or compounded. Operations composed from the basic operations include the following examples:

$$x \rightarrow y = \neg x \vee y$$

$$x \oplus y = (x \vee y) \wedge \neg(x \wedge y)$$

$$x \equiv y = \neg(x \oplus y)$$

These definitions give rise to the following truth tables giving the values of these operations for all four possible inputs.

x	y	$x \rightarrow y$	$x \oplus y$	$x \equiv y$
0	0	1	0	1
1	0	0	1	0
0	1	1	1	0
1	1	1	0	1

The first operation, $x \rightarrow y$, or Cxy, is called material implication. If x is true then the value of $x \rightarrow y$ is taken to be that of y. But if x is false then the value of y can be ignored; however the operation must return *some* boolean value and there are only two choices. So by definition, $x \rightarrow y$ is *true* when x is false. (Relevance logic suggests this definition by viewing an implication with a false premise as something other than either true or false.)

The second operation, $x \oplus y$, or Jxy, is called exclusive or (often abbreviated as XOR) to distinguish it from disjunction as the inclusive kind. It excludes the possibility of both x and y. Defined in terms of arithmetic it is addition mod 2 where $1 + 1 = 0$.

The third operation, the complement of exclusive or, is equivalence or Boolean equality: $x \equiv y$, or Exy, is true just when x and y have the same value. Hence $x \oplus y$ as its complement can be understood as $x \neq y$, being true just when x and y are different. Equivalence's counterpart in arithmetic mod 2 is $x + y + 1$.

Given two operands, each with two possible values, there are $2^2 = 4$ possible combinations of inputs. Because each output can have two possible values, there are a total of $2^4 = 16$ possible binary Boolean operations.

Laws

A law of Boolean algebra is an identity such as $x\vee(y\vee z) = (x\vee y)\vee z$ between two Boolean terms, where a Boolean term is defined as an expression built up from variables and the constants 0 and 1 using the operations $\wedge$, $\vee$, and $\neg$. The concept can be extended to terms involving other Boolean operations such as $\oplus$, $\rightarrow$, and $\equiv$, but such extensions are unnecessary for the purposes to which the laws are put. Such purposes include the definition of a Boolean algebra as any model of the Boolean laws, and as a means for deriving new laws from old as in the derivation of $x\vee(y\wedge z) = x\vee(z\wedge y)$ from $y\wedge z = z\wedge y$ as treated in the section on axiomatization.

Monotone Laws

Boolean algebra satisfies many of the same laws as ordinary algebra when one matches up $\vee$ with addition and $\wedge$ with multiplication. In particular the following laws are common to both kinds of algebra:

Associativity of $\vee$:	$x \vee (y \vee z) = (x \vee y) \vee z$
Associativity of $\wedge$:	$x \wedge (y \wedge z) = (x \wedge y) \wedge z$
Commutativity of $\vee$:	$x \vee y = y \vee x$
Commutativity of $\wedge$:	$x \wedge y = y \wedge x$
Distributivity of $\wedge$ over $\vee$:	$x \wedge (y \vee z) = (x \wedge y) \vee (x \wedge z)$
Identity for $\vee$:	$x \vee 0 = x$
Identity for $\wedge$:	$x \wedge 1 = x$
Annihilator for $\wedge$:	$x \wedge 0 = 0$

The following laws hold in Boolean Algebra, but not in ordinary algebra:

Annihilator for $\vee$:	$x \vee 1 = 1$
Idempotence of $\vee$:	$x \vee x = x$
Idempotence of $\wedge$:	$x \wedge x = x$
Absorption 1:	$x \wedge (x \vee y) = x$
Absorption 2:	$x \vee (x \wedge y) = x$
Distributivity of $\vee$ over $\wedge$:	$x \vee (y \wedge z) = (x \vee y) \wedge (x \vee z)$

Taking $x = 2$ in the third law above shows that it is not an ordinary algebra law, since $2\times2 = 4$. The remaining five laws can be falsified in ordinary algebra by taking all variables to be 1, for example in Absorption Law 1 the left hand side would be $1(1+1) = 2$ while the right hand side would be 1, and so on.

All of the laws treated so far have been for conjunction and disjunction. These operations have the property that changing either argument either leaves the output unchanged or the output changes in the same way as the input. Equivalently, changing any variable from 0 to 1 never results in the output changing from 1 to 0. Operations with this property are said to be monotone. Thus the axioms so far have all been for monotonic Boolean logic. Nonmonotonicity enters via complement $\neg$ as follows.

Nonmonotone Laws

The complement operation is defined by the following two laws.

Complementation 1	$x \wedge \neg x = 0$
Complementation 2	$x \vee \neg x = 1$

All properties of negation including the laws below follow from the above two laws alone.

In both ordinary and Boolean algebra, negation works by exchanging pairs of elements, whence in both algebras it satisfies the double negation law (also called involution law)

Double negation $\quad \neg(\neg x) = x$

But whereas *ordinary algebra* satisfies the two laws

$$(-x)(-y) = xy$$
$$(-x)+(-y) = -(x+y)$$

Boolean algebra satisfies De Morgan's laws:

De Morgan1	$\neg x \wedge \neg y = \neg(x \vee y)$
De Morgan 2	$\neg x \vee \neg y = \neg(x \wedge y)$

Completeness

The laws listed above define Boolean algebra, in the sense that they entail the rest of the subject. The laws *Complementation* 1 and 2, together with the monotone laws, suffice for this purpose and can therefore be taken as one possible *complete* set of laws or axiomatization of Boolean algebra. Every law of Boolean algebra follows logically from these axioms. Furthermore, Boolean algebras can then be defined as the models of these axioms as treated in the section thereon.

To clarify, writing down further laws of Boolean algebra cannot give rise to any new consequences of these axioms, nor can it rule out any model of them. In contrast, in a list of some but not all of the same laws, there could have been Boolean laws that did not follow from those on the list, and moreover there would have been models of the listed laws that were not Boolean algebras.

This axiomatization is by no means the only one, or even necessarily the most natural given that we did not pay attention to whether some of the axioms followed from others but simply chose to stop when we noticed we had enough laws, treated further in the section on axiomatizations. Or the intermediate notion of axiom can be sidestepped altogether by defining a Boolean law directly as any tautology, understood as an equation that holds for all values of its variables over 0 and 1. All these definitions of Boolean algebra can be shown to be equivalent.

Duality Principle

Principle: If {X, R} is a poset, then {X, R(inverse)} is also a poset.

There is nothing magical about the choice of symbols for the values of Boolean algebra. We could rename 0 and 1 to say α and β, and as long as we did so consistently throughout it would still be Boolean algebra, albeit with some obvious cosmetic differences.

But suppose we rename 0 and 1 to 1 and 0 respectively. Then it would still be Boolean algebra, and moreover operating on the same values. However it would not be identical to our original Boolean algebra because now we find ∨ behaving the way ∧ used to do and vice versa. So there are still some cosmetic differences to show that we've been fiddling with the notation, despite the fact that we're still using 0s and 1s.

But if in addition to interchanging the names of the values we also interchange the names of the two binary operations, *now* there is no trace of what we have done. The end product is completely indistinguishable from what we started with. We might notice that the columns for $x \wedge y$ and $x \vee y$ in the truth tables had changed places, but that switch is immaterial.

When values and operations can be paired up in a way that leaves everything important unchanged when all pairs are switched simultaneously, we call the members of each pair dual to each other. Thus 0 and 1 are dual, and ∧ and ∨ are dual. The Duality Principle, also called De Morgan duality, asserts that Boolean algebra is unchanged when all dual pairs are interchanged.

One change we did not need to make as part of this interchange was to complement. We say that complement is a self-dual operation. The identity or do-nothing operation x (copy the input to the output) is also self-dual. A more complicated example of a self-dual operation is $(x \wedge y) \vee (y \wedge z) \vee (z \wedge x)$. There is no self-dual binary operation that depends on both its arguments. A composition of self-dual operations is a self-dual operation. For example, if $f(x, y, z) = (x \wedge y) \vee (y \wedge z) \vee (z \wedge x)$, then $f(f(x, y, z), x, t)$ is a self-dual operation of four arguments x,y,z,t.

The principle of duality can be explained from a group theory perspective by the fact that there are exactly four functions that are one-to-one mappings (automorphisms) of the set of Boolean polynomials back to itself: the identity function, the complement function, the dual function and the contradual function (complemented dual). These four functions form a group under function composition, isomorphic to the Klein four-group, acting on the set of Boolean polynomials. Walter Gottschalk remarked that consequently a more appropriate name for the phenomenon would be the *principle* (or *square*) *of quaternality*.

Diagrammatic Representations

Venn Diagrams

A Venn diagram is a representation of a Boolean operation using shaded overlapping regions. There is one region for each variable, all circular in the examples here. The interior and exterior of region x corresponds respectively to the values 1 (true) and 0 (false) for variable x. The shading indicates the value of the operation for each combination of regions, with dark denoting 1 and light 0 (some authors use the opposite convention).

The three Venn diagrams in the figure below represent respectively conjunction $x \wedge y$, disjunction $x \vee y$, and complement $\neg x$.

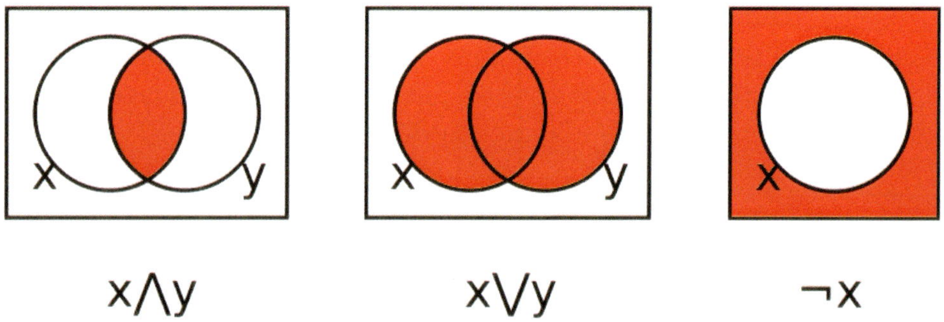

Figure. Venn diagrams for conjunction, disjunction, and complement

For conjunction, the region inside both circles is shaded to indicate that $x \wedge y$ is 1 when both variables are 1. The other regions are left unshaded to indicate that $x \wedge y$ is 0 for the other three combinations.

The second diagram represents disjunction $x \vee y$ by shading those regions that lie inside either or both circles. The third diagram represents complement $\neg x$ by shading the region *not* inside the circle.

While we have not shown the Venn diagrams for the constants 0 and 1, they are trivial, being respectively a white box and a dark box, neither one containing a circle. However we could put a circle for x in those boxes, in which case each would denote a function of one argument, x, which returns the same value independently of x, called a constant function. As far as their outputs are concerned, constants and constant functions are indistinguishable; the difference is that a constant takes no arguments, called a *zeroary* or *nullary* operation, while a constant function takes one argument, which it ignores, and is a *unary* operation.

Venn diagrams are helpful in visualizing laws. The commutativity laws for $\wedge$ and $\vee$ can be seen from the symmetry of the diagrams: a binary operation that was not commutative would not have a symmetric diagram because interchanging x and y would have the effect of reflecting the diagram horizontally and any failure of commutativity would then appear as a failure of symmetry.

Idempotence of $\wedge$ and $\vee$ can be visualized by sliding the two circles together and noting that the shaded area then becomes the whole circle, for both $\wedge$ and $\vee$.

To see the first absorption law, $x \wedge (x \vee y) = x$, start with the diagram in the middle for $x \vee y$ and note that the portion of the shaded area in common with the x circle is the whole of the x circle. For the second absorption law, $x\lor(x\land y) = x$, start with the left diagram for $x\land y$ and note that shading the whole of the x circle results in just the x circle being shaded, since the previous shading was inside the x circle.

The double negation law can be seen by complementing the shading in the third diagram for $\neg x$, which shades the x circle.

To visualize the first De Morgan's law, $(\neg x)\land(\neg y) = \neg(x\lor y)$, start with the middle diagram for $x\lor y$ and complement its shading so that only the region outside both circles is shaded, which is what the right hand side of the law describes. The result is the same as if we shaded that region which is both outside the x circle *and* outside the y circle, i.e. the conjunction of their exteriors, which is what the left hand side of the law describes.

The second De Morgan's law, $(\neg x)\lor(\neg y) = \neg(x\land y)$, works the same way with the two diagrams interchanged.

The first complement law, $x\land\neg x = 0$, says that the interior and exterior of the x circle have no overlap. The second complement law, $x\lor\neg x = 1$, says that everything is either inside or outside the x circle.

Digital Logic Gates

Digital logic is the application of the Boolean algebra of 0 and 1 to electronic hardware consisting of logic gates connected to form a circuit diagram. Each gate implements a Boolean operation, and is depicted schematically by a shape indicating the operation. The shapes associated with the gates for conjunction (AND-gates), disjunction (OR-gates), and complement (inverters) are as follows.

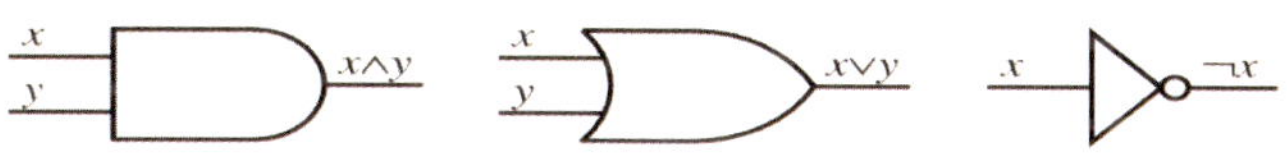

Logic gates

The lines on the left of each gate represent input wires or *ports*. The value of the input is represented by a voltage on the lead. For so-called "active-high" logic, 0 is represented by a voltage close to zero or "ground", while 1 is represented by a voltage close to the supply voltage; active-low reverses this. The line on the right of each gate represents the output port, which normally follows the same voltage conventions as the input ports.

Complement is implemented with an inverter gate. The triangle denotes the operation that simply copies the input to the output; the small circle on the output denotes the actual inversion complementing the input. The convention of putting such a circle on any port means that the signal passing through this port is complemented on the way through, whether it is an input or output port.

The Duality Principle, or De Morgan's laws, can be understood as asserting that complementing all three ports of an AND gate converts it to an OR gate and vice versa, as shown in Figure below. Complementing both ports of an inverter however leaves the operation unchanged.

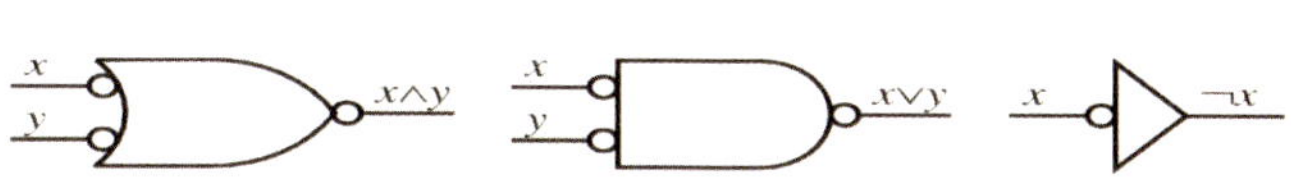

De Morgan equivalents

More generally one may complement any of the eight subsets of the three ports of either an AND or OR gate. The resulting sixteen possibilities give rise to only eight Boolean operations, namely those with an odd number of 1's in their truth table. There are eight such because the "odd-bit-out" can be either 0 or 1 and can go in any of four positions in the truth table. There being sixteen binary Boolean operations, this must leave eight operations with an even number of 1's in their truth tables. Two of these are the constants 0 and 1 (as binary operations that ignore both their inputs); four are the operations that depend nontrivially on exactly one of their two inputs, namely x, y, $\neg x$, and $\neg y$; and the remaining two are $x \oplus y$ (XOR) and its complement $x \equiv y$.

Propositional Logic

Propositional logic is a logical system that is intimately connected to Boolean algebra. Many syntactic concepts of Boolean algebra carry over to propositional logic with only minor changes in notation and terminology, while the semantics of propositional logic are defined via Boolean algebras in a way that the tautologies (theorems) of propositional logic correspond to equational theorems of Boolean algebra.

Syntactically, every Boolean term corresponds to a propositional formula of propositional logic. In this translation between Boolean algebra and propositional logic, Boolean variables x, y... become propositional variables (or atoms) P, Q,..., Boolean terms such as $x \vee y$ become propositional formulas $P \vee Q$, 0 becomes *false* or $\perp$, and 1 becomes *true* or T. It is convenient when referring to generic propositions to use Greek letters Φ, Ψ,... as metavariables (variables outside the language of propositional calculus, used when talking *about* propositional calculus) to denote propositions.

The semantics of propositional logic rely on truth assignments. The essential idea of a truth assignment is that the propositional variables are mapped to elements of a fixed Boolean algebra, and then the truth value of a propositional formula using these letters is the element of the Boolean algebra that is obtained by computing the value of the Boolean term corresponding to the formula. In classical semantics, only the two-element Boolean algebra is used, while in Boolean-valued semantics arbitrary Boolean algebras are considered. A tautology is a propositional formula that is assigned truth value *1* by every truth assignment of its propositional variables to an arbitrary Boolean algebra (or, equivalently, every truth assignment to the two element Boolean algebra).

These semantics permit a translation between tautologies of propositional logic and equational theorems of Boolean algebra. Every tautology Φ of propositional logic can be expressed as the Boolean equation $\Phi = 1$, which will be a theorem of Boolean algebra. Conversely every theorem $\Phi = \Psi$ of Boolean algebra corresponds to the tautologies $(\Phi \vee \neg\Psi) \wedge (\neg\Phi \vee \Psi)$ and $(\Phi \wedge \Psi) \vee (\neg\Phi \wedge \neg\Psi)$. If $\rightarrow$ is in the language these last tautologies can also be written as $(\Phi \rightarrow \Psi) \wedge (\Psi \rightarrow \Phi)$, or as two separate theorems $\Phi \rightarrow \Psi$ and $\Psi \rightarrow \Phi$; if $\equiv$ is available then the single tautology $\Phi \equiv \Psi$ can be used.

Applications

One motivating application of propositional calculus is the analysis of propositions and deductive arguments in natural language. Whereas the proposition "if $x = 3$ then $x+1 = 4$" depends on the meanings of such symbols as + and 1, the proposition "if $x = 3$ then $x = 3$" does not; it is true merely by virtue of its structure, and remains true whether "$x = 3$" is replaced by "$x = 4$" or "the moon is made of green cheese." The generic or abstract form of this tautology is "if P then P", or in the language of Boolean algebra, "$P \to P$".

Replacing P by $x = 3$ or any other proposition is called instantiation of P by that proposition. The result of instantiating P in an abstract proposition is called an instance of the proposition. Thus "$x = 3 \to x = 3$" is a tautology by virtue of being an instance of the abstract tautology "$P \to P$". All occurrences of the instantiated variable must be instantiated with the same proposition, to avoid such nonsense as $P \to x = 3$ or $x = 3 \to x = 4$.

Propositional calculus restricts attention to abstract propositions, those built up from propositional variables using Boolean operations. Instantiation is still possible within propositional calculus, but only by instantiating propositional variables by abstract propositions, such as instantiating Q by $Q \to P$ in $P \to (Q \to P)$ to yield the instance $P \to ((Q \to P) \to P)$.

(The availability of instantiation as part of the machinery of propositional calculus avoids the need for metavariables within the language of propositional calculus, since ordinary propositional variables can be considered within the language to denote arbitrary propositions. The metavariables themselves are outside the reach of instantiation, not being part of the language of propositional calculus but rather part of the same language for talking about it that this sentence is written in, where we need to be able to distinguish propositional variables and their instantiations as being distinct syntactic entities.)

Deductive Systems for Propositional Logic

An axiomatization of propositional calculus is a set of tautologies called axioms and one or more inference rules for producing new tautologies from old. A *proof* in an axiom system A is a finite nonempty sequence of propositions each of which is either an instance of an axiom of A or follows by some rule of A from propositions appearing earlier in the proof (thereby disallowing circular reasoning). The last proposition is the theorem proved by the proof. Every nonempty initial segment of a proof is itself a proof, whence every proposition in a proof is itself a theorem. An axiomatization is sound when every theorem is a tautology, and complete when every tautology is a theorem.

Sequent Calculus

Propositional calculus is commonly organized as a Hilbert system, whose operations are just those of Boolean algebra and whose theorems are Boolean tautologies, those Boolean terms equal to the Boolean constant 1. Another form is sequent calculus, which has two sorts, propositions as in ordinary propositional calculus, and pairs of lists of propositions called sequents, such as $A \vee B, A \wedge C, \ldots \vdash A, B \to C, \ldots$. The two halves of a sequent are called the antecedent and the succedent respectively. The customary metavariable denoting an antecedent or part thereof is Γ, and for a succedent Δ; thus $\Gamma, A \vdash \Delta$ would denote a sequent whose succedent is a list Δ and whose antecedent is a list Γ with an additional proposition A appended after it. The antecedent is interpreted as the conjunction of its propositions, the succedent as the disjunction of its propositions, and the sequent itself as the entailment of the succedent by the antecedent.

Entailment differs from implication in that whereas the latter is a binary *operation* that returns a value in a Boolean algebra, the former is a binary *relation* which either holds or does not hold. In this sense entailment is an *external* form of implication, meaning external to the Boolean algebra, thinking of the reader of the sequent as also being external and interpreting and comparing antecedents and succedents in some Boolean algebra. The natural interpretation of $\vdash$ is as $\leq$ in the partial order of the Boolean algebra defined by $x \leq y$ just when $x \vee y = y$. This ability to mix external implication $\vdash$ and internal implication $\rightarrow$ in the one logic is among the essential differences between sequent calculus and propositional calculus.

Applications

Boolean algebra as the calculus of two values is fundamental to computer circuits, computer programming, and mathematical logic, and is also used in other areas of mathematics such as set theory and statistics.

Computers

In the early 20th century, several electrical engineers intuitively recognized that Boolean algebra was analogous to the behavior of certain types of electrical circuits. Claude Shannon formally proved such behavior was logically equivalent to Boolean algebra in his 1937 master's thesis, *A Symbolic Analysis of Relay and Switching Circuits.*

Today, all modern general purpose computers perform their functions using two-value Boolean logic; that is, their electrical circuits are a physical manifestation of two-value Boolean logic. They achieve this in various ways: as voltages on wires in high-speed circuits and capacitive storage devices, as orientations of a magnetic domain in ferromagnetic storage devices, as holes in punched cards or paper tape, and so on. (Some early computers used decimal circuits or mechanisms instead of two-valued logic circuits.)

Of course, it is possible to code more than two symbols in any given medium. For example, one might use respectively 0, 1, 2, and 3 volts to code a four-symbol alphabet on a wire, or holes of different sizes in a punched card. In practice, the tight constraints of high speed, small size, and low power combine to make noise a major factor. This makes it hard to distinguish between symbols when there are several possible symbols that could occur at a single site. Rather than attempting to distinguish between four voltages on one wire, digital designers have settled on two voltages per wire, high and low.

Computers use two-value Boolean circuits for the above reasons. The most common computer architectures use ordered sequences of Boolean values, called bits, of 32 or 64 values, e.g. 011010001 101011001010101001011. When programming in machine code, assembly language, and certain other programming languages, programmers work with the low-level digital structure of the data registers. These registers operate on voltages, where zero volts represents Boolean 0, and a reference voltage (often +5V, +3.3V, +1.8V) represents Boolean 1. Such languages support both numeric operations and logical operations. In this context, "numeric" means that the computer treats sequences of bits as binary numbers (base two numbers) and executes arithmetic operations like add, subtract, multiply, or divide. "Logical" refers to the Boolean logical operations of disjunction, conjunction, and negation between two sequences of bits, in which each bit in one sequence is simply compared to its counterpart in the other sequence. Programmers therefore have the option of working in and applying the rules of either numeric algebra or Boolean algebra as needed. A core differentiating feature between these families of operations is the existence of the carry operation in the first but not the second.

Two-valued Logic

Other areas where two values is a good choice are the law and mathematics. In everyday relaxed conversation, nuanced or complex answers such as "maybe" or "only on the weekend" are acceptable. In more focused situations such as a court of law or theorem-based mathematics however it is deemed advantageous to frame questions so as to admit a simple yes-or-no answer—is the defendant guilty or not guilty, is the proposition true or false—and to disallow any other answer. However much of a straitjacket this might prove in practice for the respondent, the principle of the simple yes-no question has become a central feature of both judicial and mathematical logic, making two-valued logic deserving of organization and study in its own right.

A central concept of set theory is membership. Now an organization may permit multiple degrees of membership, such as novice, associate, and full. With sets however an element is either in or out. The candidates for membership in a set work just like the wires in a digital computer: each candidate is either a member or a nonmember, just as each wire is either high or low.

Algebra being a fundamental tool in any area amenable to mathematical treatment, these considerations combine to make the algebra of two values of fundamental importance to computer hardware, mathematical logic, and set theory.

Two-valued logic can be extended to multi-valued logic, notably by replacing the Boolean domain {0, 1} with the unit interval [0,1], in which case rather than only taking values 0 or 1, any value between and including 0 and 1 can be assumed. Algebraically, negation (NOT) is replaced with $1 - x$, conjunction (AND) is replaced with multiplication (xy), and disjunction (OR) is defined via De Morgan's law. Interpreting these values as logical truth values yields a multi-valued logic, which forms the basis for fuzzy logic and probabilistic logic. In these interpretations, a value is interpreted as the "degree" of truth – to what extent a proposition is true, or the probability that the proposition is true.

Boolean Operations

The original application for Boolean operations was mathematical logic, where it combines the truth values, true or false, of individual formulas.

Natural languages such as English have words for several Boolean operations, in particular conjunction (*and*), disjunction (*or*), negation (*not*), and implication (*implies*). *But not* is synonymous with *and not*. When used to combine situational assertions such as "the block is on the table" and "cats drink milk," which naively are either true or false, the meanings of these logical connectives often have the meaning of their logical counterparts. However, with descriptions of behavior such as "Jim walked through the door", one starts to notice differences such as failure of commutativity, for example the conjunction of "Jim opened the door" with "Jim walked through the door" in that order is not equivalent to their conjunction in the other order, since *and* usually means *and then* in such cases. Questions can be similar: the order "Is the sky blue, and why is the sky blue?" makes more sense than the reverse order. Conjunctive commands about behavior are like behavioral assertions, as in *get dressed and go to school.* Disjunctive commands such *love me or leave me* or *fish or cut bait* tend to be asymmetric via the implication that one alternative is less preferable. Conjoined nouns such as *tea and milk* generally describe aggregation as with set union while *tea or milk* is a choice. However context can reverse these senses, as in *your choices are coffee and tea* which usually means the same as *your choices are coffee or tea* (alternatives). Double negation as in "I don't not like milk" rarely means literally "I do like milk" but rather conveys some sort of hedging, as though to imply that there is a third possibility. "Not not P" can be loosely interpreted as "surely P", and although P necessarily implies "not not P" the converse is suspect in English, much as with intuitionistic logic. In view of the highly idiosyncratic usage of conjunctions in natural languages, Boolean algebra cannot be considered a reliable framework for interpreting them.

Boolean operations are used in digital logic to combine the bits carried on individual wires, thereby interpreting them over {0,1}. When a vector of n identical binary gates are used to combine two bit vectors each of n bits, the individual bit operations can be understood collectively as a single operation on values from a Boolean algebra with 2^n elements.

Naive set theory interprets Boolean operations as acting on subsets of a given set X. As we saw earlier this behavior exactly parallels the coordinate-wise combinations of bit vectors, with the union of two sets corresponding to the disjunction of two bit vectors and so on.

The 256-element free Boolean algebra on three generators is deployed in computer displays based on raster graphics, which use bit blit to manipulate whole regions consisting of pixels, relying on Boolean operations to specify how the source region should be combined with the destination, typically with the help of a third region called the mask. Modern video cards offer all $2^{2^3} = 256$ ternary operations for this purpose, with the choice of operation being a one-byte (8-bit) parameter. The constants SRC = 0xaa or 10101010, DST = 0xcc or 11001100, and MSK = 0xf0 or 11110000 allow Boolean operations such as (SRC^DST)&MSK (meaning XOR the source and destination and then AND the result with the mask) to be written directly as a constant denoting a byte calculated at compile time, 0x60 in the (SRC^DST)&MSK example, 0x66 if just SRC^DST, etc. At run time the video card interprets the byte as the raster operation indicated by the original expression in a uniform way that requires remarkably little hardware and which takes time completely independent of the complexity of the expression.

Solid modeling systems for computer aided design offer a variety of methods for building objects from other objects, combination by Boolean operations being one of them. In this method the space in which objects exist is understood as a set S of voxels (the three-dimensional analogue of pixels in two-dimensional graphics) and shapes are defined as subsets of S, allowing objects to be combined as sets via union, intersection, etc. One obvious use is in building a complex shape from simple shapes simply as the union of the latter. Another use is in sculpting understood as removal of material: any grinding, milling, routing, or drilling operation that can be performed with physical machinery on physical materials can be simulated on the computer with the Boolean operation $x \wedge \neg y$ or $x - y$, which in set theory is set difference, remove the elements of y from those of x. Thus given two shapes one to be machined and the other the material to be removed, the result of machining the former to remove the latter is described simply as their set difference.

Boolean Searches

Search engine queries also employ Boolean logic. For this application, each web page on the Internet may be considered to be an "element" of a "set". The following examples use a syntax supported by Google.

- Doublequotes are used to combine whitespace-separated words into a single search term.
- Whitespace is used to specify logical AND, as it is the default operator for joining search terms:

"Search term 1" "Search term 2"

- The OR keyword is used for logical OR:

"Search term 1" OR "Search term 2"

- A prefixed minus sign is used for logical NOT:

"Search term 1" –"Search term 2"

Boolean Algebra Examples

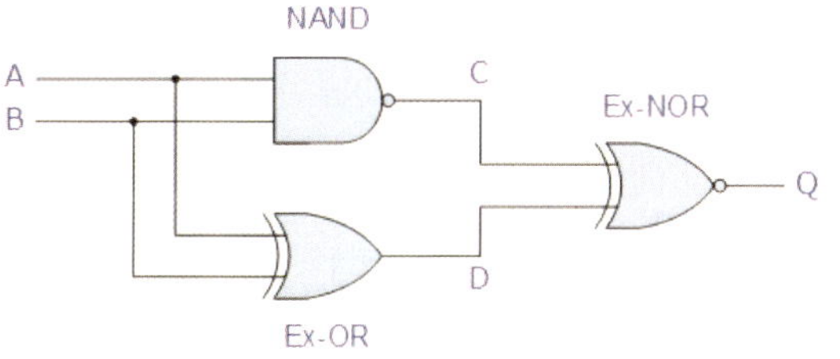

Boolean Algebra can be used to identify unnecessary logic gates within a digital logic design reducing the number of gates required.

Boolean Algebra Example 1

Construct a Truth Table for the logical functions at points C, D and Q in the following circuit and identify a single logic gate that can be used to replace the whole circuit.

First observations tell us that the circuit consists of a 2-input NAND gate, a 2-input EX-OR gate and finally a 2-input EX-NOR gate at the output. As there are only 2 inputs to the circuit labelled A and B, there can only be 4 possible combinations of the input (2^2) and these are: 0-0, 0-1, 1-0 and finally 1-1. Plotting the logical functions from each gate in tabular form will give us the following truth table for the whole of the logic circuit below.

Inputs		Output at		
A	B	C	D	Q
0	0	1	0	0
0	1	1	1	1
1	0	1	1	1
1	1	0	0	1

From the truth table above, column C represents the output function generated by the NAND gate, while column D represents the output function from the Ex-OR gate. Both of these two output expressions then become the input condition for the Ex-NOR gate at the output.

It can be seen from the truth table that an output at Q is present when any of the two inputs A or B are at logic 1. The only truth table that satisfies this condition is that of an OR Gate. Therefore, the whole of the above circuit can be replaced by just one single 2-input OR Gate.

Boolean Algebra Example 2

Find the Boolean algebra expression for the following system.

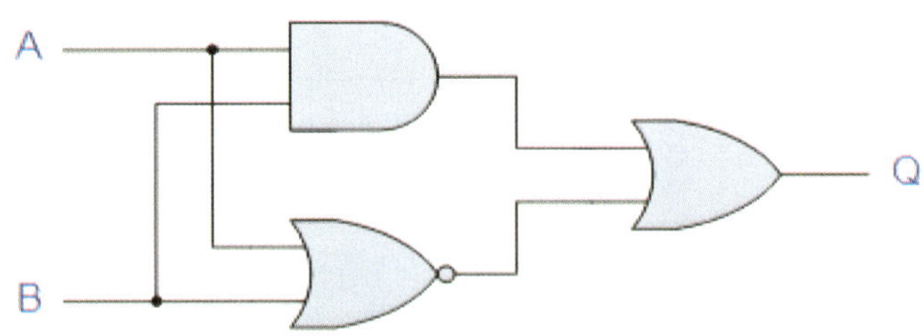

The system consists of an AND Gate, a NOR Gate and finally an OR Gate. The expression for the AND gate is A.B, and the expression for the NOR gate is $\overline{A+B}$. Both these expressions are also separate inputs to the OR gate which is defined as $\overline{A+B}$. Thus the final output expression is given as:

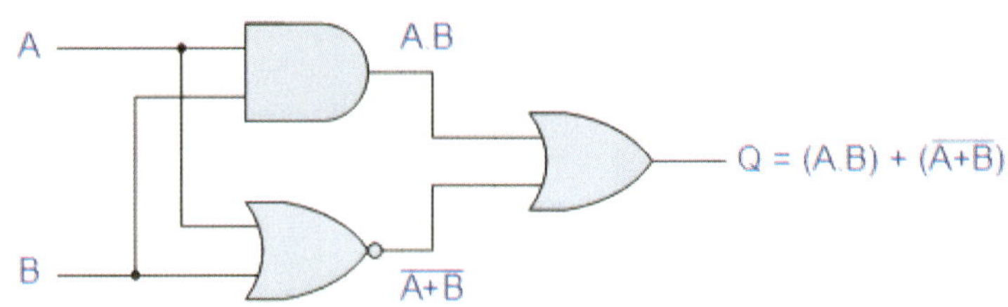

The output of the system is given as Q = (A.B) + ($\overline{A+B}$), but the notation $\overline{A+B}$ is the same as the De Morgan´s notation $\overline{A}.\overline{B}$, Then substituting $\overline{A}.\overline{B}$ into the output expression gives us a final output notation of Q = (A.B)+($\overline{A}.\overline{B}$), which is the Boolean notation for an Exclusive-NOR Gate as seen in the previous topic.

Inputs		Intermediates		Output
B	A	A.B	$\overline{A+B}$	Q
0	0	0	1	1
0	1	0	0	0
1	0	0	0	0
1	1	1	0	1

Then, the whole circuit above can be replaced by just one single Exclusive-NOR Gate and indeed an Exclusive-NOR Gate is made up of these individual gate functions.

Boolean Algebra Example 3

Find the Boolean algebra expression for the following system.

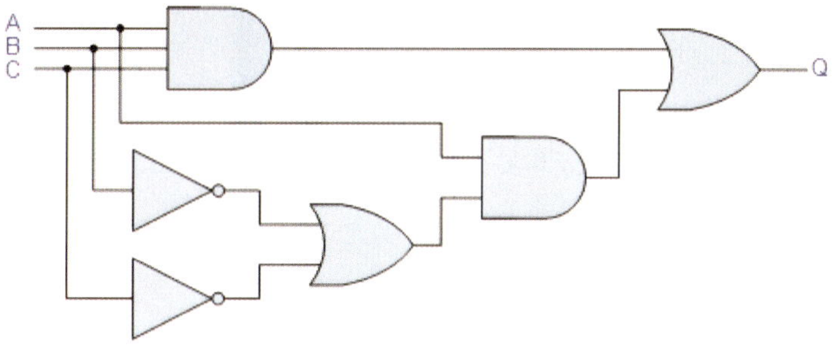

This system may look more complicated than the other two to analyse but again, the logic circuit just consists of simple AND, OR and NOT gates connected together.

As with the previous Boolean examples, we can simplify the circuit by writing down the Boolean notation for each logic gate function in turn in order to give us a final expression for the output at Q.

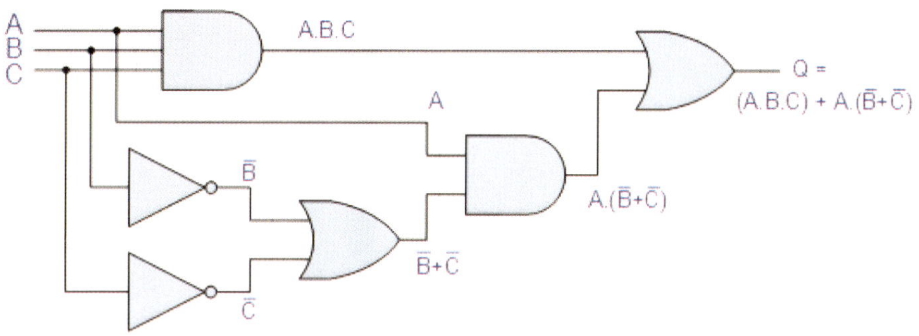

The output from the 3-input AND gate is only at logic "1" when ALL the gates inputs are HIGH at logic level "1" (A.B.C). The output from the lower OR gate is only a "1" when one or both inputs B or C are at logic level "0". The output from the 2-input AND gate is a "1" when input A is a "1" and inputs B or C are at "0". Then the output at Q is only a "1" when inputs A.B.C equal "1" or A is equal to "1" and both inputs B or C equal "0", A.($\overline{B}+\overline{C}$).

By using "de Morgan's theorem" inputs B and input C cancel out as to produce an output at Q they can be either at logic "1" or at logic "0". Then this just leaves input A as the only input needed to give an output at Q as shown in the table below.

Inputs			Intermediates					Output
C	B	A	A.B.C	B	C	B+C	A.(B+C)	Q
0	0	0	0	1	1	1	0	0
0	0	1	0	1	1	1	1	1
0	1	0	0	0	1	1	0	0
0	1	1	0	0	1	1	1	1
1	0	0	0	1	0	1	0	0
1	0	1	0	1	0	1	1	1
1	1	0	0	0	0	0	0	0
1	1	1	1	0	0	0	0	1

Then we can see that the entire logic circuit above can be replaced by just one single input labelled A thereby reducing a circuit of six individual logic gates to just one single piece of wire, (or Buffer). This type of circuit analysis using Boolean Algebra can be very powerful and quickly identify any unnecessary logic gates within a digital logic design thereby reducing the number of gates required, the power consumption of the circuit and of course the cost.

References

- Robson, Eleanor; Stedall, Jacqueline, eds. (2009), "Myth No. 2: the Horus eye fractions", The Oxford Handbook of the History of Mathematics, Oxford University Press, p. 790, ISBN 9780199213122
- Bender, Andrea; Beller, Sieghard (16 December 2013). "Mangarevan invention of binary steps for easier calculation". Proceedings of the National Academy of Sciences. 111: 1322–1327. doi:10.1073/pnas.1309160110. PMC 3910603. PMID 24344278
- Digital-Electronics-Introduction-Number-Systems-Number-Codes: engineersgarage.com, Retrieved 28 June 2018
- "National Inventors Hall of Fame – George R. Stibitz". 20 August 2008. Archived from the original on 9 July 2010. Retrieved 5 July 2010
- J.E.H. Smith (2008). Leibniz: What Kind of Rationalist?: What Kind of Rationalist?. Springer. p. 415. ISBN 978-1-4020-8668-7
- Write-your-name-in-binary-code, educational-resources: sciencefriday.com, Retrieved 05 July 2018
- Shirley, John W. (1951). "Binary numeration before Leibniz". American Journal of Physics. 19 (8): 452–454. doi:10.1119/1.1933042
- "Pioneers – The people and ideas that made a difference – George Stibitz (1904–1995)". Kerry Redshaw. 20 February 2006. Retrieved 5 July 2010
- Binary-codes, computer-logical-organization: tutorialspoint.com, Retrieved 15 April 2018
- J. Michael Dunn; Gary M. Hardegree (2001). Algebraic methods in philosophical logic. Oxford University Press US. p. 2. ISBN 978-0-19-853192-0
- Küveler, Gerd; Schwoch, Dietrich (2013) [1996]. Arbeitsbuch Informatik - eine praxisorientierte Einführung in die Datenverarbeitung mit Projektaufgabe (in German). Vieweg-Verlag, reprint: Springer-Verlag. doi:10.1007/978-3-322-92907-5. ISBN 978-3-528-04952-2. 9783322929075. Retrieved 2015-08-05
- Binary-coded-decimal: electronics-tutorials.ws, Retrieved 25 May 2018
- Shannon, Claude (1949). «The Synthesis of Two-Terminal Switching Circuits». Bell System Technical Journal. 28: 59–98. doi:10.1002/j.1538-7305.1949.tb03624.x
- Boole, George (2009) [1854]. An Investigation of the Laws of Thought on Which are Founded the Mathematical Theories of Logic and Probabilities (Macmillan, Dover Publications, reprinted with corrections [1958] ed.). New York: Cambridge University Press. ISBN 978-1-108-00153-3

4

Digital Circuits

Circuits that operate using digital signals are called digital circuits. These operate using logic gates that are printed on integrated circuits. This chapter closely examines some of the devices used in digital circuits, such as binary adder, subtractor, multiplier, encoder, decoder, shift register, etc. The principles of combinational logic circuits, sequential logic circuits, asynchronous and synchronous sequential circuits, etc. which are crucial to the understanding of digital circuit operations have also been carefully dealt in this chapter.

Circuit

A digital circuit is a circuit that functions on a number of different logic gates. The logic gates differentiate power signals. The power signals are then transferred to different parts of the digital circuit through other gates to create an output signal directly pertinent to the energy level at the moment of signal input.

Most digital circuits are comprised mainly of smaller analog components that, because of the logic gate occurrence, only operate within a certain frame of voltages. Usually these operate at extremely low voltage signals. Digital circuits also require that the analog components located throughout the circuit not be placed in a manner that will allow them to perform analog functions. This usually means there is a logic gate both before and after the analog component.

Digital circuits are becoming more and more popular as technology requires the electronic devices used every day to become smaller and smaller, making the items more readily accessible regardless of location or circumstance. A digital circuit reacts quickly as power or signal is delivered to them, as long as the signal it receives is within the correct frame of voltage or energy. Digital circuits usually work best with low voltage signals, as they are more capable of handling these signals than higher voltages. Analog circuits typically require much more voltage behind the input signal to perform with the speed that digital circuits provide.

Digital circuits have become the first option in supplying power signals to a number of devices large or small in function, purpose, or size, because of the speed and accuracy they are able to provide. Digital circuits are also capable of greatly reducing the amount of electrical noise generated by many circuits as voltage or power signal is applied to them. This makes digital circuits a first choice when the stability of the signal is essential to the accuracy or function of the device.

As digital circuits are a bit more complex in most cases than analog circuitry, the use of a trained engineer or electrician is necessary whenever maintenance of a digital circuit is required. Another of the drawbacks of a digital circuit is that digital circuits require more energy to be supplied from the low voltage power or signal input in order to create the same results as an analog circuit. As a result, many digital circuit layouts have problems related to the temperature they generate during their functioning.

A digital circuit is a circuit where the signal should be one of two discrete levels. Each level is interpreted as one of two different states (for instance, 0 or 1). These circuits built with transistors to make logic gates in order to execute Boolean logic operation. This logic is the base of digital electronics & computer processing. Digital circuits are less vulnerable to degradation in excellence than analog circuits. It is also simpler to execute error detection and rectification with digital signals. To make the routine process of designing these circuits, designers use EDA (electronic design automation) tools, a kind of software that develops the logic in a digital circuit.

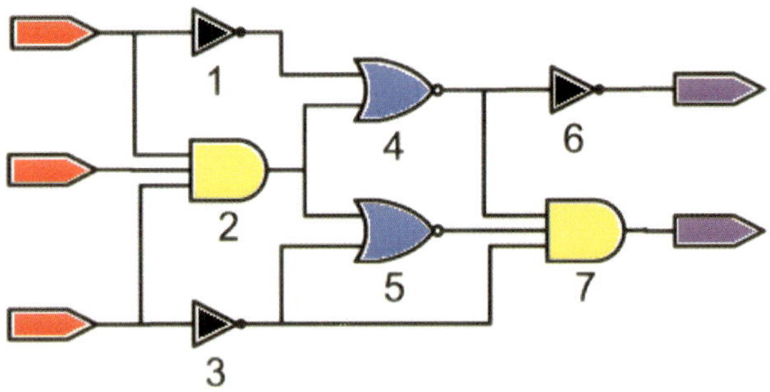

Digital Circuit

A digital circuit is used to alter the waves into pulse waves. It measures a waveform thousand of times every second and stores the data in binary form. For instance, after 12msecs, a signal might be 2.4 volts high & after 14msecs, it might be at 2.6 volts. This circuit changes the volts and times into binary data, and the waves become a series of 1's and 0's. When the circuit has to generate sound from a speaker, it produces an o/p signal which is at 2.4 V after 12msec and at 2.6 V after 14msec, similar to the original wave.

Digital circuits generate an estimate of the natural waveform, so their quality of o/p is incomplete by the number of curve size they carry out.

Efficiency of a Circuit

An efficiency of a circuit mainly depends on how rapidly it can produce results and how much power it utilizes. These circuits have to wait for a wave to complete a cycle before they can carry out their tasks. And they require sufficient power to generate the peak of the wave. The digital circuit's speed is incomplete only by the speed of the gears that make up the circuit and not by the indications they are processing. To work with tiny pulses needs moderately little power. In most of the applications, analog circuits are slower & require more power than digital circuits.

Digital circuits just have to maintain path of their pulses. Even if numerous pulses get vanished, it influences only a few of the thousands of sizes. As a result, these circuits are more accurate and can replicate their i/ps more precisely.

Construction

A digital circuit is typically constructed from small electronic circuits called logic gates that can be used to create combinational logic. Each logic gate is designed to perform a function of boolean logic when acting on logic signals. A logic gate is generally created from one or more electrically controlled switches, usually transistors but thermionic valves have seen historic use. The output of a logic gate can, in turn, control or feed into more logic gates.

A binary clock, hand-wired on breadboards

Integrated circuits consist of multiple transistors on one silicon chip, and are the least expensive way to make large number of interconnected logic gates. Integrated circuits are usually designed by engineers using electronic design automation software to perform some type of function.

Integrated circuits are usually interconnected on a printed circuit board which is a board which holds electrical components, and connects them together with copper traces.

Design

Each logic symbol is represented by a different shape. The actual set of shapes was introduced in 1984 under IEEE/ANSI standard 91-1984. "The logic symbol given under this standard are being increasingly used now and have even started appearing in the literature published by manufacturers of digital integrated circuits."

Another form of digital circuit is constructed from lookup tables, (many sold as "programmable logic devices", though other kinds of PLDs exist). Lookup tables can perform the same functions as machines based on logic gates, but can be easily reprogrammed without changing the wiring. This means that a designer can often repair design errors without changing the arrangement of wires. Therefore, in small volume products, programmable logic devices are often the preferred solution. They are usually designed by engineers using electronic design automation software.

When the volumes are medium to large, and the logic can be slow, or involves complex algorithms or sequences, often a small microcontroller is programmed to make an embedded system. These are usually programmed by software engineers.

When only one digital circuit is needed, and its design is totally customized, as for a factory production line controller, the conventional solution is a programmable logic controller, or PLC. These are usually programmed by electricians, using ladder logic.

Structure of Digital Systems

Engineers use many methods to minimize logic functions, in order to reduce the circuit's complexity. When the complexity is less, the circuit also has fewer errors and less electronics, and is therefore less expensive.

The most widely used simplification is a minimization algorithm like the Espresso heuristic logic minimizer within a CAD system, although historically, binary decision diagrams, an automated Quine–McCluskey algorithm, truth tables, Karnaugh maps, and Boolean algebra have been used.

Representation

Representations are crucial to an engineer's design of digital circuits. Some analysis methods only work with particular representations.

The classical way to represent a digital circuit is with an equivalent set of logic gates. Another way, often with the least electronics, is to construct an equivalent system of electronic switches (usually transistors). One of the easiest ways is to simply have a memory containing a truth table. The inputs are fed into the address of the memory, and the data outputs of the memory become the outputs.

For automated analysis, these representations have digital file formats that can be processed by computer programs. Most digital engineers are very careful to select computer programs ("tools") with compatible file formats.

Combinational vs. Sequential

To choose representations, engineers consider types of digital systems. Most digital systems divide into "combinational systems" and "sequential systems." A combinational system always presents the same output when given the same inputs. It is basically a representation of a set of logic functions, as already discussed.

A sequential system is a combinational system with some of the outputs fed back as inputs. This makes the digital machine perform a "sequence" of operations. The simplest sequential system is probably a flip flop, a mechanism that represents a binary digit or "bit".

Sequential systems are often designed as state machines. In this way, engineers can design a system's gross behavior, and even test it in a simulation, without considering all the details of the logic functions.

Sequential systems divide into two further subcategories. "Synchronous" sequential systems change state all at once, when a "clock" signal changes state. "Asynchronous" sequential systems propagate changes whenever inputs change. Synchronous sequential systems are made of well-characterized asynchronous circuits such as flip-flops, that change only when the clock changes, and which have carefully designed timing margins.

Synchronous Systems

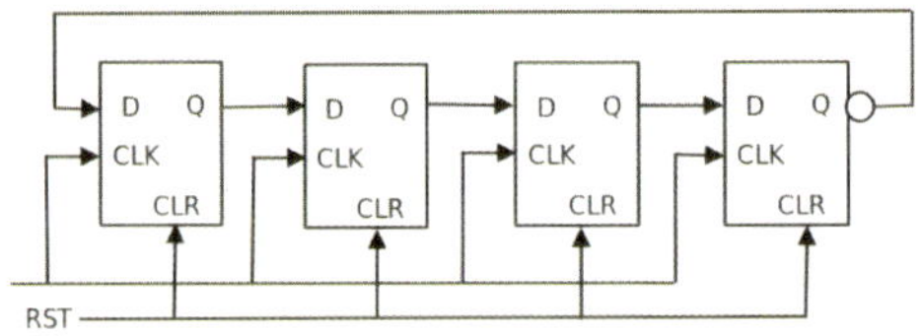

A 4-bit ring counter using D-type flip flops is an example of synchronous logic. Each device is connected to the clock signal, and update together.

The usual way to implement a synchronous sequential state machine is to divide it into a piece of combinational logic and a set of flip flops called a "state register." Each time a clock signal ticks, the state register captures the feedback generated from the previous state of the combinational logic, and feeds it back as an unchanging input to the combinational part of the state machine. The fastest rate of the clock is set by the most time-consuming logic calculation in the combinational logic.

The state register is just a representation of a binary number. If the states in the state machine are numbered (easy to arrange), the logic function is some combinational logic that produces the number of the next state.

Asynchronous Systems

As of 2014, most digital logic is synchronous because it is easier to create and verify a synchronous design. However, asynchronous logic is thought can be superior because its speed is not constrained by an arbitrary clock; instead, it runs at the maximum speed of its logic gates. Building an asynchronous system using faster parts makes the circuit faster.

Nevertherless, most systems need circuits that allow external unsynchronized signals to enter synchronous logic circuits. These are inherently asynchronous in their design and must be analyzed as such. Examples of widely used asynchronous circuits include synchronizer flip-flops, switch debouncers and arbiters.

Asynchronous logic components can be hard to design because all possible states, in all possible timings must be considered. The usual method is to construct a table of the minimum and maximum time that each such state can exist, and then adjust the circuit to minimize the number of such states. Then the designer must force the circuit to periodically wait for all of its parts to enter a compatible state (this is called "self-resynchronization"). Without such careful design, it is easy to accidentally produce asynchronous logic that is "unstable," that is, real electronics will have unpredictable results because of the cumulative delays caused by small variations in the values of the electronic components.

Register Transfer Systems

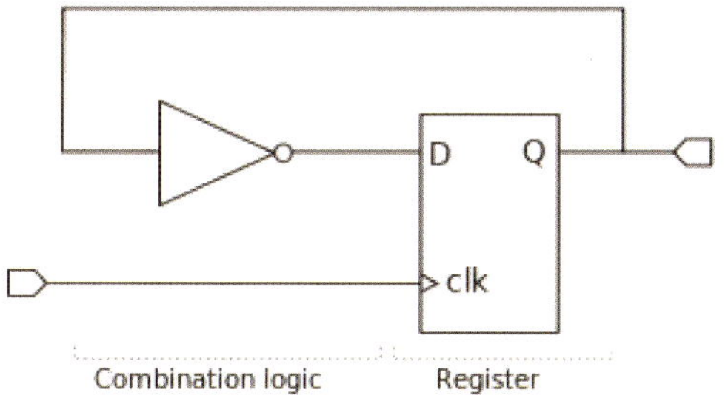

Example of a simple circuit with a toggling output. The inverter forms the combinational logic in this circuit, and the register holds the state.

Many digital systems are data flow machines. These are usually designed using synchronous register transfer logic, using hardware description languages such as VHDL or Verilog.

In register transfer logic, binary numbers are stored in groups of flip flops called registers. The outputs of each register are a bundle of wires called a "bus" that carries that number to other calculations. A calculation is simply a piece of combinational logic. Each calculation also has an output bus, and these may be connected to the inputs of several registers. Sometimes a register will have a multiplexer on its input, so that it can store a number from any one of several buses. Alternatively, the outputs of several items may be connected to a bus through buffers that can turn off the output of all of the devices except one. A sequential state machine controls when each register accepts new data from its input.

Asynchronous register-transfer systems (such as computers) have a general solution. In the 1980s, some researchers discovered that almost all synchronous register-transfer machines could be converted to asynchronous designs by using first-in-first-out synchronization logic. In this scheme, the digital machine is characterized as a set of data flows. In each step of the flow, an asynchronous "synchronization circuit" determines when the outputs of that step are valid, and presents a signal that says, "grab the data" to the stages that use that stage's inputs. It turns out that just a few relatively simple synchronization circuits are needed.

Computer Design

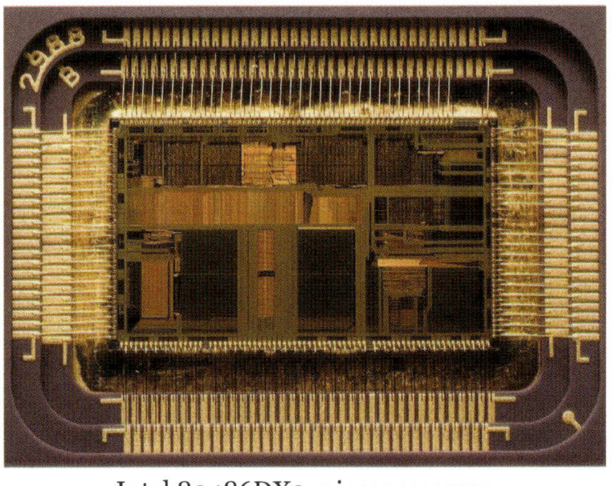

Intel 80486DX2 microprocessor

The most general-purpose register-transfer logic machine is a computer. This is basically an automatic binary abacus. The control unit of a computer is usually designed as a microprogram run by a microsequencer. A microprogram is much like a player-piano roll. Each table entry or "word" of the microprogram commands the state of every bit that controls the computer. The sequencer then counts, and the count addresses the memory or combinational logic machine that contains the microprogram. The bits from the microprogram control the arithmetic logic unit, memory and other parts of the computer, including the microsequencer itself. A "specialized computer" is usually a conventional computer with special-purpose control logic or microprogram.

In this way, the complex task of designing the controls of a computer is reduced to a simpler task of programming a collection of much simpler logic machines.

Almost all computers are synchronous. However, true asynchronous computers have also been designed. One example is the Aspida DLX core. Another was offered by ARM Holdings. Speed advantages have not materialized, because modern computer designs already run at the speed of their slowest component, usually memory. These do use somewhat less power because a clock distribution network is not needed. An unexpected advantage is that asynchronous computers do not produce spectrally-pure radio noise, so they are used in some mobile-phone base-station controllers. They may be more secure in cryptographic applications because their electrical and radio emissions can be more difficult to decode.

Computer Architecture

Computer architecture is a specialized engineering activity that tries to arrange the registers, calculation logic, buses and other parts of the computer in the best way for some purpose. Computer architects have applied large amounts of ingenuity to computer design to reduce the cost and increase the speed and immunity to programming errors of computers. An increasingly common goal is to reduce the power used in a battery-powered computer system, such as a cell-phone. Many computer architects serve an extended apprenticeship as microprogrammers.

Design Issues in Digital Circuits

Digital circuits are made from analog components. The design must assure that the analog nature of the components doesn't dominate the desired digital behavior. Digital systems must manage noise and timing margins, parasitic inductances and capacitances, and filter power connections.

Bad designs have intermittent problems such as "glitches", vanishingly fast pulses that may trigger some logic but not others, "runt pulses" that do not reach valid "threshold" voltages, or unexpected ("undecoded") combinations of logic states.

Additionally, where clocked digital systems interface to analog systems or systems that are driven from a different clock, the digital system can be subject to metastability where a change to the input violates the set-up time for a digital input latch. This situation will self-resolve, but will take a random time, and while it persists can result in invalid signals being propagated within the digital system for a short time.

Since digital circuits are made from analog components, digital circuits calculate more slowly than low-precision analog circuits that use a similar amount of space and power. However, the digital circuit will calculate more repeatably, because of its high noise immunity. On the other hand, in the high-precision domain (for example, where 14 or more bits of precision are needed), analog circuits require much more power and area than digital equivalents.

Automated Design Tools

To save costly engineering effort, much of the effort of designing large logic machines has been automated. The computer programs are called "electronic design automation tools" or just "EDA."

Simple truth table-style descriptions of logic are often optimized with EDA that automatically produces reduced systems of logic gates or smaller lookup tables that still produce the desired outputs. The most common example of this kind of software is the Espresso heuristic logic minimizer.

Most practical algorithms for optimizing large logic systems use algebraic manipulations or binary decision diagrams, and there are promising experiments with genetic algorithms and annealing optimizations.

To automate costly engineering processes, some EDA can take state tables that describe state machines and automatically produce a truth table or a function table for the combinational logic of a state machine. The state table is a piece of text that lists each state, together with the conditions controlling the transitions between them and the belonging output signals.

It is common for the function tables of such computer-generated state-machines to be optimized with logic-minimization software such as Minilog.

Often, real logic systems are designed as a series of sub-projects, which are combined using a "tool flow." The tool flow is usually a "script," a simplified computer language that can invoke the software design tools in the right order.

Tool flows for large logic systems such as microprocessors can be thousands of commands long, and combine the work of hundreds of engineers.

Writing and debugging tool flows is an established engineering specialty in companies that produce digital designs. The tool flow usually terminates in a detailed computer file or set of files that describe how to physically construct the logic. Often it consists of instructions to draw the transistors and wires on an integrated circuit or a printed circuit board.

Parts of tool flows are "debugged" by verifying the outputs of simulated logic against expected inputs. The test tools take computer files with sets of inputs and outputs, and highlight discrepancies between the simulated behavior and the expected behavior.

Once the input data is believed correct, the design itself must still be verified for correctness. Some tool flows verify designs by first producing a design, and then scanning the design to produce compatible input data for the tool flow. If the scanned data matches the input data, then the tool flow has probably not introduced errors.

The functional verification data are usually called "test vectors". The functional test vectors may be preserved and used in the factory to test that newly constructed logic works correctly. However, functional test patterns don't discover common fabrication faults. Production tests are often designed by software tools called "test pattern generators". These generate test vectors by examining the structure of the logic and systematically generating tests for particular faults. This way the fault coverage can closely approach 100%, provided the design is properly made testable.

Once a design exists, and is verified and testable, it often needs to be processed to be manufacturable as well. Modern integrated circuits have features smaller than the wavelength of the light used to expose the photoresist. Manufacturability software adds interference patterns to the exposure masks to eliminate open-circuits, and enhance the masks' contrast.

Design for Testability

There are several reasons for testing a logic circuit. When the circuit is first developed, it is necessary to verify that the design circuit meets the required functional and timing specifications. When multiple copies of a correctly designed circuit are being manufactured, it is essential to test each copy to ensure that the manufacturing process has not introduced any flaws.

A large logic machine (say, with more than a hundred logical variables) can have an astronomical number of possible states. Obviously, in the factory, testing every state is impractical if testing each state takes a microsecond, and there are more states than the number of microseconds since the universe began. Unfortunately, this ridiculous-sounding case is typical.

Fortunately, large logic machines are almost always designed as assemblies of smaller logic machines. To save time, the smaller sub-machines are isolated by permanently installed "design for test" circuitry, and are tested independently.

One common test scheme known as "scan design" moves test bits serially (one after another) from external test equipment through one or more serial shift registers known as "scan chains". Serial scans have only one or two wires to carry the data, and minimize the physical size and expense of the infrequently used test logic.

After all the test data bits are in place, the design is reconfigured to be in "normal mode" and one or more clock pulses are applied, to test for faults (e.g. stuck-at low or stuck-at high) and capture the test result into flip-flops and/or latches in the scan shift register(s). Finally, the result of the test is shifted out to the block boundary and compared against the predicted "good machine" result.

In a board-test environment, serial to parallel testing has been formalized with a standard called "JTAG" (named after the "Joint Test Action Group" that made it).

Another common testing scheme provides a test mode that forces some part of the logic machine to enter a "test cycle." The test cycle usually exercises large independent parts of the machine.

Trade-offs

Several numbers determine the practicality of a system of digital logic: cost, reliability, fanout and speed. Engineers explored numerous electronic devices to get a favourable combination of these personalities.

Cost

The cost of a logic gate is crucial, primarily because very many gates are needed to build a computer or other advanced digital system and because the more gates can be used, the more able and/or respondent the machine can become. Since the bulk of a digital computer is simply an interconnected network of logic gates, the overall cost of building a computer correlates strongly with the price per logic gate. In the 1930s, the earliest digital logic systems were constructed from telephone relays because these were inexpensive and relatively reliable. After that, electrical engineers always used the cheapest available electronic switches that could still fulfill the requirements.

The earliest integrated circuits were a happy accident. They were constructed not to save money, but to save weight, and permit the Apollo Guidance Computer to control an inertial guidance system for a spacecraft. The first integrated circuit logic gates cost nearly $50 (in 1960 dollars, when an engineer earned $10,000/year). To everyone's surprise, by the time the circuits were mass-produced, they had become the least-expensive method of constructing digital logic. Improvements in this technology have driven all subsequent improvements in cost.

With the rise of integrated circuits, reducing the absolute number of chips used represented another way to save costs. The goal of a designer is not just to make the simplest circuit, but to keep the component count down. Sometimes this results in more complicated designs with respect to the underlying digital logic but nevertheless reduces the number of components, board size, and even power consumption. A major motive for reducing component count on printed circuit boards is to reduce the manufacturing defect rate and increase reliability, as every soldered connection is a potentially bad one, so the defect and failure rates tend to increase along with the total number of component pins.

For example, in some logic families, NAND gates are the simplest digital gate to build. All other logical operations can be implemented by NAND gates. If a circuit already required a single NAND gate, and a single chip normally carried four NAND gates, then the remaining gates could be used to implement other logical operations like logical and. This could eliminate the need for a separate chip containing those different types of gates.

Reliability

The "reliability" of a logic gate describes its mean time between failure (MTBF). Digital machines often have millions of logic gates. Also, most digital machines are "optimized" to reduce their cost. The result is that often, the failure of a single logic gate will cause a digital machine to stop working. It is possible to design machines to be more reliable by using redundant logic which will not malfunction as a result of the failure of any single gate (or even any two, three, or four gates), but this necessarily entails using more components, which raises the financial cost and also usually increases the weight of the machine and may increase the power it consumes.

Digital machines first became useful when the MTBF for a switch got above a few hundred hours. Even so, many of these machines had complex, well-rehearsed repair procedures, and would be nonfunctional for hours because a tube burned-out, or a moth got stuck in a relay. Modern transistorized integrated circuit logic gates have MTBFs greater than 82 billion hours ($8.2 \cdot 10^{10}$ hours), and need them because they have so many logic gates.

Fanout

Fanout describes how many logic inputs can be controlled by a single logic output without exceeding the electrical current ratings of the gate outputs. The minimum practical fanout is about five. Modern electronic logic gates using CMOS transistors for switches have fanouts near fifty, and can sometimes go much higher.

Speed

The "switching speed" describes how many times per second an inverter (an electronic representation of a "logical not" function) can change from true to false and back. Faster logic can accomplish more operations in less time. Digital logic first became useful when switching speeds got above 50 Hz, because that was faster than a team of humans operating mechanical calculators. Modern electronic digital logic routinely switches at 5 GHz ($5 \cdot 10^9$ Hz), and some laboratory systems switch at more than 1 THz ($1 \cdot 10^{12}$ Hz).

Logic Families

Design started with relays. Relay logic was relatively inexpensive and reliable, but slow. Occasionally a mechanical failure would occur. Fanouts were typically about 10, limited by the resistance of the coils and arcing on the contacts from high voltages.

Later, vacuum tubes were used. These were very fast, but generated heat, and were unreliable because the filaments would burn out. Fanouts were typically 5...7, limited by the heating from the tubes' current. In the 1950s, special "computer tubes" were developed with filaments that omitted volatile elements like silicon. These ran for hundreds of thousands of hours.

The first semiconductor logic family was resistor–transistor logic. This was a thousand times more reliable than tubes, ran cooler, and used less power, but had a very low fan-in of 3. Diode–transistor logic improved the fanout up to about 7, and reduced the power. Some DTL designs used two power-supplies with alternating layers of NPN and PNP transistors to increase the fanout.

Transistor–transistor logic (TTL) was a great improvement over these. In early devices, fanout improved to 10, and later variations reliably achieved 20. TTL was also fast, with some variations achieving switching times as low as 20 ns. TTL is still used in some designs.

Emitter coupled logic is very fast but uses a lot of power. It was extensively used for high-performance computers made up of many medium-scale components (such as the Illiac IV).

By far, the most common digital integrated circuits built today use CMOS logic, which is fast, offers high circuit density and low-power per gate. This is used even in large, fast computers, such as the IBM System z.

Combinational Logic Circuit

Unlike Sequential Logic Circuits whose outputs are dependant on both their present inputs and their previous output state giving them some form of Memory.

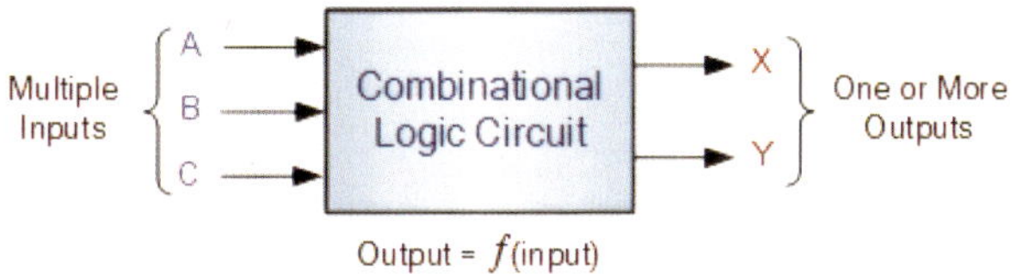

The outputs of Combinational Logic Circuits are only determined by the logical function of their current input state, logic "0" or logic "1", at any given instant in time.

The result is that combinational logic circuits have no feedback, and any changes to the signals being applied to their inputs will immediately have an effect at the output. In other words, in a Combinational Logic Circuit, the output is dependant at all times on the combination of its inputs. So if one of its inputs condition changes state, from 0-1 or 1-0, so too will the resulting output as by default combinational logic circuits have "no memory", "timing" or "feedback loops" within their design.

Combinational Logic

Combinational Logic Circuits are made up from basic logic NAND, NOR or NOT gates that are "combined" or connected together to produce more complicated switching circuits. These logic gates are the building blocks of combinational logic circuits. An example of a combinational circuit is a decoder, which converts the binary code data present at its input into a number of different output lines, one at a time producing an equivalent decimal code at its output.

Combinational logic circuits can be very simple or very complicated and any combinational circuit can be implemented with only NAND and NOR gates as these are classed as "universal" gates.

The three main ways of specifying the function of a combinational logic circuit are:

- Boolean Algebra – This forms the algebraic expression showing the operation of the logic circuit for each input variable either True or False that results in a logic "1" output.
- Truth Table – A truth table defines the function of a logic gate by providing a concise list that shows all the output states in tabular form for each possible combination of input variable that the gate could encounter.
- Logic Diagram – This is a graphical representation of a logic circuit that shows the wiring and connections of each individual logic gate, represented by a specific graphical symbol, that implements the logic circuit.

and all three of these logic circuit representations are shown below.

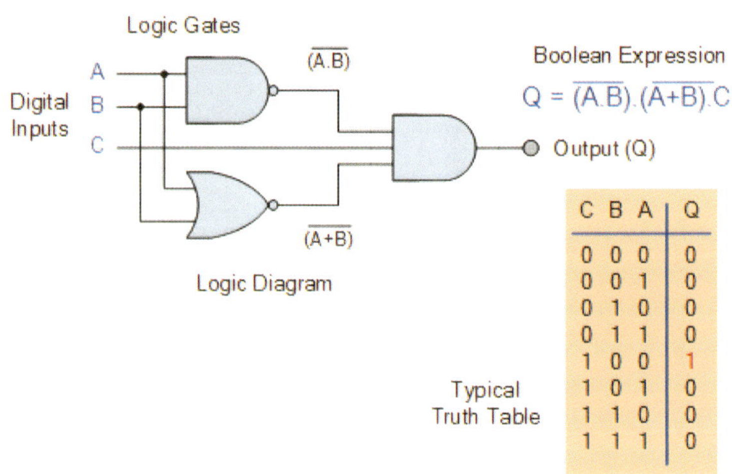

C	B	A	Q
0	0	0	0
0	0	1	0
0	1	0	0
0	1	1	0
1	0	0	1
1	0	1	0
1	1	0	0
1	1	1	0

As combinational logic circuits are made up from individual logic gates only, they can also be considered as "decision making circuits" and combinational logic is about combining logic gates together to process two or more signals in order to produce at least one output signal according to the logical function of each logic gate. Common combinational circuits made up from individual logic gates that carry out a desired application include Multiplexers, De-multiplexers, Encoders, Decoders, Full and Half Adders etc.

Classification of Combinational Logic

One of the most common uses of combinational logic is in Multiplexer and De-multiplexer type circuits. Here, multiple inputs or outputs are connected to a common signal line and logic gates are used to decode an address to select a single data input or output switch.

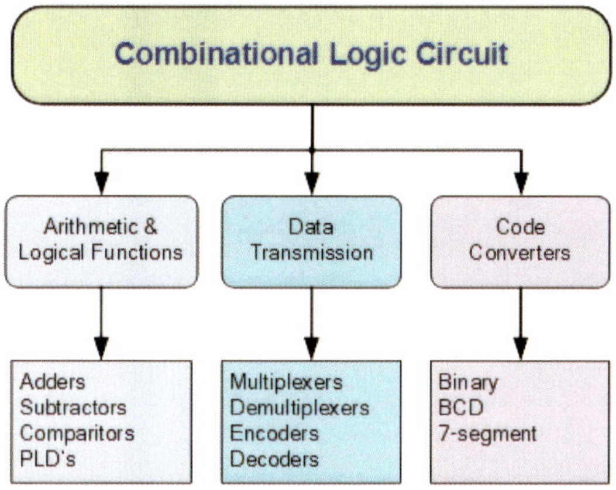

A multiplexer consist of two separate components, a logic decoder and some solid state switches, but before we can discuss multiplexers, decoders and de-multiplexers in more detail we first need to understand how these devices use these "solid state switches" in their design.

Solid State Switches

Standard TTL logic devices made up from Transistors can only pass signal currents in one direction only making them "uni-directional" devices and poor imitations of conventional electro-mechanical switches or relays. However, some CMOS switching devices made up from FET's act as near perfect "bi-directional" switches making them ideal for use as solid state switches.

Solid State Switch Applications

- Analogue Switches – Used in Data Switching and Communications, Video and Audio Signal Switching, Instrumentation and Process Control Circuits etc.
- Digital Switches – High Speed Data Transmission, Switching and Signal Routing, Ethernet, LAN's, USB and Serial Transmissions etc.
- Power Switches – Power Supplies and General "Standby Power" Switching Applications, Switching of Larger Voltages and Currents etc.

Binary Adder

A common and very useful combinational logic circuit which can be constructed using just a few basic logic gates allowing it to add together two or more binary numbers is the Binary Adder.

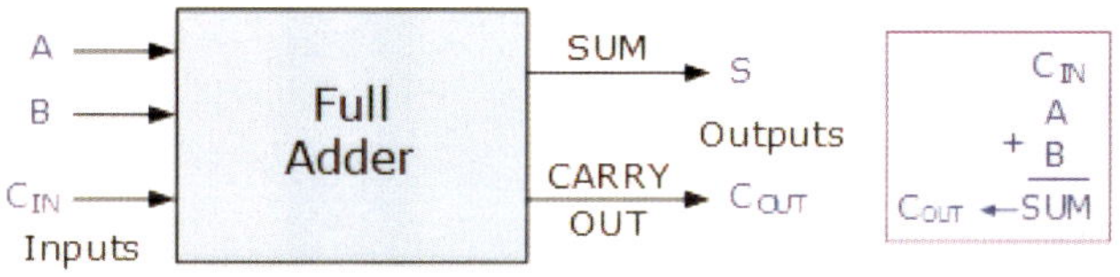

A basic Binary Adder circuit can be made from standard AND and Ex-OR gates allowing us to "add" together two single bit binary numbers, A and B.

The addition of these two digits produces an output called the SUM of the addition and a second output called the CARRY or Carry-out, (C_{OUT}) bit according to the rules for binary addition. One of the main uses for the Binary Adder is in arithmetic and counting circuits. Consider the simple addition of the two denary (base 10) numbers below.

123	A	(Augend)
+ 789	B	(Addend)
912	SUM	

From our maths lessons at school, we learnt that each number column is added together starting from the right hand side and that each digit has a weighted value depending upon its position within the columns.

When each column is added together a carry is generated if the result is greater or equal to 10, the base number. This carry is then added to the result of the addition of the next column to the left and so on, simple school math's addition, add the numbers and carry.

The adding of binary numbers is exactly the same idea as that for adding together decimal numbers but this time a carry is only generated when the result in any column is greater or equal to "2", the base number of binary. In other words 1 + 1 creates a carry.

Binary Addition

Binary Addition follows these same basic rules as for the denary addition above except in binary there are only two digits with the largest digit being "1". So when adding binary numbers, a carry out is generated when the "SUM" equals or is greater than two (1+1) and this becomes a "CARRY" bit for any subsequent addition being passed over to the next column for addition and so on. Consider the single bit addition below.

Binary Addition of Two Bits

0	0	1	1
+ 0	+ 1	+ 0	+ 1
0	1	1	(carry) 1←0

When the two single bits, A and B are added together, the addition of "0 + 0", "0 + 1" and "1 + 0" results in either a "0" or a "1" until you get to the final column of "1 + 1" then the sum is equal to "2". But the number two does not exists in binary however, 2 in binary is equal to 10, in other words a zero for the sum plus an extra carry bit.

Then the operation of a simple adder requires two data inputs producing two outputs, the Sum (S) of the equation and a Carry (C) bit as shown.

Binary Adder Block Diagram

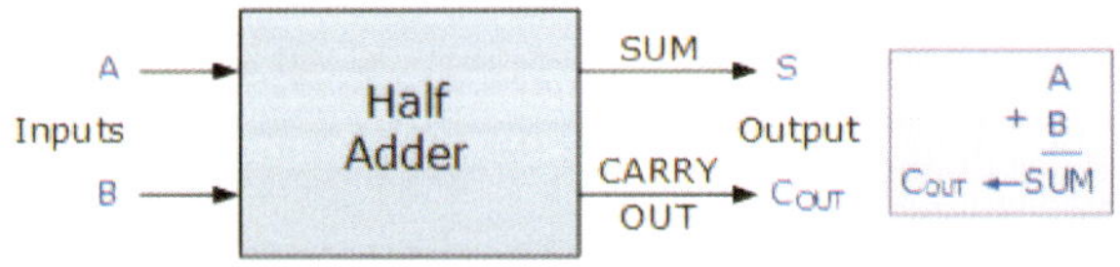

For the simple 1-bit addition problem above, the resulting carry bit could be ignored but you may have noticed something else with regards to the addition of these two bits, the sum of their binary addition resembles that of an Exclusive-OR Gate. If we label the two bits as A and B then the resulting truth table is the sum of the two bits but without the final carry.

2-input Exclusive-OR Gate

Symbol	Truth Table		
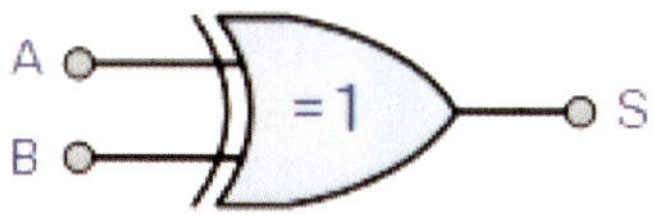	B	A	S
	0	0	0
	0	1	1
2-input Ex-OR Gate	1	0	1
	1	1	0

We can see from the truth table above, that an Exclusive-OR gate only produces an output "1" when either input is at logic "1", but not both the same as for the binary addition of the previous two bits. However in order to perform the addition of two numbers, microprocessors and electronic calculators require the extra carry bit to correctly calculate the equations so we need to rewrite the previous summation to include two-bits of output data as shown below.

00	00	01	01
+ 00	+ 01	+ 00	+ 01
00	01	01	10

From the above equations we now know that an Exclusive-OR gate will only produce an output "1" when "EITHER" input is at logic "1", so we need an additional output to produce the carry bit when "BOTH" inputs A and B are at logic "1". One digital gate that fits the bill perfectly producing an output "1" when both of its inputs A and B are "1" (HIGH) is the standard AND Gate.

2-input AND Gate

Symbol	Truth Table		
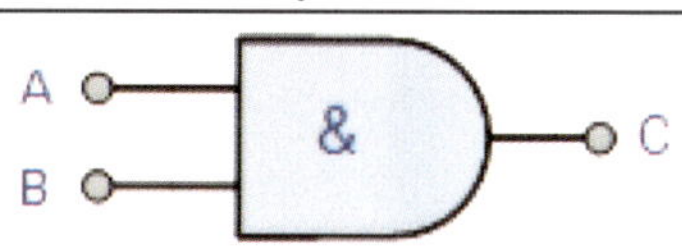	B	A	C
	0	0	0
	0	1	0
2-input AND Gate	1	0	0
	1	1	1

By combining the Exclusive-OR gate with the AND gate results in a simple digital binary adder circuit known commonly as the "Half Adder" circuit.

Half Adder

A half adder is a type of adder, an electronic circuit that performs the addition of numbers. The half adder is able to add two single binary digits and provide the output plus a carry value. It has two inputs, called A and B, and two outputs S (sum) and C (carry). The common representation uses a XOR logic gate and an AND logic gate.

Half Adder Truth Table with Carry-out

<table>
<tr><th rowspan="2">Symbol</th><th colspan="4">Truth Table</th></tr>
<tr><th>B</th><th>A</th><th>SUM</th><th>CARRY</th></tr>
<tr><td rowspan="4">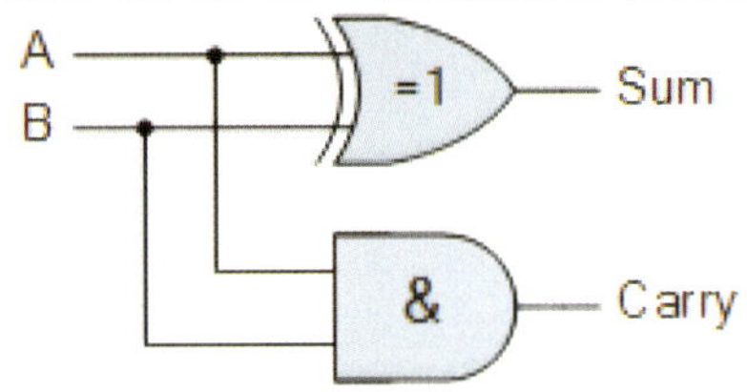
</td><td>0</td><td>0</td><td>0</td><td>0</td></tr>
<tr><td>0</td><td>1</td><td>1</td><td>0</td></tr>
<tr><td>1</td><td>0</td><td>1</td><td>0</td></tr>
<tr><td>1</td><td>1</td><td>0</td><td>1</td></tr>
</table>

\From the truth table of the half adder we can see that the SUM (S) output is the result of the Exclusive-OR gate and the Carry-out (Cout) is the result of the AND gate. Then the Boolean expression for a half adder is as follows.

For the SUM bit:

$$SUM = A \text{ XOR } B = A \oplus B$$

For the CARRY bit:

$$CARRY = A \text{ AND } B = A.B$$

One major disadvantage of the Half Adder circuit when used as a binary adder, is that there is no provision for a "Carry-in" from the previous circuit when adding together multiple data bits.

For example, suppose we want to add together two 8-bit bytes of data, any resulting carry bit would need to be able to "ripple" or move across the bit patterns starting from the least significant bit (LSB). The most complicated operation the half adder can do is "1 + 1" but as the half adder has no carry input the resultant added value would be incorrect. One simple way to overcome this problem is to use a Full Adder type binary adder circuit.

Full Adder

A full adder is a digital circuit that performs addition. Full adders are implemented with logic gates in hardware. A full adder adds three one-bit binary numbers, two operands and a carry bit. The adder outputs two numbers, a sum and a carry bit. The term is contrasted with a half adder, which adds two binary digits.

This adder is difficult to implement than a half-adder. The difference between a half-adder and a full-adder is that the full-adder has three inputs and two outputs, whereas half adder has only two inputs and two outputs. The first two inputs are A and B and the third input is an input carry as C-IN. When a full-adder logic is designed, you string eight of them together to create a byte-wide adder and cascade the carry bit from one adder to the next.

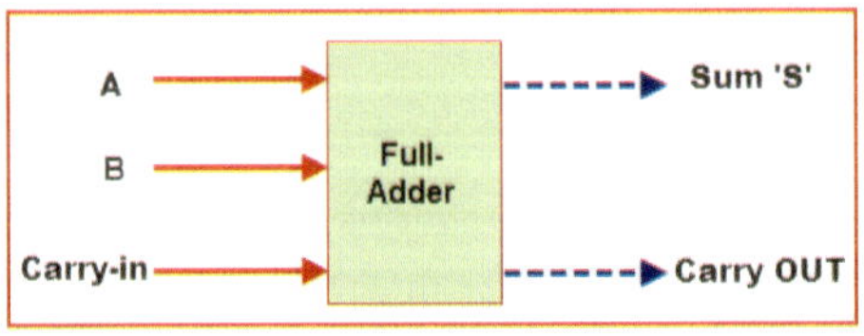

Full Adder

The output carry is designated as C-OUT and the normal output is designated as S.

Full Adder Truth Table:

INPUTS			OUTPUT	
A	B	C-IN	C-OUT	S
0	0	0	0	0
0	0	1	0	1
0	1	0	0	1
0	1	1	1	0
1	0	0	0	1
1	0	1	1	0
1	1	0	1	0
1	1	1	1	1

Full Adder Truth Table

With the truth-table, the full adder logic can be implemented. You can see that the output S is an XOR between the input A and the half-adder, SUM output with B and C-IN inputs. We take C-OUT will only be true if any of the two inputs out of the three are HIGH.

So, we can implement a full adder circuit with the help of two half adder circuits. At first, half adder will be used to add A and B to produce a partial Sum and a second half adder logic can be used to add C-IN to the Sum produced by the first half adder to get the final S output.

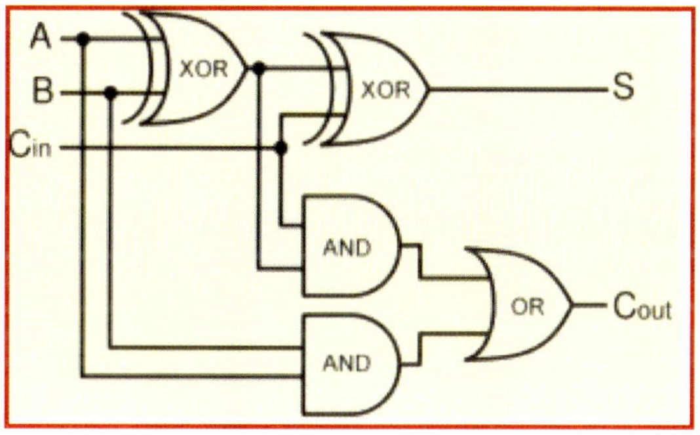

Full Adder Logic Circuit

If any of the half adder logic produces a carry, there will be an output carry. So, COUT will be an OR function of the half-adder Carry outputs. Take a look at the implementation of the full adder circuit shown below.

The implementation of larger logic diagrams is possible with the above full adder logic a simpler symbol is mostly used to represent the operation. Given below is a simpler schematic representation of a one-bit full adder.

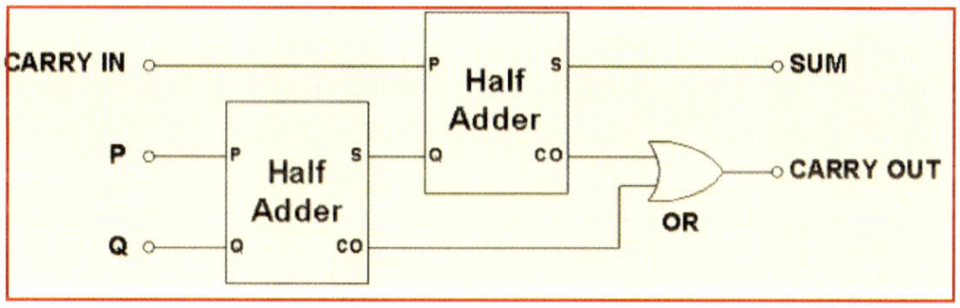

Full Adder Design Using Half Adders

With this type of symbol, we can add two bits together, taking a carry from the next lower order of magnitude, and sending a carry to the next higher order of magnitude. In a computer, for a multi-bit operation, each bit must be represented by a full adder and must be added simultaneously. Thus, to add two 8-bit numbers, you will need 8 full adders which can be formed by cascading two of the 4-bit blocks.

Then the Boolean expression for a half adder is as follows.

For the SUM bit:

$$SUM = A\ XOR\ B = A \oplus B$$

For the CARRY bit:

$$CARRY = A\ AND\ B = A.B$$

One major disadvantage of the Half Adder circuit when used as a binary adder, is that there is no provision for a "Carry-in" from the previous circuit when adding together multiple data bits.

For example, suppose we want to add together two 8-bit bytes of data, any resulting carry bit would need to be able to "ripple" or move across the bit patterns starting from the least significant bit (LSB). The most complicated operation the half adder can do is "1 + 1" but as the half adder has no carry input the resultant added value would be incorrect. One simple way to overcome this problem is to use a Full Adder type binary adder circuit.

A Full Adder Circuit

The main difference between the Full Adder and the previous Half Adder is that a full adder has three inputs. The same two single bit data inputs A and B as before plus an additional Carry-in (C-in) input to receive the carry from a previous stage as shown below.

Full Adder Block Diagram

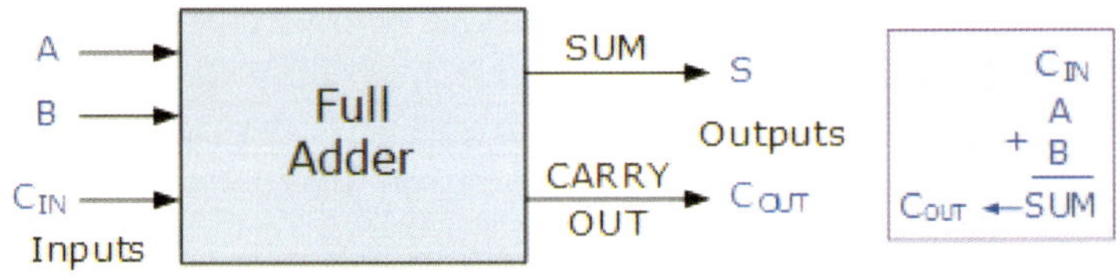

Then the full adder is a logical circuit that performs an addition operation on three binary digits and just like the half adder, it also generates a carry out to the next addition column. Then a Carry-in is a possible carry from a less significant digit, while a Carry-out represents a carry to a more significant digit.

In many ways, the full adder can be thought of as two half adders connected together, with the first half adder passing its carry to the second half adder as shown.

Full Adder Logic Diagram

As the full adder circuit above is basically two half adders connected together, the truth table for the full adder includes an additional column to take into account the Carry-in, C_{IN} input as well as the summed output, S and the Carry-out, C_{OUT} bit.

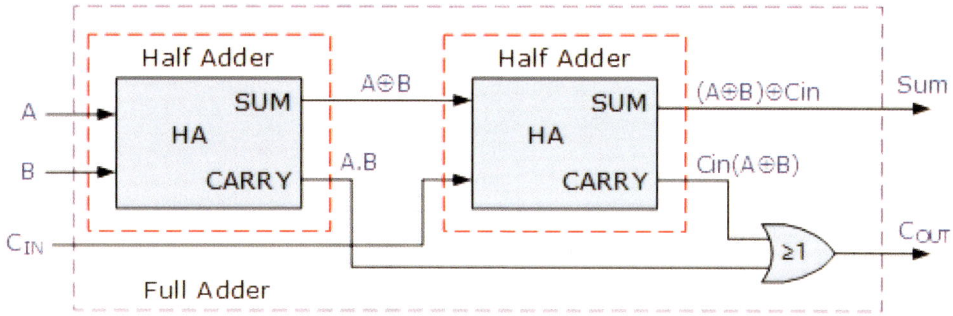

Full Adder Truth Table with Carry

Symbol	Truth Table				
	C-in	B	A	Sum	C-out
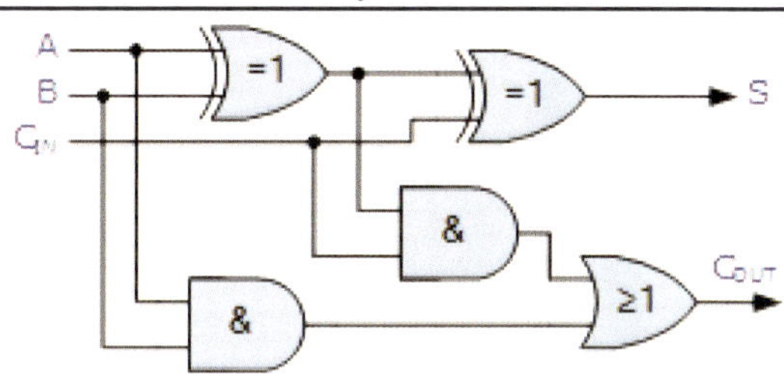	0	0	0	0	0
	0	0	1	1	0
	0	1	0	1	0
	0	1	1	0	1
	1	0	0	1	0
	1	0	1	0	1
	1	1	0	0	1
	1	1	1	1	1

Then the Boolean expression for a full adder is as follows.

For the SUM (S) bit:

$$SUM = (A\ XOR\ B)\ XOR\ Cin = (A \oplus B) \oplus Cin$$

For the CARRY-OUT (Cout) bit:

$$CARRY\text{-}OUT = A\ AND\ B\ OR\ Cin(A\ XOR\ B) = A.B + Cin(A \oplus B)$$

The relationship between the Full-Adder and the Half-Adder is half adder produces results and full adder uses half adder to produce some other result. Similarly, while the Full-Adder is of two Half-Adders, the Full-Adder is the actual block that we use to create the arithmetic circuits.

Kogge–Stone adder

KSA is a parallel prefix form carry look ahead adder. It generates carry in O (log n) time and is widely considered as the fastest adder and is widely used in the industry for high performance arithmetic circuits.

In KSA, carries are computed fast by computing them in parallel at the cost of increased area. A tree structure of 3 bit Kogge stone adder (KSA) is shown fig. It has three processing stages for calculating the sum bits.

Complete Functionality of KSA

I. Preprocessing:

This step involves computation of generate and propagate signals corresponding too each pair of bits in A and B. These signals are given by the logic equations below:

pi = Ai xor Bi

gi = Ai and Bi

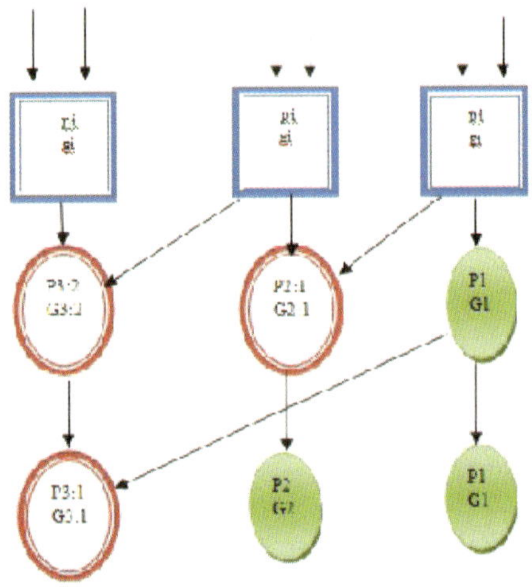

Fig: Tree Structure of 3bit Kogge Stone Adder

II. Carry look ahead network:

This block differentiates KSA from other adders and is the main force behind its high performance. This step involves computation of carries corresponding to each bit. It uses group propagate and generate as intermediate signals which are given by the logic equations below:

Pi:j = Pi:k+1 and Pk:j

Gi:j = Gi:k+1 or (Pi:k+1 and Gk:j)

III. Post processing:

This is the final step and is common to all adders of this family (carry look ahead). It involves computation of sum bits. Sum bits are computed by the logic given below:

Si = pi xor Ci-1

Illustration of KSA using an example:

A = 110 and B = 101

I Step: pi = Ai xor Bi and gi = Ai and Bi

p3 = 1 xor 1=0;p2 = 1 xor 0 = 1;p1 = 0 xor 1 = 1;

g3 = 1 and 1=1;g2 = 1 and 0 =0;g1 = 0 and 1 = 0;

II Step: Pi:j = Pi:k+1 and Pk:j and Gi:j = Gi:k+1 or (Pi:k+1 and Gk:j)

P3:2 = p3 and p2; 0 and 1 = 0;

G3:2 = g3 or (p3 and g2); 1or (0 and 0) = 1;

P2:1 = p2 and p1; 1 and 1 = 1;

G2:1 = g2 or (p2 and g1); 0 or (1 and 0) = 0;

P3:1 = P3:2 and p1; 0 and 1= 0;

G3:1 = G3:2 or (P3:2 and g1); 1 or (0 and 0) = 1;

III Step: Si = pi xor Ci-1

S1 = p1 xor 0; 1 xor 0 = 1;

S2 = p2 xor g1; 1 xor 0 = 1;

S3 = p3 xor G2; 0 xor 0 = 0;

Cout = G3:1; 1;

Sum Bits : Cout S3 S2 S1 : 1011;

Enhancements

Enhancements to the original implementation include increasing the radix and sparsity of the adder. The *radix* of the adder refers to how many results from the previous level of computation are used to generate the next one. The original implementation uses radix-2, although it's possible to create radix-4 and higher. Doing so increases the power and delay of each stage, but reduces the number of required stages. In the so called sparse Kogge–Stone adder (SKA) the *sparsity* of the adder refers to how many carry bits are generated by the carry-tree. Generating every carry bit is called sparsity-1, whereas generating every other is sparsity-2 and every fourth is sparsity-4. The resulting carries are then used as the carry-in inputs for much shorter ripple carry adders or some other adder design, which generates the final sum bits. Increasing sparsity reduces the total needed computation and can reduce the amount of routing congestion.

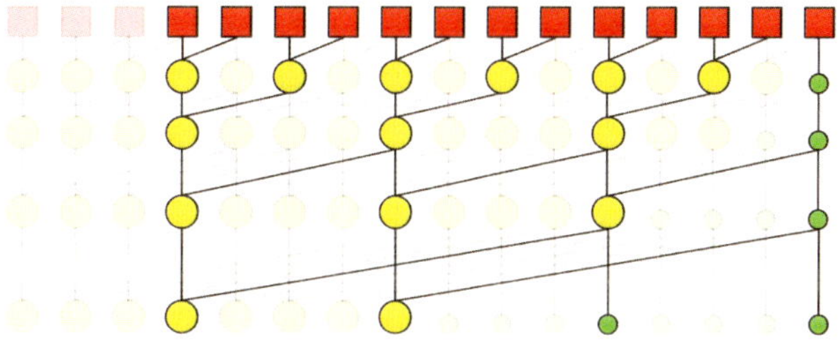

Above is an example of a Kogge–Stone adder with sparsity-4. Elements eliminated by sparsity shown marked with transparency. As shown, power and area of the carry generation is improved significantly, and routing congestion is substantially reduced. Each generated carry feeds a multiplexer for a carry select adder or the carry-in of a ripple carry adder.

Carry-Save Adder

There are many cases where it is desired to add more than two numbers together. The straightforward way of adding together m numbers (all n bits wide) is to add the first two, then add that sum to the next, and so on. This requires a total of m – 1 additions, for a total gate delay of O(m lg n) (assuming lookahead carry adders). Instead, a tree of adders can be formed, taking only O(lgm • lg n) gate delays.

Using carry save addition, the delay can be reduced further still. The idea is to take 3 numbers that we want to add together, x + y + z, and convert it into 2 numbers c + s such that x + y + z = c + s, and do this in O(1) time. The reason why addition can not be performed in O(1) time is because the carry information must be propagated. In carry save addition, we refrain from directly passing on the carry information until the very last step. We will first illustrate the general concept with a base 10 example.

To add three numbers by hand, we typically align the three operands, and then proceed column by column in the same fashion that we perform addition with two numbers. The three digits in a row are added, and any overflow goes into the next column. Observe that when there is some non-zero carry, we are really adding four digits (the digits of x,y and z, plus the carry).

Carry	1	1	2	1	
x:	1	2	3	4	5
y:	3	8	1	7	2
z: +	2	0	5	8	7
sum:	7	1	1	0	4

The carry save approach breaks this process down into two steps. The first is to compute the sum ignoring any carries:

x :	1	2	3	4	5
y :	3	8	1	7	2
z : +	2	0	5	8	7
s:	6	0	9	9	4

Each s_i is equal to the sum of $x_i + y_i + z_i$ modulo 10. Now, separately, we can compute the carry on a column by column basis:

x :	1	2	3	4	5
y :	3	8	1	7	2
z : +	2	0	5	8	7
c:	1	0	1	1	

In this case, each ci is the sum of the bits from the previous column divided by 10 (ignoring any remainder). Another way to look at it is that any carry over from one column gets put into the next column. Now, we can add together c and s, and we'll verify that it indeed is equal to x + y + z:

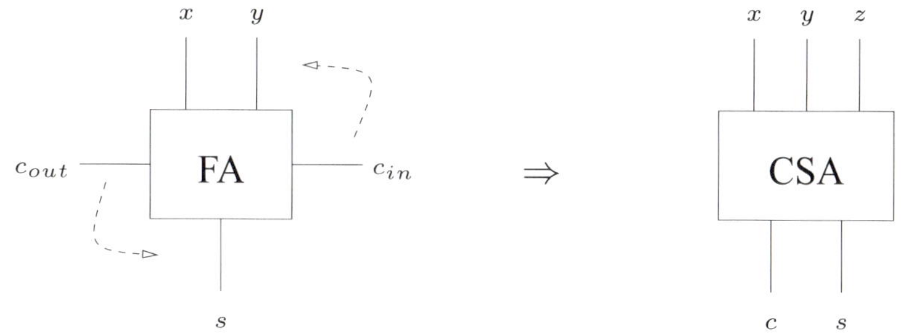

Figure : The carry save adder block is the same circuit as the full adder.

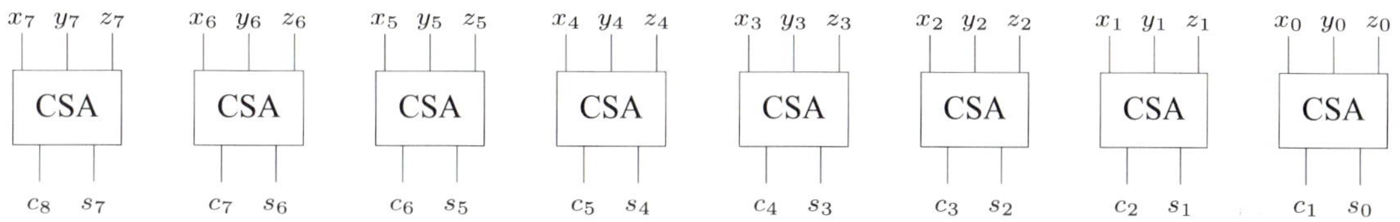

Figure: One CSA block is used for each bit. This circuit adds three n = 8 bit numbers together into two new numbers.

s:	6	0	9	9	4
c: +	1	0	1	1	
sum:	7	1	1	0	4

The important point is that c and s can be computed independently, and furthermore, each c_i (and s_i) can be computed independently from all of the other c's (and s's). This achieves our original goal of converting three numbers that we wish to add into two numbers that add up to the same sum, and in O(1) time.

The same concept can be applied to binary numbers. As a quick example:

x:	1		0	0	1	1
y:		1	1	0	0	1
z: +		0	1	0	1	1
s:		0	0	0	0	1
c: +	1	1	0	1	1	
sum:	1	1	0	1	1	1

What does the circuit to compute s and c look like? It is actually identical to the full adder, but with some of the signals renamed. Figure shows a full adder and a carry save adder. A carry save adder simply is a full adder with the c_{in} input renamed to z, the z output (the original "answer" output) renamed to s, and the c_{out} output renamed to c. Figure shows how n carry save adders are arranged to add three n bit numbers x,y and z into two numbers c and s. Note that the CSA block in bit position zero generates c_1, not c_0. Similar to the least significant column when adding numbers by hand (the "blank"), c_0 is equal to zero. Note that all of the CSA blocks are independent, thus the entire circuit takes only O(1) time. To get the final sum, we still need a LCA, which will cost us O(lg n) delay. The asymptotic gate delay to add three n-bit numbers is thus the same as adding only two n-bit numbers.

So how long does it take us to add m different n-bit numbers together? The simple approach is just to repeat this trick approximately m times over. This is illustrated in figure below. There are m−2 CSA blocks (each block in the figure actually represents many one-bit CSA blocks in parallel) that we have to go through, and then the final LCA. Note that every time we pass through a CSA block, our number increases in size by one bit. Therefore, the numbers that go to the LCA will be at most n + m − 2 bits long. So the final LCA will have a gate delay of O(lg (n + m)). Therefore the total gate delay is O(m + lg (n + m))

Instead of arranging the CSA blocks in a chain, a tree formation can actually be used. This is slightly awkward because of the odd ratio of 3 to 2. Figure shows how to build a tree of CSAs. This circuit is called a Wallace tree.

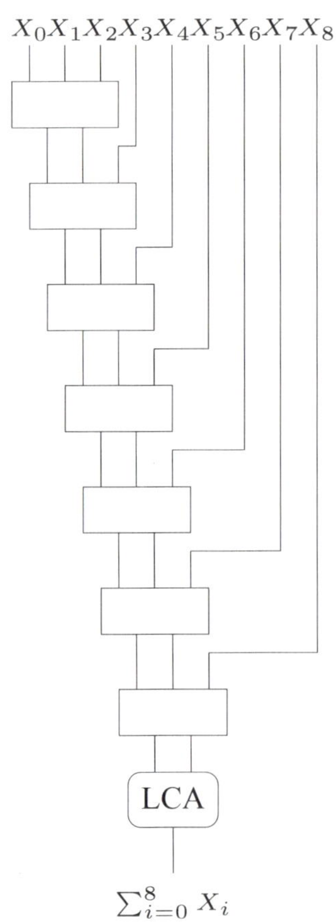

Figure: Adding m n-bit numbers with a chain of CSA's.

The depth of the tree is $\log_{\frac{3}{2}} m$. Like before, the width of the numbers will increase as we get deeper in the tree. At the end of the tree, the numbers will be O(n + log m) bits wide, and therefore the LCA will have a O(lg (n + log m)) gate delay. The total gate delay of the Wallace tree is thus O(log m + lg (n + log m)).

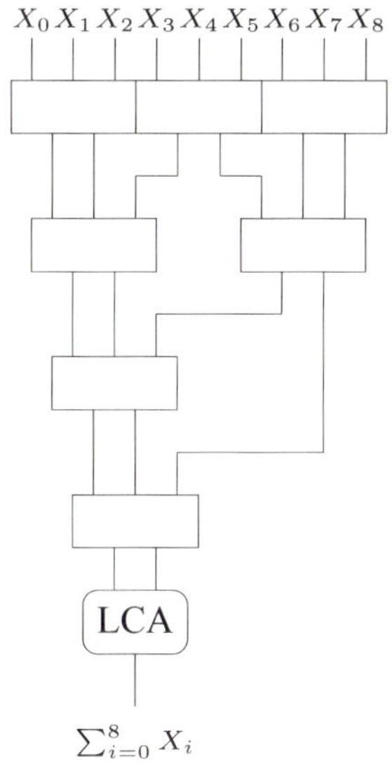

Figure: A Wallace tree for adding m n-bit numbers.

Carry-save Accumulators

Supposing that we have two bits of storage per digit, we can use a redundant binary representation, storing the values 0, 1, 2, or 3 in each digit position. It is therefore obvious that one more binary number can be added to our carry-save result without overflowing our storage capacity.

The key to success is that at the moment of each partial addition we add three bits:

- 0 or 1, from the number we are adding.
- 0 if the digit in our store is 0 or 2, or 1 if it is 1 or 3.
- 0 if the digit to its right is 0 or 1, or 1 if it is 2 or 3.

To put it another way, we are taking a carry digit from the position on our right, and passing a carry digit to the left, just as in conventional addition; but the carry digit we pass to the left is the result of the *previous* calculation and not the current one. In each clock cycle, carries only have to move one step along, and not n steps as in conventional addition.

Because signals don't have to move as far, the clock can tick much faster.

There is still a need to convert the result to binary at the end of a calculation, which effectively just means letting the carries travel all the way through the number just as in a conventional adder. But if we have done 512 additions in the process of performing a 512-bit multiplication, the cost of that final conversion is effectively split across those 512 additions, so each addition bears 1/512 of the cost of that final "conventional" addition.

Drawbacks

At each stage of a carry-save addition,

- We know the result of the addition at once.

- We *still do not know* whether the result of the addition is larger or smaller than a given number (for instance, we do not know whether it is positive or negative).

This latter point is a drawback when using carry-save adders to implement modular multiplication (multiplication followed by division, keeping the remainder only). If we cannot know whether the intermediate result is greater or less than the modulus.

Montgomery multiplication, which depends on the rightmost digit of the result, is one solution; though rather like carry-save addition itself, it carries a fixed overhead, so that a sequence of Montgomery multiplications saves time but a single one does not. Fortunately exponentiation, which is effectively a sequence of multiplications, is the most common operation in public-key cryptography.

Careful error analysis allows a choice to be made about subtracting the modulus even though we don't know for certain whether the result of the addition is big enough to warrant the subtraction. For this to work, it is necessary for the circuit design to be able to add −2, −1, 0, +1 or +2 times the modulus. The advantage over Montgomery multiplication is that there is no fixed overhead attached to each sequence of multiplications.

Technical Details

The carry-save unit consists of n full adders, each of which computes a single sum and carry bit based solely on the corresponding bits of the three input numbers. Given the three n-bit numbers a, b, and c, it produces a partial sum ps and a shift-carry sc:

$$ps_i = a_i \oplus b_i \oplus c_i,$$
$$sc_i = (a_i \wedge b_i) \vee (a_i \wedge c_i) \vee (b_i \wedge c_i).$$

The entire sum can then be computed by:

1. Shifting the carry sequence sc left by one place.
2. Appending a 0 to the front (most significant bit) of the partial sum sequence ps.
3. Using a ripple carry adder to add these two together and produce the resulting $(n + 1)$-bit value.

Carry-lookahead Adder

In case of parallel adders, the binary addition of two numbers is initiated when all the bits of the augend and the addend must be available at the same time to perform the computation. In a parallel adder circuit, the carry output of each full adder stage is connected to the carry input of the next higher-order stage, hence it is also called as ripple carry type adder.

In such adder circuits, it is not possible to produce the sum and carry outputs of any stage until the input carry occurs. So there will be a considerable time delay in the addition process , which is known as , carry propagation delay. In any combinational circuit , signal must propagate through the gates before the correct output sum is available in the output terminals.

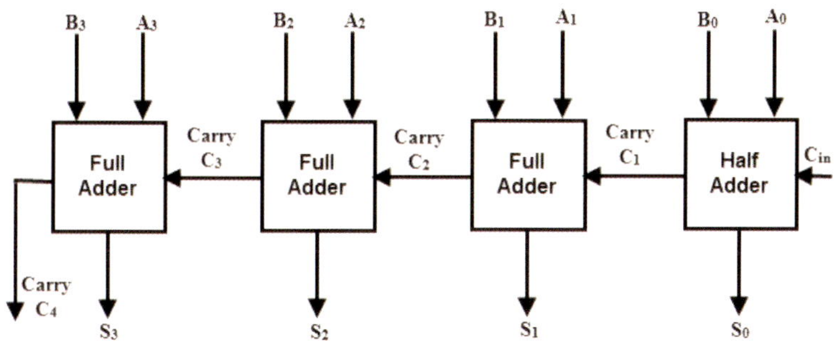

Consider the above figure, in which the sum S4 is produced by the corresponding full adder as soon as the input signals are applied to it. But the carry input C4 is not available on its final steady state value until carry c3 is available at its steady state value. Similarly C3 depends on C2 and C2 on C1. Therefore, carry must propagate to all the stages in order that output S4 and carry C5 settle their final steady-state value.

The propagation time is equal to the propagation delay of the typical gate times the number of gate levels in the circuit. For example, if each full adder stage has a propagation delay of 20n seconds, then S4 will reach its final correct value after 80n (20 × 4) seconds. If we extend the number of stages for adding more number of bits then this situation becomes much worse.

So the speed at which the number of bits added in the parallel adder depends on the carry propagation time. However, signals must be propagated through the gates at a given enough time to produce the correct or desired output.

The following are the methods to get the high speed in the parallel adder to produce the binary addition.

- By employing faster gates with reduced delays, we can reduce the propagation delay. But there will be a capability limit for every physical logic gate.
- Another way is to increase the circuit complexity in order to reduce the carry delay time. There are several methods available to speeding up the parallel adder, one commonly used method employs the principle of look ahead-carry addition by eliminating inter stage carry logic.

A carry-Lookahead adder is a fast parallel adder as it reduces the propagation delay by more complex hardware, hence it is costlier. In this design, the carry logic over fixed groups of bits of the adder is reduced to two-level logic, which is nothing but a transformation of the ripple carry design.

This method makes use of logic gates so as to look at the lower order bits of the augend and addend to see whether a higher order carry is to be generated or not. Let us discuss in detail.

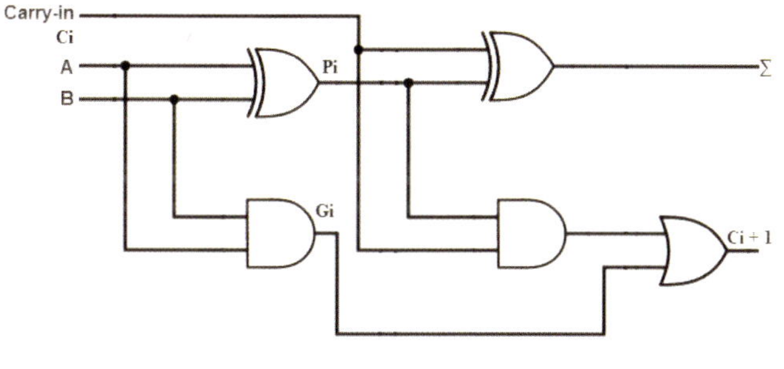

A	B	Ci	C i +1	Condition
0	0	0	0	No carry generate
0	0	1	0	
0	1	0	0	
0	1	1	1	No carry propagate
1	0	0	0	
1	0	1	1	
1	1	0	1	Carry generate
1	1	1	1	

Consider the full adder circuit shown above with corresponding truth table. If we define two variables as carry generate Gi and carry propagate Pi then,

$P_i = A_i \oplus B_i$

$Gi = Ai\ Bi$

The sum output and carry output can be expressed as

$Si = Pi \oplus Ci$

$C\,i+1 = Gi + Pi\ Ci$

Where Gi is a carry generate which produces the carry when both Ai, Bi are one regardless of the input carry. Pi is a carry propagate and it is associate with the propagation of carry from Ci to Ci +1.

The carry output Boolean function of each stage in a 4 stage carry-Lookahead adder can be expressed as

$C1 = G0 + P0\ Cin$

$C2 = G1 + P1\ C1$

$= G1 + P1\ G0 + P1\ P0\ Cin$

$C3 = G2 + P2\ C2$

$= G2 + P2\ G1 + P2\ P1\ G0 + P2\ P1\ P0\ Cin$

$C4 = G3 + P3\ C3$

$= G3 + P3\ G2 + P3\ P2\ G1 + P3\ P2\ P1\ G0 + P3\ P2\ P1\ P0\ Cin$

From the above Boolean equations we can observe that C4 does not have to wait for C3 and C2 to propagate but actually C4 is propagated at the same time as C3 and C2. Since the Boolean expression for each carry output is the sum of products so these can be implemented with one level of AND gates followed by an OR gate.

The implementation of three Boolean functions for each carry output (C2, C3 and C4) for a carry-Lookahead carry generator shown in below figure.

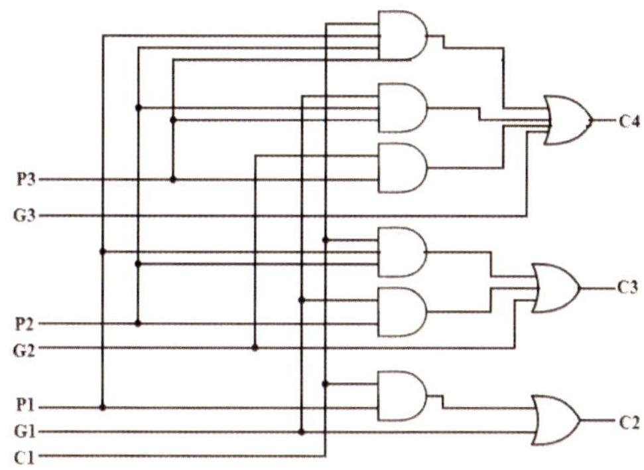

Therefore, a 4 bit parallel adder can be implemented with the carry-Lookahead scheme to increase the speed of binary addition as shown in below figure. In this, two Ex-OR gates are required by each sum output. The first Ex-OR gate generates Pi variable output and the AND gate generates Gi variable.

Hence, in two gates levels all these P's and G's are generated. The carry-Lookahead generators allows all these P and G signals to propagate after they settle into their steady state values and produces the output carriers at a delay of two levels of gates. Therefore, the sum outputs S2 to S4 have equal propagation delay times.

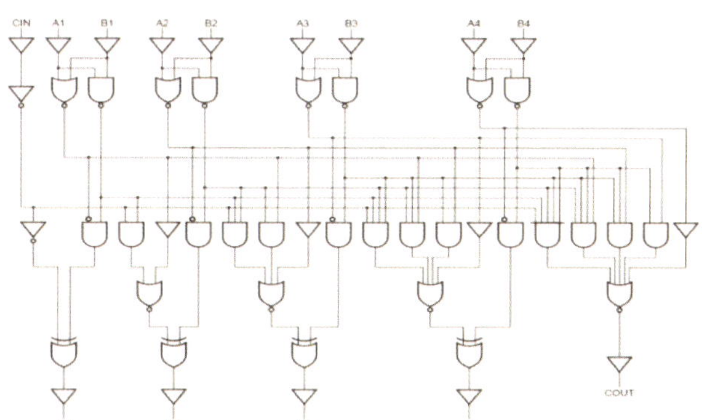

It is also possible to construct 16 bit and 32 bit parallel adders by cascading the number of 4 bit adders with carry logic. A 16 bit carry-Lookahead adder is constructed by cascading the four 4 bit adders with two more gate delays, whereas the 32 bit carry-Lookahead adder is formed by cascading of two 16 bit adders.

In a 16 bit carry-Lookahead adder, 5 and 8 gate delays are required to get C16 and S15 respectively, which are less as compared to the 9 and 10 gate delay for C16 and S15 respectively in cascaded four bit carry-Lookahead adder blocks. Similarly, in 32 bit adder, 7 and 10 gate delays are required by C32 and S31 which are less compared to 18 and 17 gate delays for the same outputs if the 32 bit adder is implemented by eight 4 bit adders.

Carry-Lookahead Adder ICs

The high speed carry-Lookahead adders are integrated on integrated circuits in different bit configurations by several manufacturers. There are several individual carry generator ICs are available so that we have to make connection with logic gates to perform the addition operation.

A typical carry-Lookahead generator IC is 74182 which accept four pairs of active low carry propagate (as Po, P1, P2 and P3) and carry generate (Go, G1, G2 and G3) signals and an active high input (Cn).

It provides active high carriers (Cn+x, Cn+y, Cn+z) across the four groups of binary adders. This IC also facilitates the other levels of look ahead by active low propagate and carry generate outputs.

The logic expressions provided by the IC 74182 are

$$C_{n+x} = G_o + P_o C_n$$

$$C_{n+y} = G_1 + P_1 G_o + P_1 P_o C_n$$

$$C_{n+z} = G_2 + P_2 G_1 + P_2 P_1 G_o$$

$$\overline{G} = \overline{G3 + P3\ G2 + P3\ P2\ G1 + P3\ P2\ P1\ Go}$$

$$\overline{P} = \overline{P3\ P2\ P1\ Po}$$

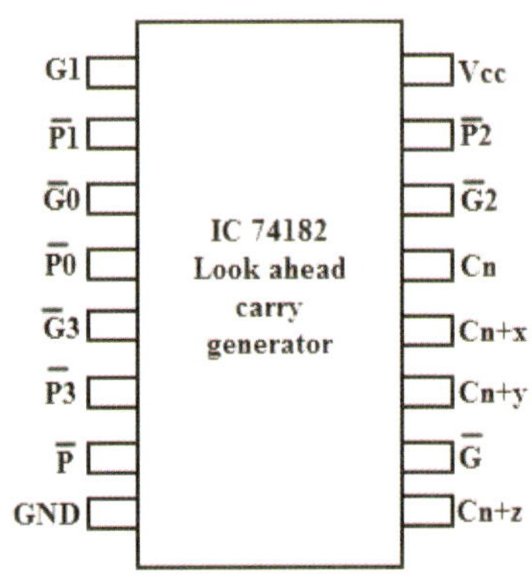

On the other hand, there are many high speed adder ICs which combine a set of full adders with carry-Lookahead circuitry. The most popular form of such IC is 74LS83/74S283 which is a 4 bit parallel adder high speed IC that contains four interconnected full adders with a carry-Lookahead circuitry.

The functional symbol for this type of IC is shown in below figure. It accepts the two 4 bit numbers as A3A2A1Ao and B3B2B1Bo and input carry Cino into the LSB position. This IC produce output sum bits as S3S2S1So and the carry output Cout3 into the MSB position.

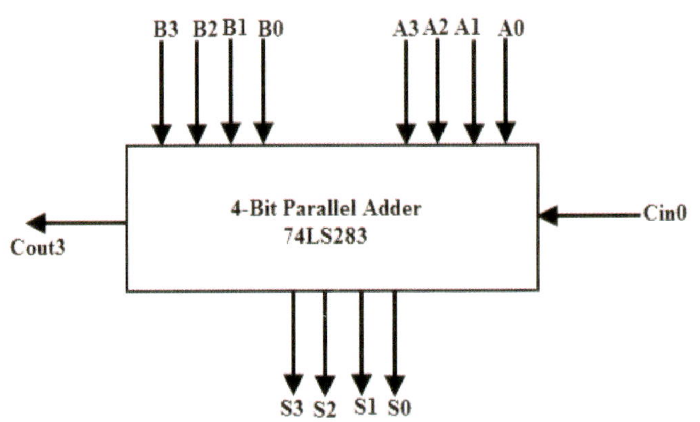

By cascading two or more parallel adder ICs we can perform the addition of larger binary numbers such as 8 bit, 24-bit and 32 bit addition.

Binary Subtractor

The Binary Subtractor is another type of combinational arithmetic circuit that is the opposite of the Binary Adder.

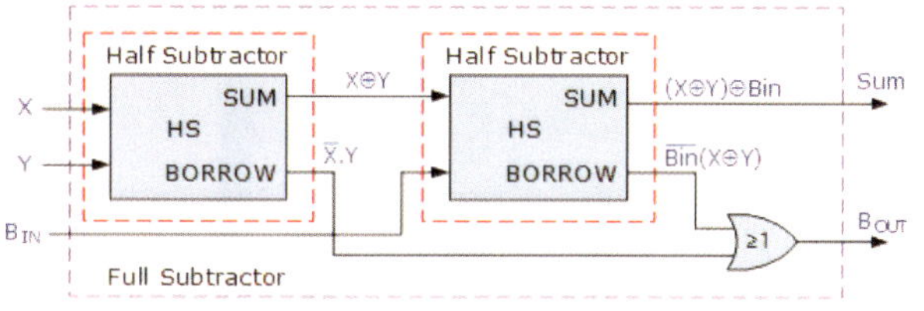

As their name implies, a Binary Subtractor is a decision making circuit that subtracts two binary numbers from each other, for example, X – Y to find the resulting difference between the two numbers.

Unlike the Binary Adder which produces a SUM and a CARRY bit when two binary numbers are added together, the binary subtractor produces a DIFFERENCE, D by using a BORROW bit, B from the previous column. Then obviously, the operation of subtraction is the opposite to that of addition.

We learnt from our maths lessons at school that the minus sign, "–" is used for a subtraction calculation, and when one number is subtracted from another, a borrow is required if the subtrahend is greater than the minuend. Consider the simple subtraction of the two denary (base) numbers below.

123	X	
– 78	Y	(Subtrahend)
45	DIFFERENCE	

We can not directly subtract 8 from 3 in the first column as 8 is greater than 3, so we have to borrow a 10, the base number, from the next column and add it to the minuend to produce 13 minus 8. This "borrowed" 10 is then return back to the subtrahend of the next column once the difference is found. Simple school math's, borrow a 10 if needed, find the difference and return the borrow.

The subtraction of one binary number from another is exactly the same idea as that for subtracting two decimal numbers but as the binary number system is a Base-2 numbering system which uses "0" and "1" as its two independent digits, large binary numbers which are to be subtracted from each other are therefore represented in terms of "0's" and "1's".

Binary Subtraction

Binary Subtraction can take many forms but the rules for subtraction are the same whichever process you use. As binary notation only has two digits, subtracting a "0" from a "0" or a "1" leaves the result unchanged as 0-0 = 0 and 1-0 = 1. Subtracting a "1" from a "1" results in a "0", but subtracting a "1" from a "0" requires a borrow. In other words 0 – 1 requires a borrow.

Binary Subtraction of Two Bits

0	1	1	(borrow) 1→ 0
– 0	– 0	– 1	– 1
0	1	0	1

For the simple 1-bit subtraction problem above, if the borrow bit is ignored the result of their binary subtraction resembles that of an Exclusive-OR Gate. To prevent any confusion in this chapter between a binary subtractor input labelled, B and the resulting borrow bit output from the binary subtractor also being labelled, B, we will label the two input bits as X for the minuend and Y for the subtrahend. Then the resulting truth table is the difference between the two input bits of a single binary subtractor is given as:

2-input Exclusive-OR Gate

Symbol	Truth Table		
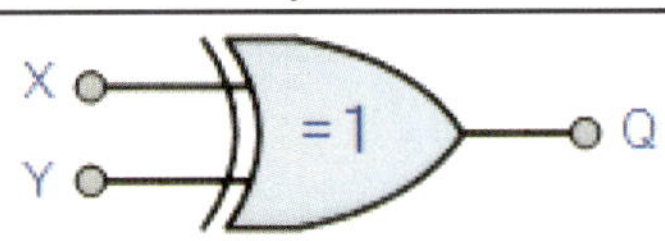 2-input Ex-OR Gate	Y	X	Q
	0	0	0
	0	1	1
	1	0	1
	1	1	0

As with the Binary Adder, the difference between the two digits is only a "1" when these two inputs are not equal as given by the Ex-OR expression. However, we need an additional output to produce the borrow bit when input X = 0 and Y = 1. Unfortunately there are no standard logic gates that will produce an output for this particular combination of X and Y inputs.

But we know that an AND Gate produces an output "1" when both of its inputs X and Y are "1" (HIGH) so if we use an inverter or NOT Gate to complement the input X before it is fed to the AND gate, we can produce the required borrow output when X = 0 and Y = 1 as shown below.

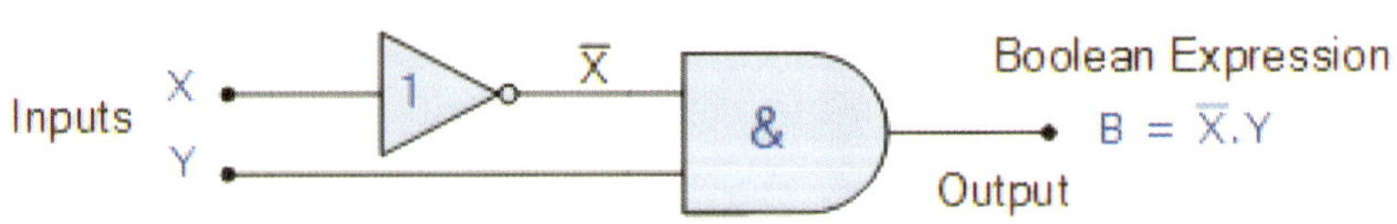

Then by combining the Exclusive-OR gate with the NOT-AND combination results in a simple digital binary subtractor circuit known commonly as the Half Subtractor

Half Subtractor

A Half Subtractor Circuit

A half subtractor is a logical circuit that performs a subtraction operation on two binary digits. The half subtractor produces a sum and a borrow bit for the next stage.

Half Subtractor with Borrow-out

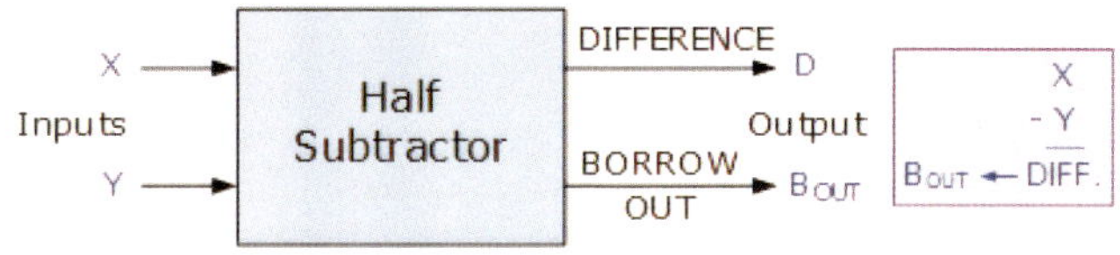

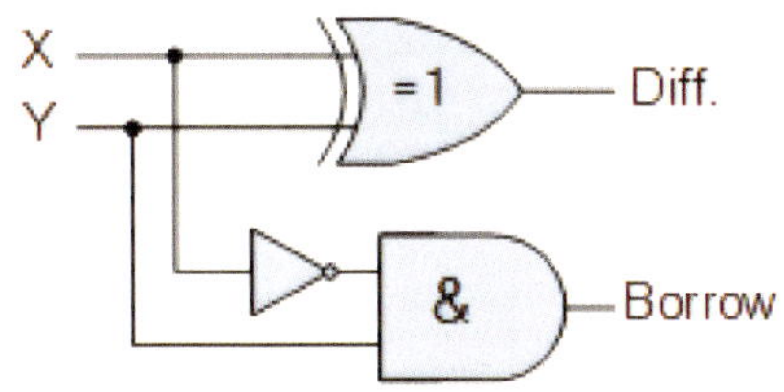

Symbol	Truth Table			
	Y	X	DIFFERENCE	BORROW
	0	0	0	0
	0	1	1	0
	1	0	1	1
	1	1	0	0

From the truth table of the half subtractor we can see that the DIFFERENCE (D) output is the result of the Exclusive-OR gate and the Borrow-out (Bout) is the result of the NOT-AND combination. Then the Boolean expression for a half subtractor is as follows.

For the DIFFERENCE bit:

$$D = X \text{ XOR } Y = X \oplus Y$$

For the BORROW bit:

$$B = \text{not-X AND } Y = \overline{X}.Y$$

If we compare the Boolean expressions of the half subtractor with a half adder, we can see that the two expressions for the SUM (adder) and DIFFERENCE (subtractor) are exactly the same and so they should be because of the Exclusive-OR gate function. The two Boolean expressions for the binary subtractor BORROW is also very similar to that for the adders CARRY. Then all that is needed to convert a half adder to a half subtractor is the inversion of the minuend input X.

One major disadvantage of the Half Subtractor circuit when used as a binary subtractor, is that there is no provision for a "Borrow-in" from the previous circuit when subtracting multiple data bits from each other. Then we need to produce what is called a "full binary subtractor" circuit to take into account this borrow-in input from a previous circuit.

Full Subtractor

A Full Binary Subtractor Circuit

The main difference between the Full Subtractor and the previous Half Subtractor circuit is that a full subtractor has three inputs. The two single bit data inputs X (minuend) and Y (subtrahend) the same as before plus an additional Borrow-in (B-in) input to receive the borrow generated by the subtraction process from a previous stage as shown below.

Full Subtractor Block Diagram

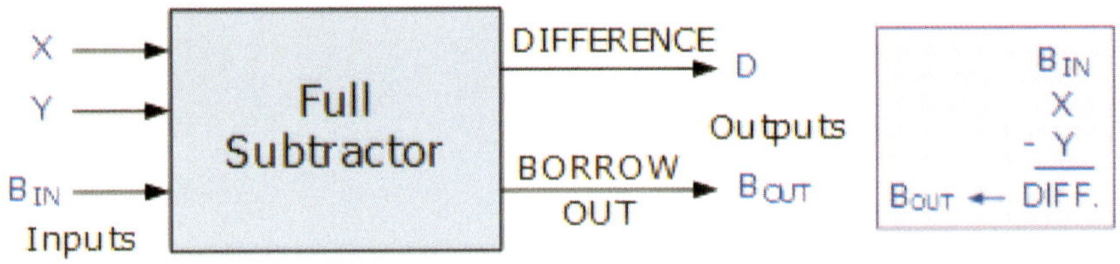

Then the combinational circuit of a "full subtractor" performs the operation of subtraction on three binary bits producing outputs for the difference D and borrow B-out. Just like the binary adder circuit, the full subtractor can also be thought of as two half subtractors connected together, with the first half subtractor passing its borrow to the second half subtractor as follows.

Full Subtractor Logic Diagram

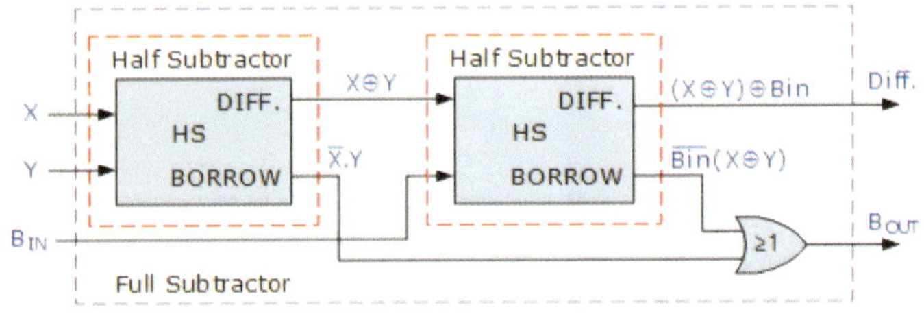

As the full subtractor circuit above represents two half subtractors cascaded together, the truth table for the full subtractor will have eight different input combinations as there are three input variables, the data bits and the Borrow-in, B_{IN} input. Also includes the difference output, D and the Borrow-out, B_{OUT} bit.

Full Subtractor Truth Table

<table>
<tr><th>Symbol</th><th colspan="5">Truth Table</th></tr>
<tr><td rowspan="9">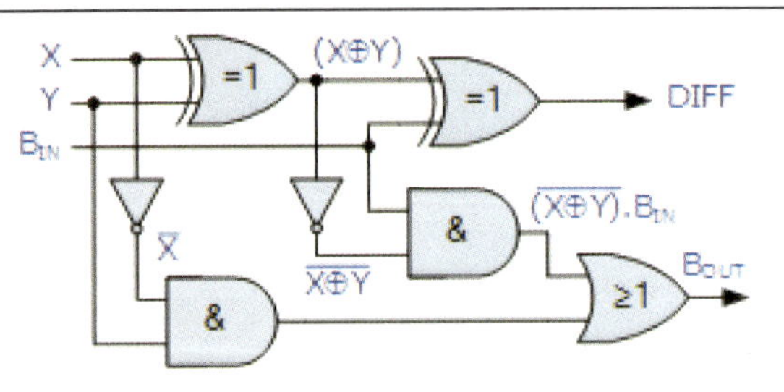
</td><td>B-in</td><td>Y</td><td>X</td><td>Diff.</td><td>B-out</td></tr>
<tr><td>0</td><td>0</td><td>0</td><td>0</td><td>0</td></tr>
<tr><td>0</td><td>0</td><td>1</td><td>1</td><td>0</td></tr>
<tr><td>0</td><td>1</td><td>0</td><td>1</td><td>1</td></tr>
<tr><td>0</td><td>1</td><td>1</td><td>0</td><td>0</td></tr>
<tr><td>1</td><td>0</td><td>0</td><td>1</td><td>1</td></tr>
<tr><td>1</td><td>0</td><td>1</td><td>0</td><td>0</td></tr>
<tr><td>1</td><td>1</td><td>0</td><td>0</td><td>1</td></tr>
<tr><td>1</td><td>1</td><td>1</td><td>1</td><td>1</td></tr>
</table>

Then the Boolean expression for a full subtractor is as follows.

For the DIFFERENCE (D) bit:

$$D=(\overline{X}.\overline{Y}.B_{IN})+\left(\overline{X}.Y.\overline{B_{IN}}\right)+\left(X.Y.B_{IN}\right)$$

which can be simplified too:

$$D=(X \text{ XOR } Y) \text{ XOR } B_{IN}=(X \oplus Y) \oplus B_{IN}$$

For the BORROW OUT (B_{OUT}) bit:

$$B_{OUT}=\left(\overline{X}.\overline{Y}.B_{IN}\right)+\left(\overline{X}.Y.\overline{B_{IN}}\right)+\left(\overline{X}.Y.B_{IN}\right)+\left(X.Y.B_{IN}\right)$$

which will also simplify too:

$$B_{OUT}=\overline{X} \text{ AND } Y \text{ OR}\left(\overline{X \text{XOR } Y}\right)B_{IN}=\overline{X}.Y+\overline{(X \oplus Y)}B_{IN}$$

An n-bit Binary Subtractor

As with the binary adder, we can also have n number of 1-bit full binary subtractor connected or "cascaded" together to subtract two parallel n-bit numbers from each other. For example two 4-bit binary numbers. We said before that the only difference between a full adder and a full subtractor was the inversion of one of the inputs.

So by using an n-bit adder and n number of inverters (NOT Gates), the process of subtraction becomes an addition as we can use two's complement notation on all the bits in the subtrahend and setting the carry input of the least significant bit to a logic "1" (HIGH).

Binary Subtractor using 2's Complement

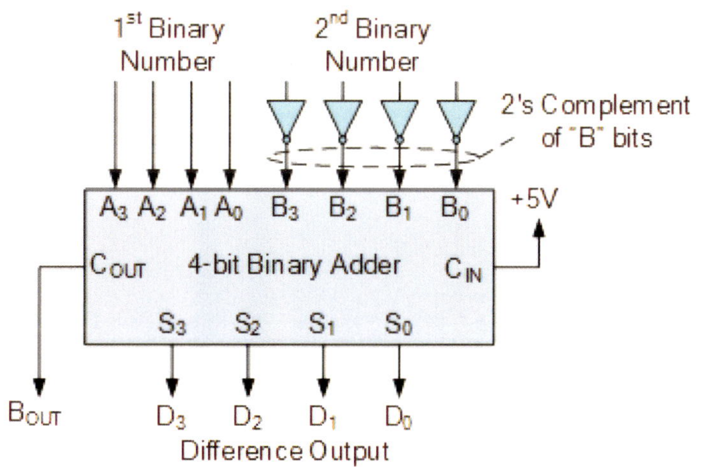

Then we can use a 4-bit full-adder ICs such as the 74LS283 and CD4008 to perform subtraction simply by using two's complement on the subtrahend, B inputs as X – Y is the same as saying, X + (-Y) which equals X plus the two's complement of Y.

If we wanted to use the 4-bit adder for addition once again, all we would need to do is set the carry-in (C_{IN}) input LOW at logic "0". Because we can use the 4-bit adder IC such as the 74LS83 or 74LS283 as a full-adder or a full-subtractor they are available as a single adder/subtractor circuit with a single control input for selecting between the two operations.

Binary Multiplier

A binary multiplier is a combinational logic circuit used in digital systems to perform the multiplication of two binary numbers. These are most commonly used in various applications especially in the field of digital signal processing to perform the various algorithms.

Commercial applications like computers, mobiles, high speed calculators and some general purpose processors require binary multipliers.

Compared with addition and subtraction, multiplication is a complex process. In multiplication process, the number which is to be multiplied by the other number is called as multiplicand and the number multiplied is called as multiplier.

Binary Multiplication

Similar to the multiplication of decimal numbers, binary multiplication follows the same process for producing a product result of the two binary numbers. The binary multiplication is much easier as it contains only 0s and 1s. The four fundamental rules for binary multiplication are

$0 \times 0 = 0$

$0 \times 1 = 0$

$1 \times 0 = 0$

$1 \times 1 = 1$

The multiplication of two binary numbers can be performed by using two common methods, namely partial product addition and shifting, and using parallel multipliers.

Before discussing about the types, let us look at the unsigned binary numbers multiplication process. Consider a two 4 bit binary numbers as 1010 and 1011, and its multiplication of these two is given as

```
        1 0 1 0    ----->  Multiplicand
      × 1 0 1 1    ----->  Multiplier
    -------------
        1 0 1 0    ----->  Partial product 1
      1 0 1 0      ----->  Partial product 2
    0 0 0 0        ----->  Partial product 3
  1 0 1 0          ----->  Partial product 4
    -------------
  1 1 0 1 1 1 0
    -------------
```

From the above multiplication, partial products are generated for each digit in the multiplier. Then all these partial products are added to produce the final product value. In the partial product multiplication, when the multiplier bit zero, the partial product is zero, and when the multiplier bit is 1, the resulted partial product is the multiplicand.

As similar to the decimal numbers, each successive partial product is shifted one position left relative to the preceding partial product before summing all partial products.

Therefore, this multiplication uses n-shifts and adds to multiply n-bit binary number. The combinational circuit implemented to perform such multiplication is called as an array multiplier or combinational multiplier.

Parallel Binary Multiplier Circuit

Let us consider two unsigned 2 bit binary numbers A and B to generalize the multiplication process. The multiplicand A is equal to A1A0 and the multiplier B is equal to B1B0. The figure below shows the multiplication process of two 2 bit binary numbers.

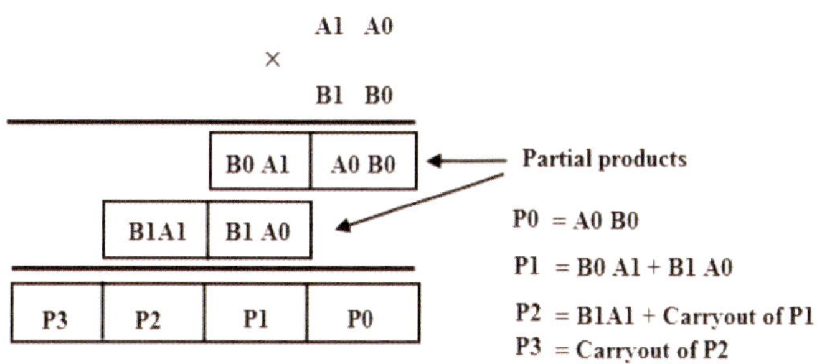

This process involves the multiplication of two digits and the addition of digits with or without carry. After the multiplication of the each bit to the multiplicand, partial products are generated, and then these products are added to produce the total sum which represents the binary multiplication value.

This multiplication is implemented by combinational circuit such that the multiplication is performed with AND gates whereas the addition is carried out by using half adders as shown in figure.

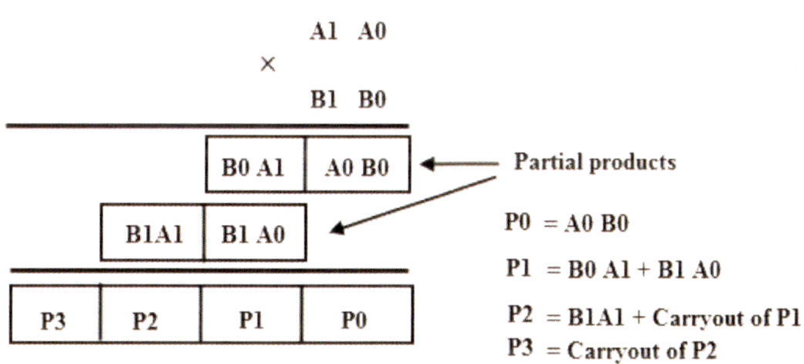

The first partial product is obtained by the AND gate which is nothing but a least significant bit of the multiplication result. Since the second partial product is shifted to the left position, the first partial second term and second partial product first term is added by half adder and produce the sum output along with the carry out.

This carry out is added at the next half adder as an input as shown in figure. Likewise, it produces the multiplication result of two binary numbers by using the simple circuit configuration. The multiplication of the two 2 bit number results a 4-bit binary number.

Let us consider two unsigned 4 bit numbers multiplication in which the multiplicand, A is equal to A3A2 A1A0 and the multiplier B is equal to B3B2B1B0. The partial products are produced depending on each multiplier bit multiplied by the multiplicand.

Each partial product consists of four product terms and these are shifted to the left relative to the previous partial product as shown in figure. All these partial products are added to produce the 8 bit product.

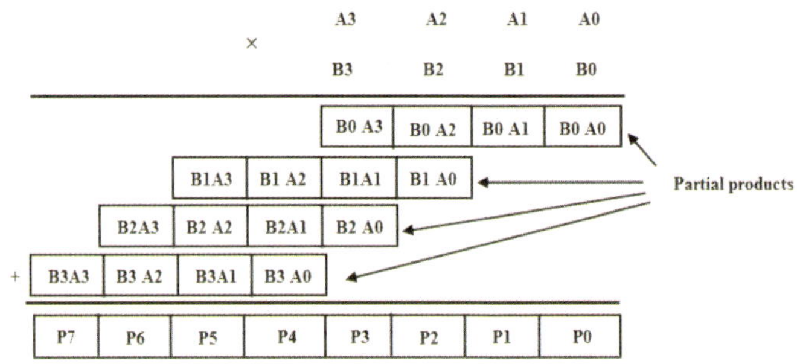

The logic circuit for the 4× 4 binary multiplication can be implemented by using three binary full adders along with AND gates.

In the above operation the first partial product is obtained by multiplying B0 with A3A2 A1A0, the second partial product is formed by multiplying B1 with A3A2 A1A0, likewise for 3rd and 4th partial products. So these partial products can be implemented with AND gates as shown in figure.

These partial products are then added by using 4 bit parallel adder. The three most significant bits of first partial product with carry (considered as zero) are added with second partial term in the first full adder.

Then the result is added to the next partial product with carry out and it goes on till the final partial product, finally it produces 8 bit sum which indicates the multiplication value of the two binary numbers.

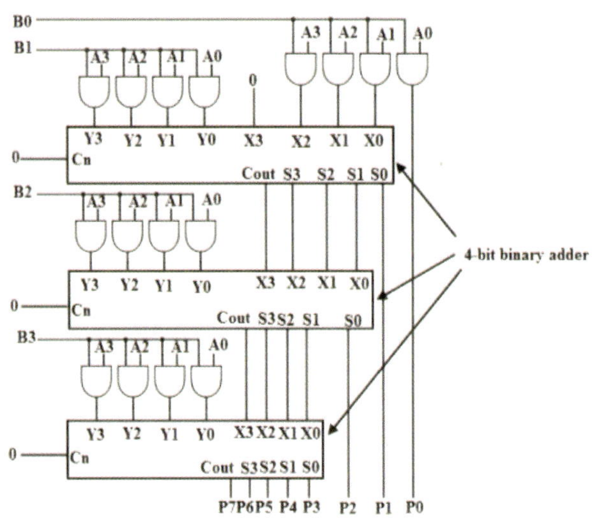

Binary Multiplier using Shift Method

Alternatively to the above automated method, manual multiplication approach can be implemented by using n-bit adder, four registers (A, B, C and Q) and shift and control logic as shown in below figure.

In this, the 4 bit multiplier is stored in Q register, the 4 bit multiplicand is stored in register B and the register A is initially cleared to zero. The multiplication process starts with checking of the least significant bit of B whether it is 0 or 1.

If the $B_0 = 1$, the number in the multiplicand (B) is added with the least significant bits of the A register and all bits of C, A and Q registers are shifted to the right one bit.

If the bit $B_0 = 0$, the combined C and Q registers are shifted to the right by one bit without performing any addition. This process is repeated for n times for n bit numbers. This method of binary multiplication is called as parallel multiplier.

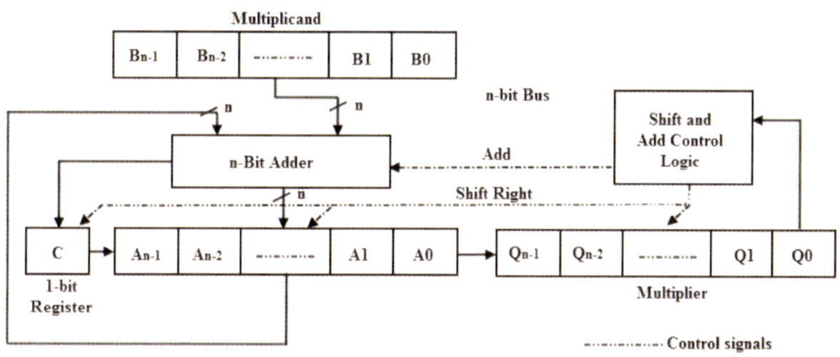

Consider the below figure in which the multiplier and multiplicand values are given as 1011 and 1101 which are loaded into the Q and A registers respectively. Initially the register C is zero and hence the A register is zero, which stores the carry in addition.

Since the $B_0 = 1$, then the number in the B is added to the bits of A and produce the addition result as 1101, and the Q and A register are shifted their values one bit right so the new values during the first cycle are 0110 and 1101 respectively.

This process has to be repeated four times to perform the 4 bit multiplication. The final multiplication result will be available in the A and Q registers as 10001111 as shown in the figure.

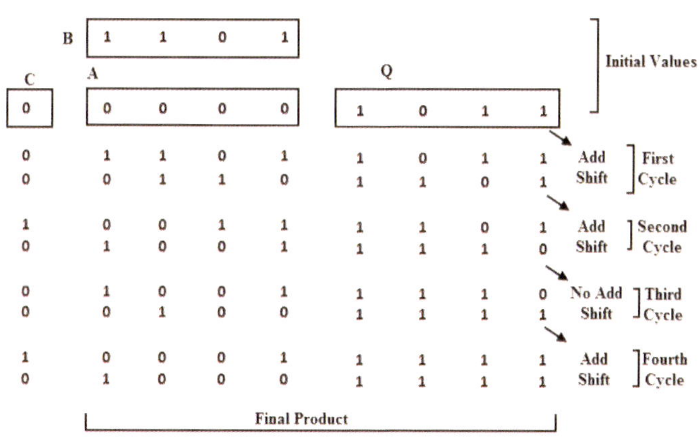

A 4 × 4 unsigned binary multiplier takes two, four bit inputs and produces an output of 8 bits. Similarly 8 × 8 multiplier accepts two 8 bit inputs and generates an output of 16 bits.

These multiplier logic circuits are implemented on integrated circuits with various pin configurations.

These ICs are used in several applications, particularly in various microprocessors used for computers, controlling devices, calculators, mobiles, digital signal processors (DSPs), etc.

Binary Encoder

An encoder is a device which converts familiar numbers or characters or symbols into a coded format. It accepts the alphabetic characters and decimal numbers as inputs and produces the outputs as a coded representation of the inputs.

It encodes the given information into a more compact form. In other words, it is a combinational circuit that performs the opposite function of a decoder.

These are mainly used to reduce the number of bits needed to represent given information. In digital systems, encoders are used for transmitting the information. Thus the transmission link uses fewer lines to transmit the encoded information.

In addition, these encoders are used for encoding the data which is to be stored for later use as it facilitates fewer bits storing over the available space.

A binary encoder has 2n input lines and n output lines, hence it encodes the information from 2n inputs into an n-bit code. From all the input lines, only one of an input line is activated at a time, and depending on the input line, it produces the n bit output code.

The figure below shows the block diagram of binary encoder which consists of 2n input lines and n output lines. It translates decimal number to binary number.

The output lines of an encoder correspond to either true binary equivalent or in BCD coded form of the binary for the input value. Some of these binary encoders include decimal to binary encoders, decimal to octal, octal to binary encoders, decimal to BCD encoders, etc.

Depending on the number of input lines, digital or binary encoders produce the output codes in the form of 2 or 3 or 4 bit codes.

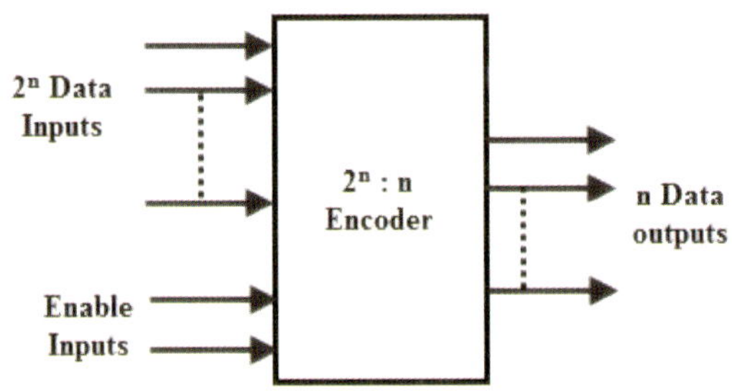

4 – to – 2 Bit Binary Encoder

The block diagram and truth table of a 4 input encoder is shown in below figure. The truth table consists of four rows , since , it is assumed that only one input is the value of 1 then the corresponding binary code associated with that enabled input is displayed at the outputs.

It is to be observed from the table is the output Yo is 1 when either input w1 or w3 is 1, also the output Y1 is set to 1 when either input w2 or w3 is 1.

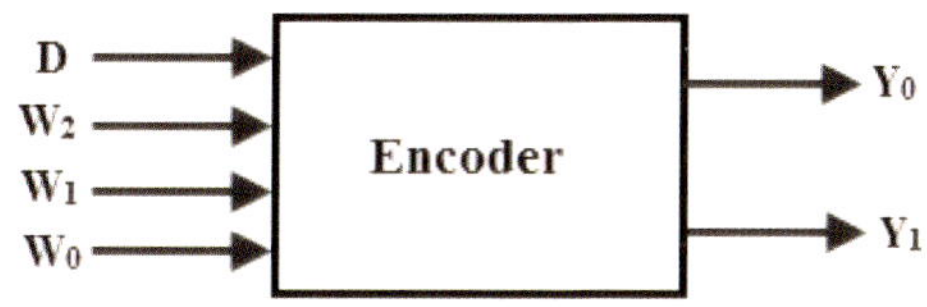

Block Diagram

D	W_2	W_1	W_0	Y_1	Y_0
0	0	0	1	0	0
0	0	1	0	0	1
0	1	0	0	1	0
1	0	0	0	1	1

Truth Table

The output from 4-to-2 encoder is generated by the logic circuit implemented by a set of OR gates as shown in below. In the figure a, the output of the encoder is same if the input activated is the Io input (Io = 1) or if no input is activated i.e., all the inputs are zero.

This causes ambiguity in the encoding output. To avoid this ambiguity, a valid encode output can be added as an additional output, so this output assumes a value 1 when Io is equal to 1.

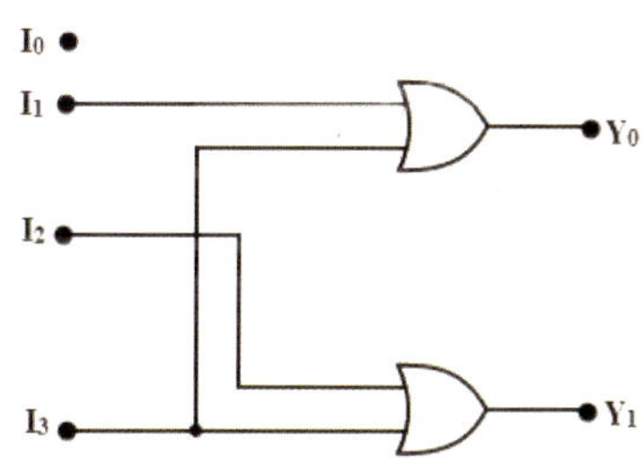

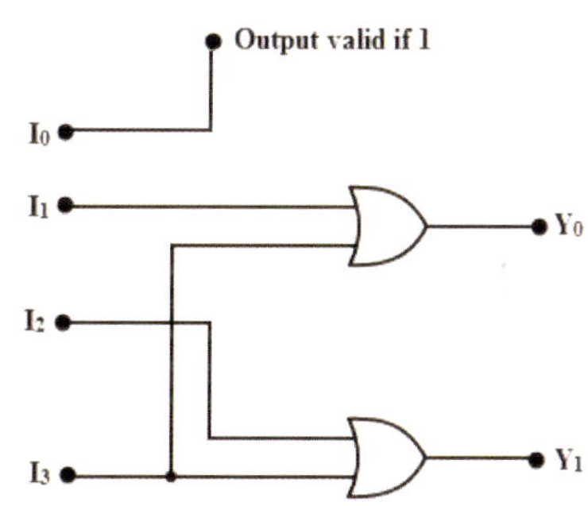

Decimal to BCD Encoder

This type of encoder usually consists of ten input lines and 4 output lines. Each input line corresponds to the each decimal digit and 4 outputs correspond to the BCD code.

This encoder accepts the decoded decimal data as an input and encodes it to the BCD output which is available on the output lines.

The figure below shows the basic logic symbol of decimal to BCD encoder along with its truth table. The truth table represents the BCD code for each decimal digit.

From this we can formulate the relationship between the BCD bit and decimal digit. It is important to note that there is no explicit input line for decimal zero. When this condition occurs, i.e., decimal inputs 1 to 9 all are zero, then the BCD output is 0000.

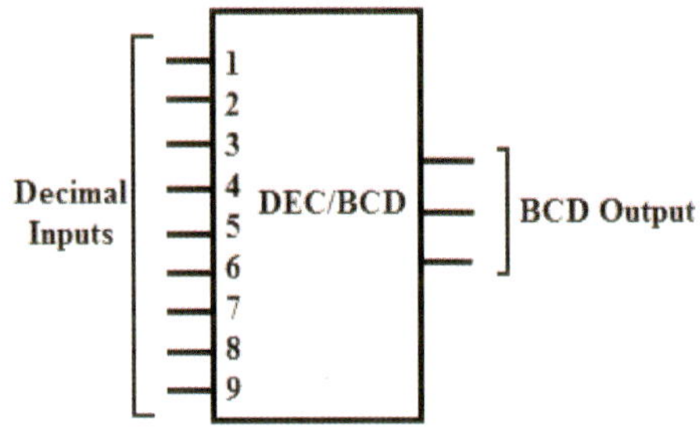

Logic Symbol

From the above table, we get the expressions as

$$Y3 = D8 + D9$$

$$Y2 = D4 + D5 + D6 + D7$$

$$Y1 = D2 + D3 + D6 + D7$$

$$Y0 = D1 + D3 + D5 + D7 + D9$$

From the above expressions, the decimal to BCD encoder logic circuit can be implemented by using set of OR gates as shown in below figure.

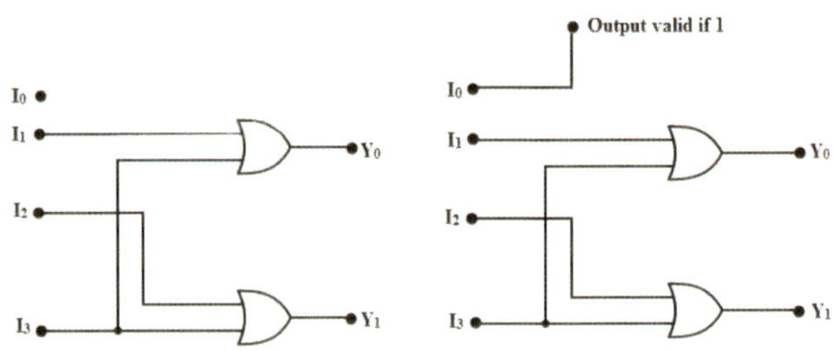

Octal to Binary Encoder

An octal to binary encoder consists of eight input lines and three output lines. Each input line corresponds to each octal digit and three outputs generate corresponding binary code.

In encoders, it is to be assumed that only one input is active or has a value 1 at any given time otherwise the circuit has no meaning. The figure below shows the logic symbol of octal to binary encoder along with its truth table.

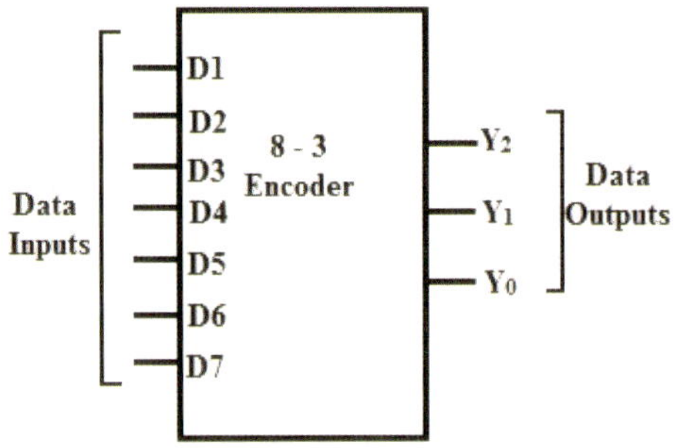

No	Inputs								Outputs		
	D_7	D_6	D_5	D_4	D_3	D_2	D_1	D_0	Y_2	Y_1	Y_0
0	0	0	0	0	0	0	0	1	0	0	0
1	0	0	0	0	0	0	1	0	0	0	1
2	0	0	0	0	0	1	0	0	0	1	0
3	0	0	0	0	1	0	0	0	0	1	1
4	0	0	0	1	0	0	0	0	1	0	0
5	0	0	1	0	0	0	0	0	1	0	1
6	0	1	0	0	0	0	0	0	1	1	0
7	1	0	0	0	0	0	0	0	1	1	1

From the above table, the output Y2 becomes 1 if any of the digits D4 or D5 or D6 or D7 is one. Thus, we can write its expression as

$$Y2 = D4 + D5 + D6 + D7$$

Similarly, $Y1 = D2 + D3 + D6 + D7$ and

$$Y0 = D1 + D3 + D5 + D7$$

Also it is to be observed that D0 does not exist in any of the expressions so it is considered as don't care. From the above expressions, we can implement the octal to binary encoder using set of OR gates as shown in figure below.

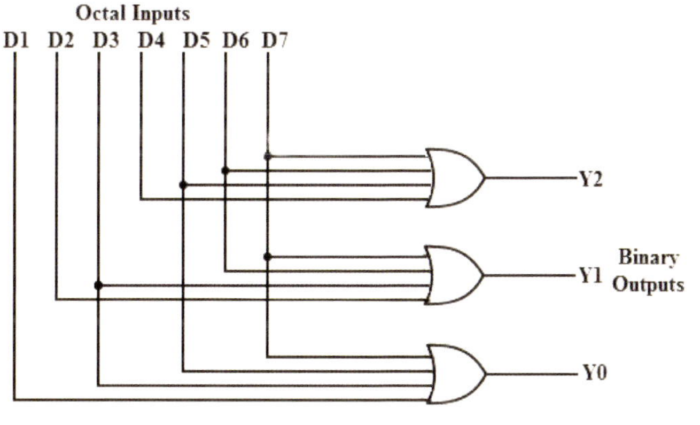

There is ambiguity in the octal to binary encoder that when all the inputs are zero, an output with all 0's is generated. Also, when D0 is 1, the output generated is zero. This is a major problem in this type of encoder. This can be resolved by specifying the condition that none of the inputs are active with an additional output.

Digital Encoder Applications

Encoders are very common electronic circuits used in all digital systems. In case of pocket calculators, these are used to translate the decimal values to the binary in order to perform the binary functions such as addition, subtraction, multiplication, etc.

These are also used to generate the digital signals in response to the movement which are classified into shaft encoders and linear encoders. Let us discuss keyboard applications of encoders in brief.

Keyboard Encoder

This type of encoder function is to generate the binary code corresponds to the alphanumeric character key depressed on a keyboard. The below figure shows a keyboard encoder used to encode the 10 decimal digits in BCD code by using a diode matrix. In this circuit, BCD data are stored in the S-R flip-flops.

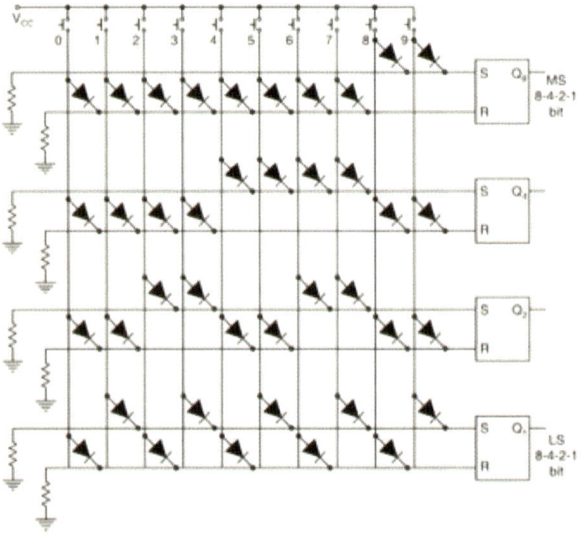

When a key corresponding to one of the decimal digits is pressed, the selected diodes are forward biased by a positive voltage, and these are connected to the set and reset terminals of S-R flip-flops. The diode arrangement is made in such a way that each flip-flop set or resets so as to produce the 4 bit BCD code.

Suppose if key corresponds to decimal digit 7 is pressed, the diodes which are connected to the S inputs of Q4, Q2 and Q1 are forward biased, also the diode which is connected to the R input of Q8. Therefore, the output BCD code is 0111.

From the logic diagram of the encoder, it is to be observed that the diode configuration at each S and R input is essentially a diode OR gate. This type of diode matrix encoders is used in printed circuit boards on many electronic devices having keyboards as user data interface.

Priority Encoder

The Priority Encoder solves the problems mentioned above by allocating a priority level to each input. The priority encoders output corresponds to the currently active input which has the highest priority. So when an input with a higher priority is present, all other inputs with a lower priority will be ignored.

The priority encoder comes in many different forms with an example of an 8-input priority encoder along with its truth table shown below.

8-to-3 Bit Priority Encoder

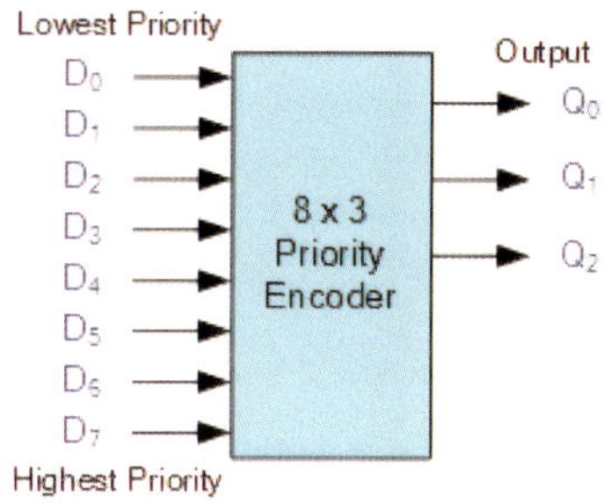

Inputs								Outputs		
D_7	D_6	D_5	D_4	D_3	D_2	D_1	D_0	Q_2	Q_1	Q_0
0	0	0	0	0	0	0	1	0	0	0
0	0	0	0	0	0	1	x	0	0	1
0	0	0	0	0	1	x	x	0	1	0
0	0	0	0	1	x	x	x	0	1	1
0	0	0	1	x	x	x	x	1	0	0
0	0	1	x	x	x	x	x	1	0	1
0	1	x	x	x	x	x	x	1	1	0
1	x	x	x	x	x	x	x	1	1	1

X = dont care

Priority encoders are available in standard IC form and the TTL 74LS148 is an 8-to-3 bit priority encoder which has eight active LOW (logic "0") inputs and provides a 3-bit code of the highest ranked input at its output.

Priority encoders output the highest order input first for example, if input lines "D2", "D3" and "D5" are applied simultaneously the output code would be for input "D5" ("101") as this has the highest order out of the 3 inputs. Once input "D5" had been removed the next highest output code would be for input "D3" ("011"), and so on.

The truth table for a 8-to-3 bit priority encoder is given as:

Digital Inputs								Binary Output		
D_7	D_6	D_5	D_4	D_3	D_2	D_1	D_0	Q_2	Q_1	Q_0
0	0	0	0	0	0	0	1	0	0	0
0	0	0	0	0	0	1	X	0	0	1
0	0	0	0	0	1	X	X	0	1	0
0	0	0	0	1	X	X	X	0	1	1
0	0	0	1	X	X	X	X	1	0	0
0	0	1	X	X	X	X	X	1	0	1
0	1	X	X	X	X	X	X	1	1	0
1	X	X	X	X	X	X	X	1	1	1

Where X equals "dont care", that is logic "0" or a logic "1".

From this truth table, the Boolean expression for the encoder above with data inputs D_0 to D_7 and outputs Q_0, Q_1, Q_2 is given as:

Output Q_0

$$Q_0 = \sum(1, 3, 5, 7)$$
$$Q_0 = \sum\left(\overline{D_7}\,\overline{D_6}\,\overline{D_5}\,\overline{D_4}\,\overline{D_3}D_1 + \overline{D_7}\,\overline{D_6}\,\overline{D_5}\,\overline{D_4}D_3 + \overline{D_7}\,\overline{D_6}D_5 + D_7\right)$$
$$Q_0 = \sum\left(\overline{D_6}\,\overline{D_4}\,\overline{D_2}D_1 + \overline{D_6}\,\overline{D_4}D_3 + \overline{D_6}D_5 + D_7\right)$$
$$Q_0 = \sum\left(\overline{D_6}\left(\overline{D_4}\,\overline{D_2}D_1 + \overline{D_4}D_3 + D_5\right) + D_7\right)$$

Output Q_1

$$Q_1 = \sum(2, 3, 6, 7)$$
$$Q_1 = \sum\left(\overline{D_7}\,\overline{D_6}\,\overline{D_5}\,\overline{D_4}\,\overline{D_3}D_2 + \overline{D_7}\,\overline{D_6}\,\overline{D_5}\,\overline{D_4}D_3 + \overline{D_7}D_6 + D_7\right)$$
$$Q_1 = \sum\left(\overline{D_5}\,\overline{D_4}D_2 + \overline{D_5}\,\overline{D_4}D_3 + D_6 + D_7\right)$$
$$Q_1 = \sum\left(\overline{D_5}\,\overline{D_4}\left(D_2 + D_3\right) + D_6 + D_7\right)$$

Output Q_2

$$Q_2 = \sum(4, 5, 6, 7)$$
$$Q_2 = \sum\left(\overline{D_7}\,\overline{D_6}\,\overline{D_5}D_4 + \overline{D_7}\,\overline{D_6}D_5 + \overline{D_7}D_6 + D_7\right)$$
$$Q_2 = \sum\left(D_4 + D_5 + D_6 + D_7\right)$$

Then the final Boolean expression for the priority encoder including the zero inputs is defined as:

$$Q_0 = \sum\left(\overline{D_6}\left(\overline{D_4}\,\overline{D_2}D_1 + \overline{D_4}D_3 + D_5\right) + D_7\right)$$
$$Q_1 = \sum\left(\overline{D_5}\,\overline{D_4}\left(D_2 + D_3\right) + D_6 + D_7\right)$$
$$Q_2 = \sum\left(D_4 + D_5 + D_6 + D_7\right)$$

In practice these zero inputs would be ignored allowing the implementation of the final Boolean expression for the outputs of the 8-to-3 priority encoder.

Priority Encoder Applications

Positional Encoders

Another more common application is in magnetic positional control as used on ships navigation or for robotic arm positioning etc. Here for example, the angular or rotary position of a compass is converted into a digital code by a 74LS148 8-to-3 line priority encoder and input to the systems computer to provide navigational data and an example of a simple 8 position to 3-bit output compass encoder is shown below. Magnets and reed switches could be used at each compass point to indicate the needles angular position.

Priority Encoder Navigation

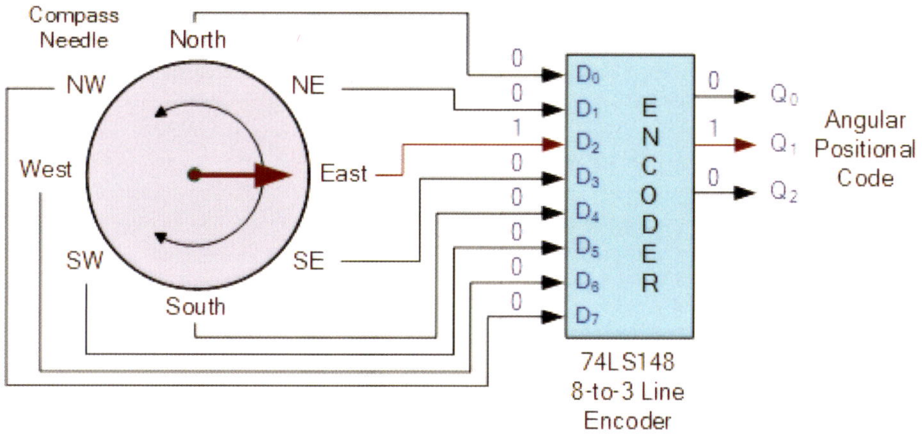

Compass Direction	Binary Output		
	Q_0	Q_1	Q_2
North	0	0	0
North-East	0	0	1
East	0	1	0
South-East	0	1	1
South	1	0	0
South-West	1	0	1
West	1	1	0
North-West	1	1	1

Interrupt Requests

Other applications especially for Priority Encoders may include detecting interrupts in microprocessor applications. Here the microprocessor uses interrupts to allow peripheral devices such as the disk drive, scanner, mouse, or printer etc, to communicate with it, but the microprocessor can only "talk" to one peripheral device at a time so needs some way of knowing when a particular peripheral device wants to communicate with it.

The processor does this by using "Interrupt Requests" or "IRQ" signals to assign priority to all the peripheral devices to ensure that the most important peripheral device is serviced first. The order of importance of the devices will depend upon their connection to the priority encoder.

IRQ Number	Typical Use	Description
IRQ 0	System timer	Internal System Timer.
IRQ 1	Keyboard	Keyboard Controller.
IRQ 3	COM2 & COM4	Second and Fourth Serial Port.
IRQ 4	COM1 & COM3	First and Third Serial Port.
IRQ 5	Sound	Sound Card.
IRQ 6	Floppy disk	Floppy Disk Controller.
IRQ 7	Parallel port	Parallel Printer.
IRQ 12	Mouse	PS/2 Mouse.
IRQ 14	Primary IDE	Primary Hard Disk Controller.
IRQ 15	Secondary IDE	Secondary Hard Disk Controller.

Because implementing such a system using priority encoders such as the standard 74LS148 priority encoder IC involves additional logic circuits, purpose built integrated circuits such as the 8259 Programmable Priority Interrupt Controller is available.

Binary Decoder

Decoder is a combinational logic circuit that converts coded input to coded outputs provided both of these are different from one another. The name decoder means translating of coded information from one format into another. So the input code generally has fewer bits than output code word.

A digital decoder converts a set of digital signals into corresponding decimal code. A decoder is also a most commonly used circuit in prior to the use of encoder. The encoded data is decoded for user interface in most of the output devices like monitors, calculator displays, printers, etc. once after the information is encoded by encoders.

A binary decoder is a multi-input, multi-output combinational circuit that converts a binary code of n input lines into a one out of 2n output code. These are used when there is need to activate exactly one of 2n output based on an n-bit input value.

The figure below shows the general structure of binary decoder in which encoded information is accepted at n input lines and the output is produced at 2n possible output lines.

Generally, decoders are provided with enable inputs so as to activate the decoded output based on data inputs. As an example, in case of BCD code, the 4 bit combinations from 0000 through 1001 are enough to represent the decimal digits 0 to 9.

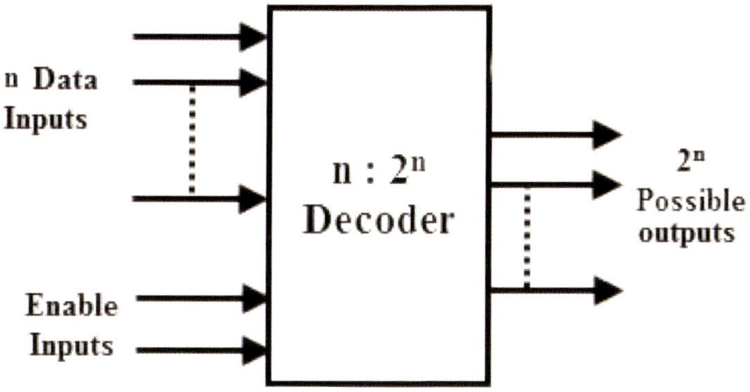

Depending on the number of input lines, the inputs of a binary code can be 2-bit or 3-bit or 4-bit codes. Upon the availability of 2n lines, it activates the one of its output by deactivating (making logic 0) all other input whenever it receives n inputs.

Usually the number of bits in output code is more than the bits in its input code. The most commonly used practical binary decoders are 2-to-4 decoder, 3-to-8 decoder and 4-to-16 line binary decoder.

2-to-4 Binary Decoder

In a 2-to-4 binary decoder, two inputs are decoded into four outputs hence it consists of two input lines and 4 output lines. Only one output is active at any time while the other outputs are maintained at logic 0 and the output which is held active or high is determined the two binary inputs A and B.

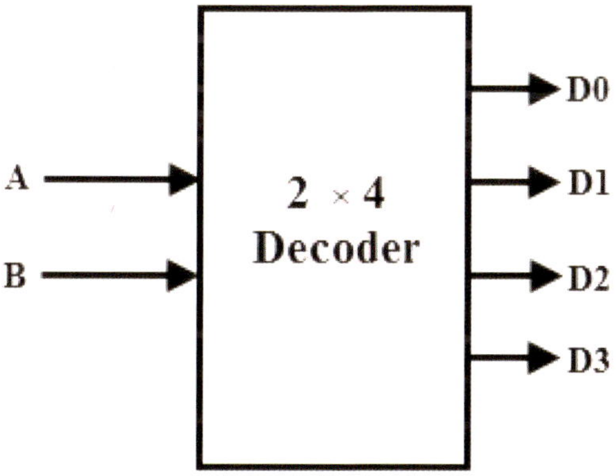

The figure below shows the truth table for a 2-to-4 decoder. For a given input, the outputs Y0 through Y3 are active high if enable input EN is active high (EN = 1). When both inputs A and B are low (or A= B= 0), the output Y0 will be active or High and all other outputs will be low.

When A = 0 and B = 1, the output Y1 will be active and when A = 1 and B = 0, then the output Y2 will be active. When both the inputs are high, then the output Y3 will be high. If the enable bit is zero then all the outputs will be set to zero. This relationship between the inputs and outputs are illustrated in below truth table clearly.

Inputs			Outputs			
EN	A	B	Y_3	Y_2	Y_1	Y_0
0	×	×	0	0	0	0
1	0	0	0	0	0	1
1	0	1	0	0	1	0
1	1	0	0	1	0	0
1	1	1	1	0	0	0

From the above truth table we can obtain Boolean expression for the each output as

$$Y0 = \overline{A}\,\overline{B}$$

$$Y1 = \overline{A}\,B$$

$$Y2 = A\,\overline{B}$$

$$Y3 = A\,B$$

These expressions can be implemented by using basic logic gates. Thus, the logic circuit design of the 2-to-4 line decoder is given below which is implemented by using NOT and AND gates. Two NOT gates or inverters provide the complement of inputs.

A common enable line is connected to each AND gate such that when EN= 0 all the outputs are zero and if EN=1, depends on the inputs A and B, outputs are produced. Each output represents one of the minterms of the 2 input variables.

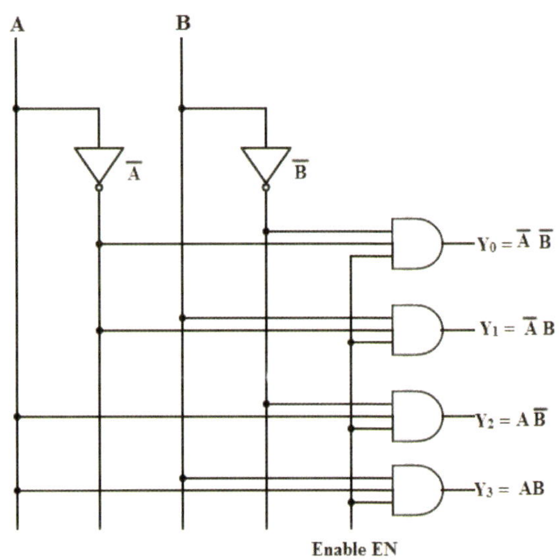

It is also possible to design 2-to-4 decoder using NAND gates as shown in figure below along with truth table. This is constructed with a principle of max terms as outputs. To generate the mint-erms, we have to use NAND gates which act as inverters. If both inputs are zero (A = B = 0), Y0 will be zero , if A = 0 and B= 1, then Y1 will be 1 and so on.

Therefore, only one output will be low for any combinations of inputs at a given time and all other outputs will be high. This type of decoders is available in IC forms so that 3 to 8, 4 to 16, and 5 to 32 decoders can also be made depends on the application requirement.

3-to-8 Decoder

In a 3-to-8 decoder, three inputs are decoded into eight outputs. It has three inputs as A, B, and C and eight output from Y0 through Y7. Based on the combinations of the three inputs, only one of the eight outputs is selected.

Inputs			Outputs			
EN	A	B	Y_3	Y_2	Y_1	Y_0
0	×	×	0	0	0	0
1	0	0	0	0	0	1
1	0	1	0	0	1	0
1	1	0	0	1	0	0
1	1	1	1	0	0	0

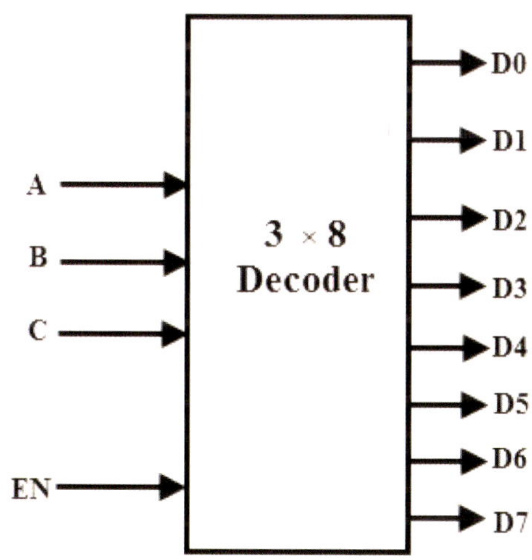

The figure below shows the truth table of a 3-to-8 decoder. Enable input is provided to activate the decoded output depends on the input combinations A, B and C. Suppose if A = B=1 and C= 0, then the output Y6 is 1 and all other outputs are zero. So from the truth table, minterms represents the each output equation and are given as:

$Y0 = \bar{A}\,\bar{B}\,\bar{C}$

$Y1 = \bar{A}\,\bar{B}\,C$

$Y2 = \bar{A}\,B\,\bar{C}$

$Y3 = \bar{A}\,B\,C$

$Y4 = A\,\bar{B}\,\bar{C}$

$Y5 = A\,\bar{B}\,C$

$Y6 = A\,B\,\bar{C}$

$Y7 = A\,B\,C$

Inputs				Outputs							
EN	A	B	C	Y_7	Y_6	Y_5	Y_4	Y_3	Y_2	Y_1	Y_0
0	×	×	×	0	0	0	0	0	0	0	0
1	0	0	0	0	0	0	0	0	0	0	1
1	0	0	1	0	0	0	0	0	0	1	0
1	0	1	0	0	0	0	0	0	1	0	0
1	0	1	1	0	0	0	0	1	0	0	0
1	1	0	0	0	0	0	1	0	0	0	0
1	1	0	1	0	0	1	0	0	0	0	0
1	1	1	0	0	1	0	0	0	0	0	0
1	1	1	1	1	0	0	0	0	0	0	0

Using the above min term expressions for each output, the circuit of 3-to-8 decoder is can be implemented by using three NOT gates and eight AND gates. Each NOT gate provides the complement of the input and AND gates generates one of the minterms.

Also enable input activate the decoded output depends on the input data. The logic diagram of this decoder is shown below.

Only one of eight outputs is high at a given time for a particular input combination, that why this decoder is also called as 1-of-8 decoder. Suppose, when ABC = 011, then only AND gate 4 has all inputs high, thus Y3 is high.

Also, 3-bit binary numbers at the input is converted to eight digits at the output (which is equivalent to octal number system), that's how; it is also called as a binary-to-octal decoder

It is also possible represent the each output equation using max terms. In such case, inversion operation is performed in the logic circuit than that of circuit with min terms. The figure below shows the truth table of 3-to-8 line decoder using NAND gates. Each output in the table gives a max term representation.

At a given time only one output is low and all other outputs will be high. For example, when A=B= 1 and C=0, then the output Y6 is zero and all other outputs are high as shown in below figure.

Inputs			Outputs							
A	B	C	Y_0	Y_1	Y_2	Y_3	Y_4	Y_5	Y_6	Y_7
0	0	0	0	1	1	1	1	1	1	1
0	0	1	1	0	1	1	1	1	1	1
0	1	0	1	1	0	1	1	1	1	1
0	1	1	1	1	1	0	1	1	1	1
1	0	0	1	1	1	1	0	1	1	1
1	0	1	1	1	1	1	1	0	1	1
1	1	0	1	1	1	1	1	1	0	1
1	1	1	1	1	1	1	1	1	1	0

From the above table , a 3-to-8 line decoder is designed by using three NAND gates and three NOT gates. NOT gates generate the complement of input while the NAND gates generate max terms of each output as shown in below figure.

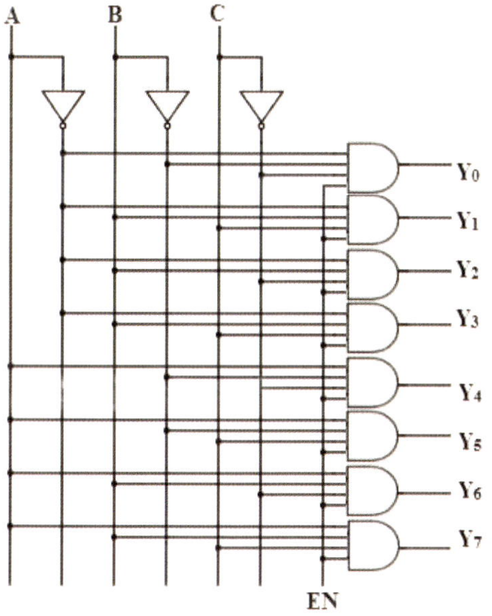

4-to-16 Decoder

A 4-to-16 decoder consists of 4 inputs and 16 outputs. Similar to all the decoders discussed above, in this also only one output will be low at a given time and all other outputs are high (using max-terms).

The truth table of this type of decoder is shown below. If the input to this decoder is 1000, then output Y8 will be low and all other outputs will be high as shown in figure. This will be so on for all the input combinations.

Inputs				Outputs															
A	B	C	D	Y_0	Y_1	Y_2	Y_3	Y_4	Y_5	Y_6	Y_7	Y_8	Y_9	Y_{10}	Y_{11}	Y_{12}	Y_{13}	Y_{14}	Y_{15}
0	0	0	0	0	1	1	1	1	1	1	1	1	1	1	1	1	1	1	1
0	0	0	1	1	0	1	1	1	1	1	1	1	1	1	1	1	1	1	1
0	0	1	0	1	1	0	1	1	1	1	1	1	1	1	1	1	1	1	1
0	0	1	1	1	1	1	0	1	1	1	1	1	1	1	1	1	1	1	1
0	1	0	0	1	1	1	1	0	1	1	1	1	1	1	1	1	1	1	1
0	1	0	1	1	1	1	1	1	0	1	1	1	1	1	1	1	1	1	1
0	1	1	0	1	1	1	1	1	1	0	1	1	1	1	1	1	1	1	1
0	1	1	1	1	1	1	1	1	1	1	0	1	1	1	1	1	1	1	1
1	0	0	0	1	1	1	1	1	1	1	1	0	1	1	1	1	1	1	1
1	0	0	1	1	1	1	1	1	1	1	1	1	0	1	1	1	1	1	1
1	0	1	0	1	1	1	1	1	1	1	1	1	1	0	1	1	1	1	1
1	0	1	1	1	1	1	1	1	1	1	1	1	1	1	0	1	1	1	1
1	1	0	0	1	1	1	1	1	1	1	1	1	1	1	1	0	1	1	1
1	1	0	1	1	1	1	1	1	1	1	1	1	1	1	1	1	0	1	1
1	1	1	0	1	1	1	1	1	1	1	1	1	1	1	1	1	1	0	1
1	1	1	1	1	1	1	1	1	1	1	1	1	1	1	1	1	1	1	0

From the above truth table, a 4-to-16 decoder can be implemented by using 4 NOT gates and 16 decoding NAND gates. For decoding all possible combinations of 4 bits input, sixteen (24 = 16) decoding gates are required.

It is important to note that all the NAND gates are implemented on this circuit produce the active low outputs as shown in figure.

Since it selects one of 16 outputs based in the particular input combination, these decoders are also called as 1-of-16 decoder. And also its output represents the sixteen digits as hexadecimal number system, this type of decoder is also called as a binary-to-hexadecimal decoder.

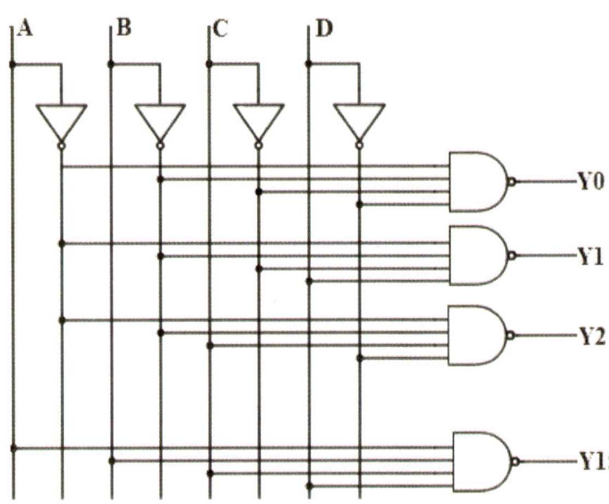

It is possible to combine or cascade two or more decoders to produce a decoder with larger number of input bits with the use of enable input of decoder. The cascade combination of two 3-to-8 line decoder is given below figure. It consists of four inputs A, B, C and Enable E and 16 outputs Y0 to Y7.

One of the input variable is used as enable input of the first 3-to-4 decoder and this same input is complemented and connected as enable input of the second decoder. The decoder to be enabled is decided by the most significant input variable and other input variables are fed to each decoder.

When enable input is zero then the top decoder is enabled while the other is disabled. Then the top decoder eight outputs generate the minterms 0000 to 0111. Likewise, when enable is 1, the lower decoder is enabled and top one is disabled. Thus the bottom decoder outputs generate minterms from 1000 to 1111.

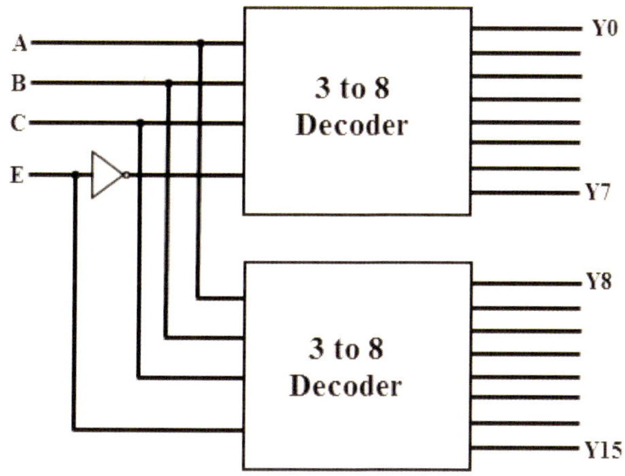

Applications of Decoders

Decoders are greatly used in applications where the particular output or group of outputs to be activated only on the occurrence of a specific combination of input levels. Very often these input levels are provided by the outputs of a register or counter.

When the counter or register continuously pulse the decoder inputs, the outputs will be activated sequentially. And these outputs can be used as sequencing signals or timing signals to switch the devices at particular times.

Binary to Decimal Decoder

Decoders are used to get the decimal digit corresponding to a specific input combination. A BCD number needs 4 binary digits to represent the 0 to 9 decimal digits, thus it consists of 4 input lines. It consists of 10 output lines corresponding to 0 to 9 decimal digits. T

This type of decoder is also called as a 1 to 10 decoder. For a specific input combination, the output will be activated corresponding to the decimal equivalent of the input combination.

Address Decoders

Amongst its many uses, a decoder is widely used to decode the particular memory location in the computer memory system. Decoders accept the address code generated by the CPU which is a combination of address bits for a specific location in the memory.

In a memory system, there are several memory ICs are combined and each one has their unique address to distinguish from other memory locations.

In such cases a decoder built in the memory ICs circuitry, is used to select a memory IC in response to a range of addresses by decoding the most significant bits of the systems address, thereby a particular memory location or IC is selected.

In a more complex memory system, the memory ICs or chips are arranged in multiple banks. When the microprocessor wants to access one or more bytes at a time, these banks must be selected simultaneously or individually.

In such cases more than one decoder must be activated. For that, cascaded decoders are used or most commonly decoders are replaced with programmable logic devices.

Instruction Decoder

Another application of the decoder can be found in the control unit of the central processing unit. This decoder is used to decode the program instructions in order to activate the specific control lines such that different operations in the ALU of the CPU are carried out.

Multiplexer

The multiplexer is a combinational logic circuit designed to switch one of several input lines to a single common output line by the application of a control logic.

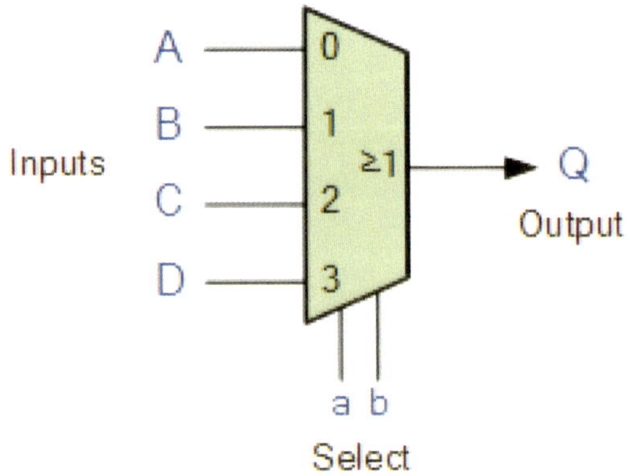

Multiplexing is the generic term used to describe the operation of sending one or more analogue or digital signals over a common transmission line at different times or speeds and as such, the device we use to do just that is called a Multiplexer.

The multiplexer, shortened to "MUX" or "MPX", is a combinational logic circuit designed to switch one of several input lines through to a single common output line by the application of a control signal. Multiplexers operate like very fast acting multiple position rotary switches connecting or controlling multiple input lines called "channels" one at a time to the output.

Multiplexers, or MUX's, can be either digital circuits made from high speed logic gates used to switch digital or binary data or they can be analogue types using transistors, MOSFET's or relays to switch one of the voltage or current inputs through to a single output.

The most basic type of multiplexer device is that of a one-way rotary switch as shown.

Basic Multiplexing Switch

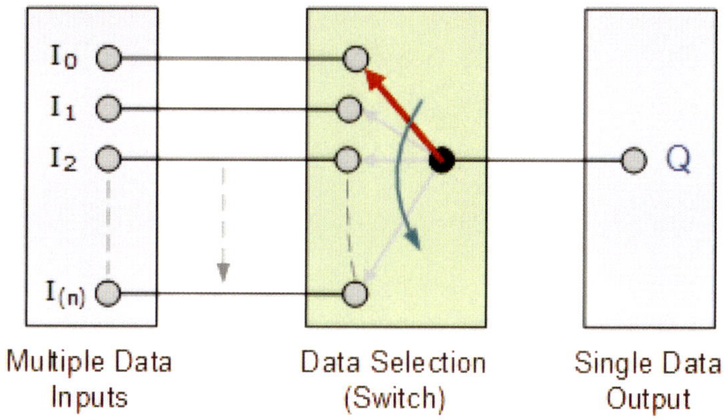

The rotary switch, also called a wafer switch as each layer of the switch is known as a wafer, is a mechanical device whose input is selected by rotating a shaft. In other words, the rotary switch is a manual switch that you can use to select individual data or signal lines simply by turning its inputs "ON" or "OFF". So how can we select each data input automatically using a digital device.

In digital electronics, multiplexers are also known as data selectors because they can "select" each input line, are constructed from individual Analogue Switches encased in a single IC package as opposed to the "mechanical" type selectors such as normal conventional switches and relays.

They are used as one method of reducing the number of logic gates required in a circuit design or when a single data line or data bus is required to carry two or more different digital signals. For example, a single 8-channel multiplexer.

Generally, the selection of each input line in a multiplexer is controlled by an additional set of inputs called control lines and according to the binary condition of these control inputs, either "HIGH" or "LOW" the appropriate data input is connected directly to the output. Normally, a multiplexer has an even number of 2^n data input lines and a number of "control" inputs that correspond with the number of data inputs.

Note that multiplexers are different in operation to Encoders. Encoders are able to switch an n-bit input pattern to multiple output lines that represent the binary coded (BCD) output equivalent of the active input.

We can build a simple 2-line to 1-line (2-to-1) multiplexer from basic logic NAND gates as shown.

2-input Multiplexer Design

The input A of this simple 2-1 line multiplexer circuit constructed from standard NAND gates acts to control which input (I_0 or I_1) gets passed to the output at Q.

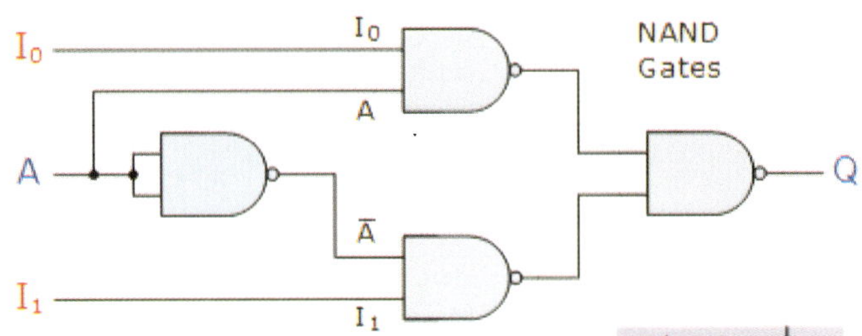

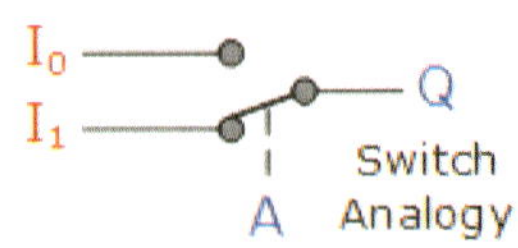

Inputs			
A	I_1	I_0	Q
0	0	0	0
0	0	1	0
0	1	0	1
0	1	1	1
1	0	0	0
1	0	1	1
1	1	0	0
1	1	1	1

Truth Table

From the truth table above, we can see that when the data select input, A is LOW at logic 0, input I_1 passes its data through the NAND gate multiplexer circuit to the output, while input I_0 is blocked. When the data select A is HIGH at logic 1, the reverse happens and now input I_0 passes data to the output Q while input I_1 is blocked.

So by the application of either a logic "0" or a logic "1" at A we can select the appropriate input, I_0 or I_1 with the circuit acting a bit like a single pole double throw (SPDT) switch.

As we only have one control line, (A) then we can only switch 2^1 inputs and in this simple example, the 2-input multiplexer connects one of two 1-bit sources to a common output, producing a 2-to-1-line multiplexer. We can confirm this in the following Boolean expression.

$$Q = \overline{A.I_0}.I + \overline{A}.\ I_0.I_1 + A.I_0.\overline{I_1} +_1 A.I_0.I_1$$

and for our 2-input multiplexer circuit above, this can be simplified too:

$$Q = \overline{A}.I_1 + A.I_0$$

We can increase the number of data inputs to be selected further simply by following the same procedure and larger multiplexer circuits can be implemented using smaller 2-to-1 multiplexers as their basic building blocks. So for a 4-input multiplexer we would therefore require two data select lines as 4-inputs represents 2^2 data control lines give a circuit with four inputs, I_0, I_1, I_2, I_3 and two data select lines A and B as shown.

4-to-1 Channel Multiplexer

The Boolean expression for this 4-to-1 Multiplexer above with inputs A to D and data select lines a, b is given as:

$$Q=\overline{a}\overline{b}A+a\overline{b}B+\overline{a}bC+abD$$

In this example at any one instant in time only ONE of the four analogue switches is closed, connecting only one of the input lines A to D to the single output at Q. As to which switch is closed depends upon the addressing input code on lines "a" and "b", so for this example to select input B to the output at Q, the binary input address would need to be "a" = logic "1" and "b" = logic "0".

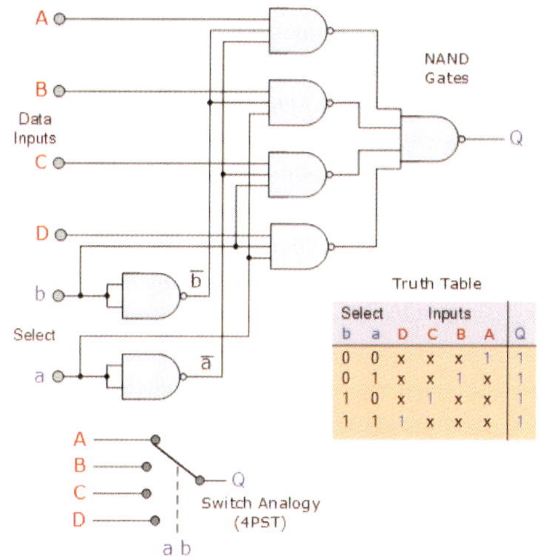

Select		Inputs				
b	a	D	C	B	A	Q
0	0	x	x	x	1	1
0	1	x	x	1	x	1
1	0	x	1	x	x	1
1	1	1	x	x	x	1

Then we can show the selection of the data through the multiplexer as a function of the data select bits as shown.

Multiplexer Input Line Selection

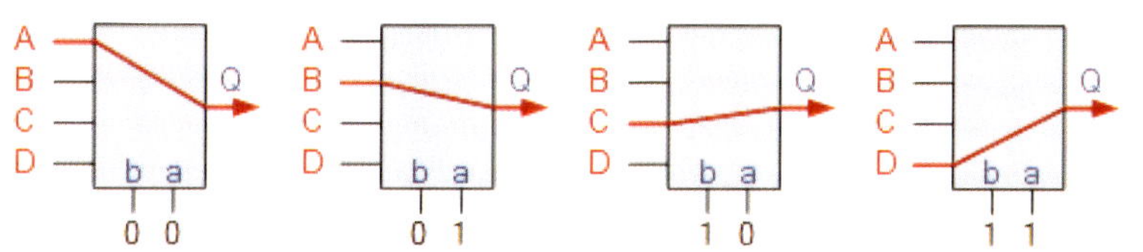

Adding more control address lines, (n) will allow the multiplexer to control more inputs as it can switch 2^n inputs but each control line configuration will connect only ONE input to the output.

Then the implementation of the Boolean expression above using individual logic gates would require the use of seven individual gates consisting of AND, OR and NOT gates as shown.

4 Channel Multiplexer using Logic Gates

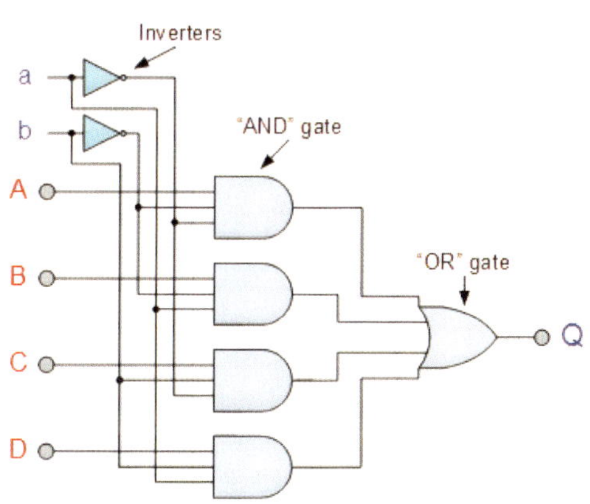

The symbol used in logic diagrams to identify a multiplexer is as follows.

Multiplexer Symbol

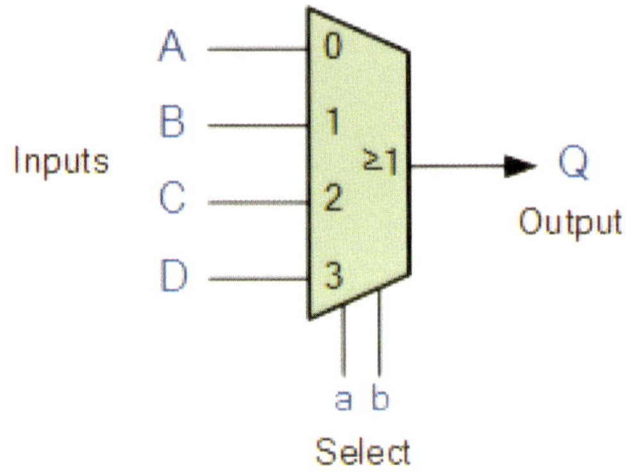

Multiplexers are not limited to just switching a number of different input lines or channels to one common single output. There are also types that can switch their inputs to multiple outputs and have arrangements or 4-to-2, 8-to-3 or even 16-to-4 etc. configurations and an example of a simple Dual channel 4 input multiplexer (4-to-2) is given below:

4-to-2 Channel Multiplexer

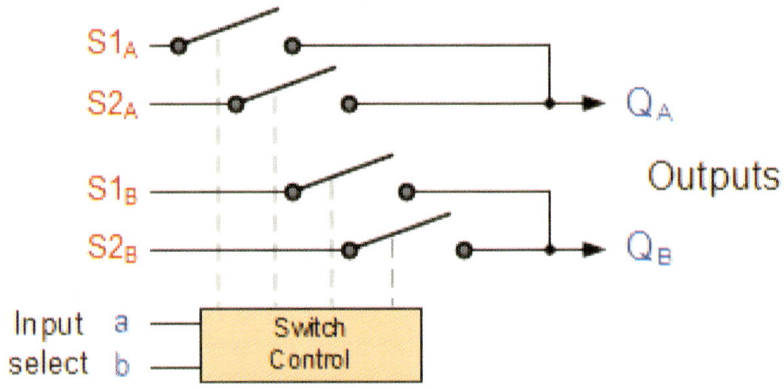

Here in this example the 4 input channels are switched to 2 individual output lines but larger arrangements are also possible. This simple 4-to-2 configuration could be used for example, to switch audio signals for stereo pre-amplifiers or mixers.

8x1 Multiplexer

Let us implement 8x1 Multiplexer using 4x1 Multiplexers and 2x1 Multiplexer. We know that 4x1 Multiplexer has 4 data inputs, 2 selection lines and one output. Whereas, 8x1 Multiplexer has 8 data inputs, 3 selection lines and one output.

So, we require two 4x1 Multiplexers in first stage in order to get the 8 data inputs. Since, each 4x1 Multiplexer produces one output, we require a 2x1 Multiplexer in second stage by considering the outputs of first stage as inputs and to produce the final output.

Let the 8x1 Multiplexer has eight data inputs I_7 to I_0, three selection lines s_2, s_1 & so and one output Y. The Truth table of 8x1 Multiplexer is shown below.

Selection Inputs			Output
S_2	S_1	S_0	Y
0	0	0	I_0
0	0	1	I_1
0	1	0	I_2
0	1	1	I_3
1	0	0	I_4
1	0	1	I_5
1	1	0	I_6
1	1	1	I_7

We can implement 8x1 Multiplexer using lower order Multiplexers easily by considering the above Truth table. The block diagram of 8x1 Multiplexer is shown in the following figure.

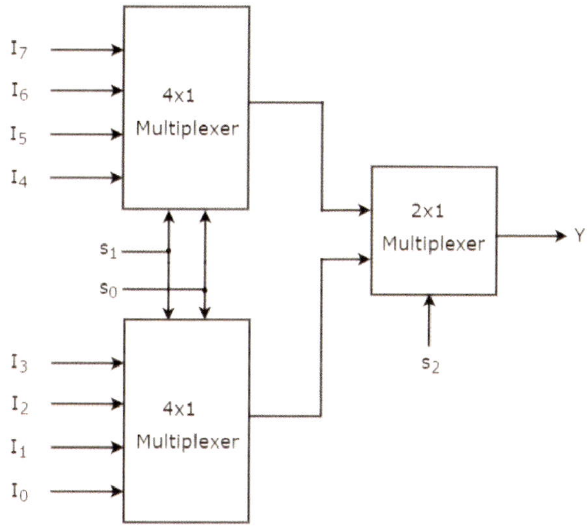

The same selection lines, s_1 & s_0 are applied to both 4x1 Multiplexers. The data inputs of upper 4x1 Multiplexer are I_7 to I_4 and the data inputs of lower 4x1 Multiplexer are I_3 to I_0. Therefore, each 4x1 Multiplexer produces an output based on the values of selection lines, s_1 & s_0.

The outputs of first stage 4x1 Multiplexers are applied as inputs of 2x1 Multiplexer that is present in second stage. The other selection line, s_2 is applied to 2x1 Multiplexer.

- If s_2 is zero, then the output of 2x1 Multiplexer will be one of the 4 inputs I_3 to I_0 based on the values of selection lines s_1 & s_0.
- If s_2 is one, then the output of 2x1 Multiplexer will be one of the 4 inputs I_7 to I_4 based on the values of selection lines s_1 & s_0.

Therefore, the overall combination of two 4x1 Multiplexers and one 2x1 Multiplexer performs as one 8x1 Multiplexer.

16x1 Multiplexer

Let us implement 16x1 Multiplexer using 8x1 Multiplexers and 2x1 Multiplexer. We know that 8x1 Multiplexer has 8 data inputs, 3 selection lines and one output. Whereas, 16x1 Multiplexer has 16 data inputs, 4 selection lines and one output.

So, we require two 8x1 Multiplexers in first stage in order to get the 16 data inputs. Since, each 8x1 Multiplexer produces one output, we require a 2x1 Multiplexer in second stage by considering the outputs of first stage as inputs and to produce the final output.

Let the 16x1 Multiplexer has sixteen data inputs I_{15} to I_0, four selection lines s_3 to s_0 and one output Y. The Truth table of 16x1 Multiplexer is shown below.

Selection Inputs				Output
S_3	S_2	S_1	S_0	Y
0	0	0	0	I_0
0	0	0	1	I_1
0	0	1	0	I_2
0	0	1	1	I_3
0	1	0	0	I_4
0	1	0	1	I_5
0	1	1	0	I_6
0	1	1	1	I_7
1	0	0	0	I_8
1	0	0	1	I_9
1	0	1	0	I_{10}
1	0	1	1	I_{11}
1	1	0	0	I_{12}
1	1	0	1	I_{13}
1	1	1	0	I_{14}
1	1	1	1	I_{15}

We can implement 16x1 Multiplexer using lower order Multiplexers easily by considering the above Truth table. The block diagram of 16x1 Multiplexer is shown in the following figure.

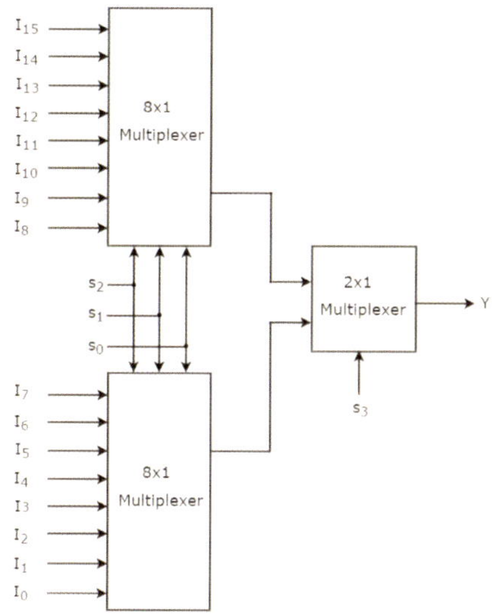

The same selection lines, s_2, s_1 & s_0 are applied to both 8x1 Multiplexers. The data inputs of upper 8x1 Multiplexer are I_{15} to I_8 and the data inputs of lower 8x1 Multiplexer are I_7 to I_0. Therefore, each 8x1 Multiplexer produces an output based on the values of selection lines, s_2, s_1 & s_0.

The outputs of first stage 8x1 Multiplexers are applied as inputs of 2x1 Multiplexer that is present in second stage. The other selection line, s_3 is applied to 2x1 Multiplexer.

- If s_3 is zero, then the output of 2x1 Multiplexer will be one of the 8 inputs Is_7 to I_0 based on the values of selection lines s_2, s_1 & s_0.
- If s_3 is one, then the output of 2x1 Multiplexer will be one of the 8 inputs I_{15} to I_8 based on the values of selection lines s_2, s_1 & s_0.

Therefore, the overall combination of two 8x1 Multiplexers and one 2x1 Multiplexer performs as one 16x1 Multiplexer.

Applications of Multiplexers

Multiplexers are used in various applications wherein multiple-data need to be transmitted by using single line.

- Communication System

A communication system has both a communication network and a transmission system. By using a multiplexer, the efficiency of the communication system can be increased by allowing the transmission of data, such as audio and video data from different channels through single lines or cables.

- Computer Memory

Multiplexers are used in computer memory to maintain a huge amount of memory in the computers, and also to reduce the number of copper lines required to connect the memory to other parts of the computer.

- Telephone Network

In telephone networks, multiple audio signals are integrated on a single line of transmission with the help of a multiplexer.

- Transmission from the Computer System of a Satellite

Multiplexer is used to transmit the data signals from the computer system of a spacecraft or a satellite to the ground system by using a GSM satellite.

Demultiplexer

The action or operation of a demultiplexer is opposite to that of the multiplexer. As inverse to the MUX , demux is a one-to-many circuit. With the use of a demultiplexer , the binary data can be bypassed to one of its many output data lines.

Demultiplexers are mainly used in Boolean function generators and decoder circuits. Different input/output configuration demultiplexers are available in the form of single integrated circuits (ICs).

Also, the facility of cascading two or more IC circuits helps to generate multiple output demultiplexers.

The process of getting information from one input and transmitting the same over one of many outputs is called demultiplexing. A demultiplexer is a combinational logic circuit that receives the information on a single input and transmits the same information over one of 2n possible output lines.

The bit combinations of the select lines control the selection of specific output line to be connected to the input at given instant. The below figure illustrates the basic idea of demultiplexer, in which the switching of the input to any one of the four outputs is possible at a given instant.

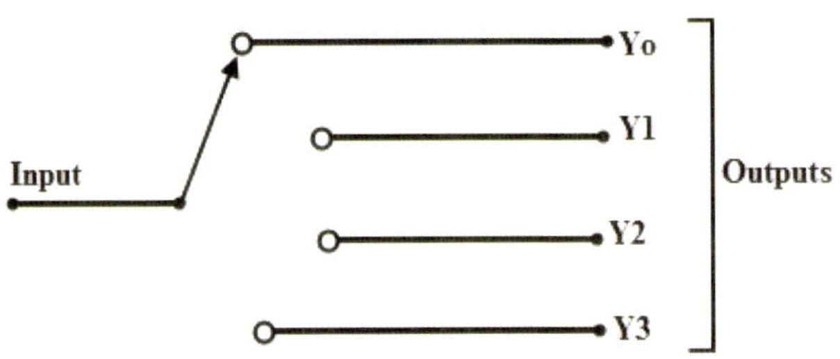

Demultiplexers are also called as data distributors, since they transmit the same data which is received at the input to different destinations.

Thus, a demultiplexer is a 1-to-N device where as the multiplexer is an N-to-1 device. The figure below shows the block diagram of a demultiplexer or simply a DEMUX.

It consists of 1 input line, n output lines and m select lines. In this, m selection lines are required to produce 2m possible output lines (consider 2m = n). For example, a 1-to-4 demultiplexer requires 2 (22) select lines to control the 4 output lines.

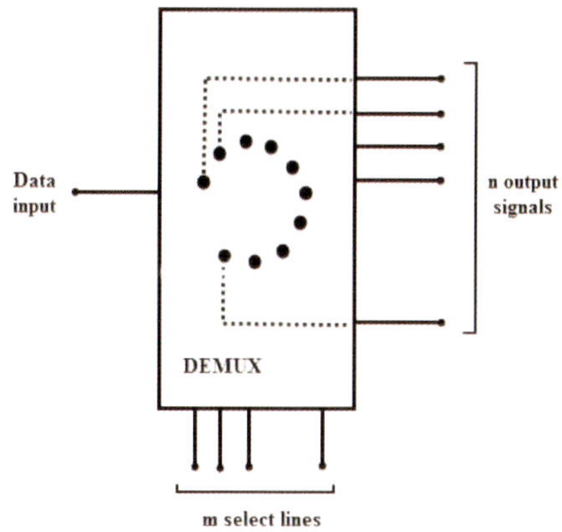

There are several types of demultiplexers based on the output configurations such as 1:4, 1:8 and 1:16.

These are available in different IC packages and some of the most commonly used demultiplexer ICs includes 74139 (dual 1:4 DEMUX), 73136 (1:8 DEMUX), 74154 (1:16 DEMUX), 74159 (1:16 DEMUX open collector type), etc.

1-to-2 Demultiplexer

A 1-to-2 demultiplexer consists of one input line, two output lines and one select line. The signal on the select line helps to switch the input to one of the two outputs. The figure below shows the block diagram of a 1-to-2 demultiplexer with additional enable input.

In the figure, there are only two possible ways to connect the input to output lines, thus only one select signal is enough to do the demultiplexing operation. When the select input is low, then the input will be passed to Y0 and if the select input is high then the input will be passed to Y1.

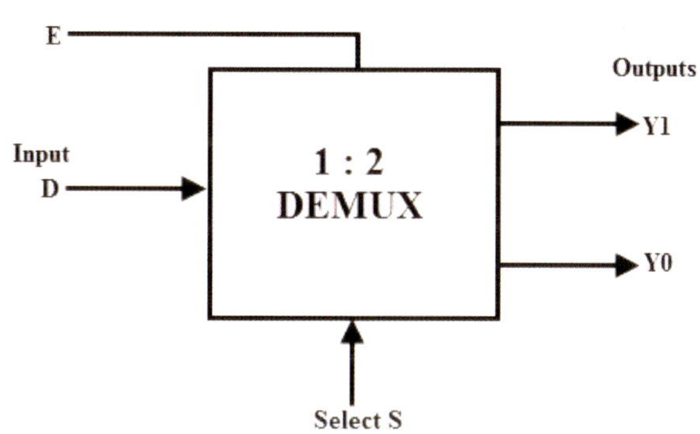

The truth table of a 1-to-2 demultiplexer is shown below in which the input is routed to Y0 and Y1 depends on the value of select input S. In the table output Y1 is active when the combination of select line and input line are active high, i.e., S F = 11.

Therefore, the output Y1 = SF and similarly the output Y0 is equal to S⁻ F.

Select	Input	Outputs	
S	D	Y_1	Y_0
0	0	0	0
0	1	0	1
1	0	0	0
1	1	1	0

From the above truth table, the logic diagram of this demultiplexer can be designed by using two AND gates and one NOT gate as shown in below figure. When the select lines S=0, AND gate A1 is enabled while A2 is disabled.

Then, the data from the input flows to the output line Y1. Similarly, when S=1, AND gate A2 is enabled and AND gate A1 is disabled, thus data is passed to the Y0 output.

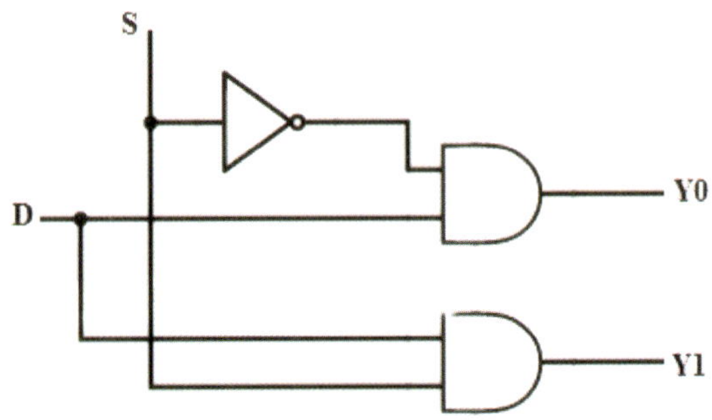

1-to-4 Demultiplexer

A 1-to-4 demultiplexer has a single input (D), two selection lines (S1 and S0) and four outputs (Y0 to Y3). The input data goes to any one of the four outputs at a given time for a particular combination of select lines.

This demultiplexer is also called as a 2-to-4 demultiplexer which means that two select lines and 4 output lines. The block diagram of 1:4 DEMUX is shown below.

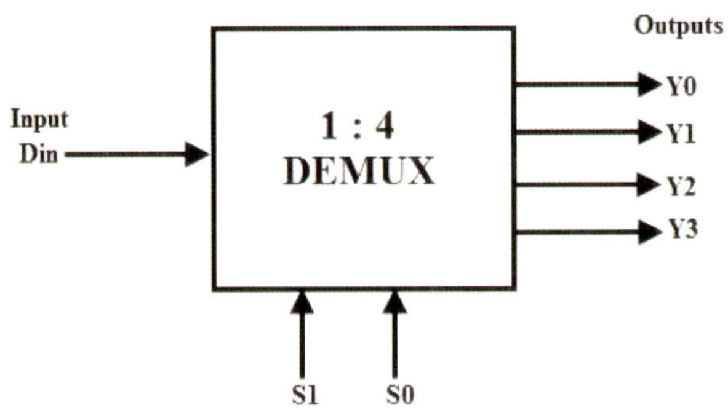

The truth table of this type of demultiplexer is given below. From the truth table it is clear that, when S1=0 and S0= 0, the data input is connected to output Y0 and when S1= 0 and s0=1, then the data input is connected to output Y1.

Similarly, other outputs are connected to the input for other two combinations of select lines.

Data Input	Select Inputs		Outputs			
D	S_1	S_0	Y_3	Y_2	Y_1	Y_0
D	0	0	0	0	0	D
D	0	1	0	0	D	0
D	1	0	0	D	0	0
D	1	1	D	0	0	0

From the table, the output logic can be expressed as min terms and are given below.

$$Y0 = \overline{S1}\ \overline{S0}\ D$$

$$Y1 = \overline{S1}\ S0\ D$$

$$Y2 = S1\ \overline{S0}\ D$$

$$Y3 = S1\ S0\ D$$

Where D is the input data, Y0 to Y3 are output lines and S0 & S1 are select lines.

From the above Boolean expressions, a 1-to-4 demultiplexer can be implemented by using four 3-input AND gates and two NOT gates as shown in figure below. The two selection lines enable the particular gate at a time.

So depends on the combination of select inputs, input data is passed through the selected gate to the associated output.

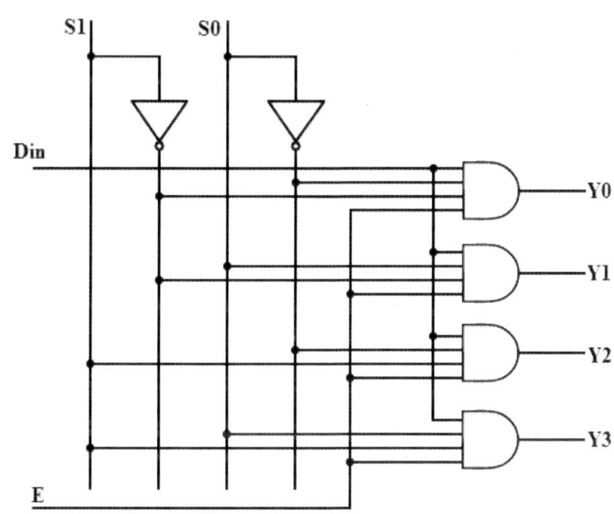

This type of demultiplexer is available in IC form and a typical IC 74139 is most commonly used dual 1-to-4 demultiplexer. It has two independent demultiplexers and each DEMUX accepts two binary inputs as select lines and four mutually exclusive active-low outputs.

Both demultiplexers share a common set of selection lines so they are selected in parallel. Also, each demultiplexer consists of enable pin or data input, for one demultiplexer it is active high data input and for other it is active low data input.

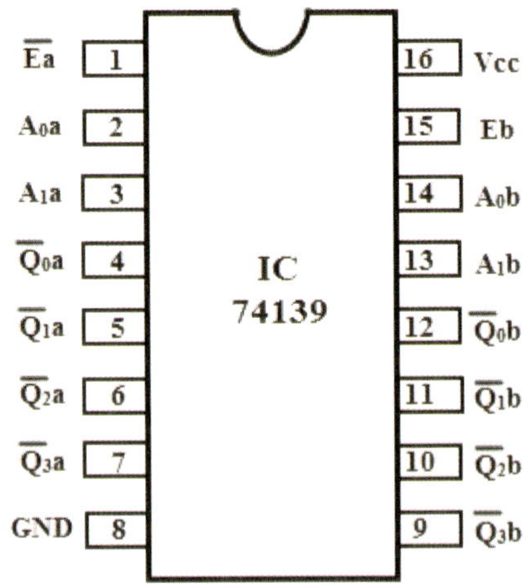

1-to-8 Demultiplexer

The below figure shows the block diagram of a 1-to-8 demultiplexer that consists of single input D, three select inputs S2, S1 and S0 and eight outputs from Y0 to Y7.

It is also called as 3-to-8 demultiplexer due to three select input lines. It distributes one input line to one of 8 output lines depending on the combination of select inputs.

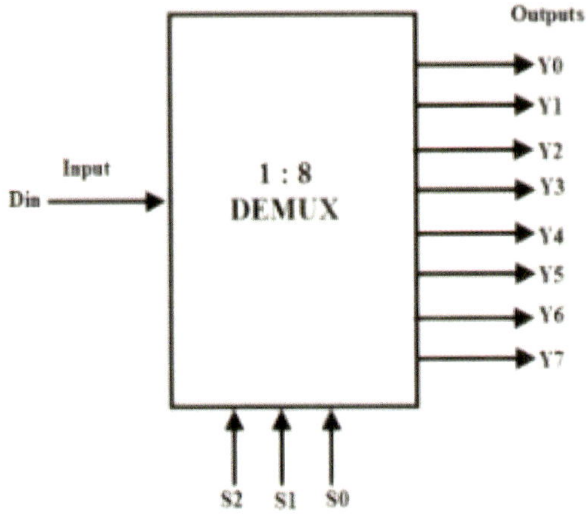

The truth table for this type of demultiplexer is shown below. The input D is connected with one of the eight outputs from Y0 to Y7 based on the select lines S2, S1 and S0.

For example, if S2S1S0=000, then the input D is connected to the output Y0 and so on.

Data Input	Select Inputs			Outputs							
D	S_2	S_1	S_0	Y_7	Y_6	Y_5	Y_4	Y_3	Y_2	Y_1	Y_0
D	0	0	0	0	0	0	0	0	0	0	D
D	0	0	1	0	0	0	0	0	0	D	0
D	0	1	0	0	0	0	0	0	D	0	0
D	0	1	1	0	0	0	0	D	0	0	0
D	1	0	0	0	0	0	D	0	0	0	0
D	1	0	1	0	0	D	0	0	0	0	0
D	1	1	0	0	D	0	0	0	0	0	0
D	1	1	1	D	0	0	0	0	0	0	0

From this truth table, the Boolean expressions for all the outputs can be written as follows.

$$Y0 = D\,\overline{S2}\,\overline{S1}\,\overline{S0}$$

$$Y1 = D\,\overline{S2}\,\overline{S1}\,S0$$

$$Y2 = D\,\overline{S2}\,S1\,\overline{S0}$$

$$Y3 = D\,\overline{S2}\,S1\,S0$$

$$Y4 = D\,S2\,\overline{S1}\,\overline{S0}$$

$$Y5 = D\,S2\,\overline{S1}\,S0$$

$$Y6 = D\,S2\,S1\,\overline{S0}$$

$$Y7 = D\,S2\,S1\,S0$$

From these obtained equations, the logic diagram of this demultiplexer can be implemented by using eight AND gates and three NOT gates as shown in below figure. The different combinations of the select lines , select one AND gate at given time , such that data input will appear at a particular output.

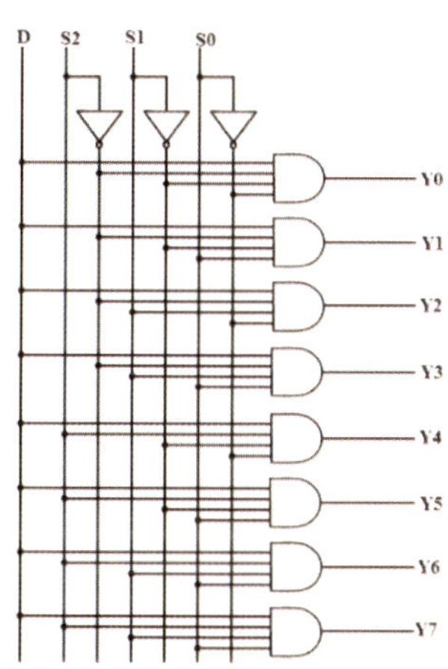

A typical IC74237 is a 1-to-8 demultiplexer that consists of latches at three select inputs. The pin out of this IC is given below.

The pins A0 to A2 are data inputs, Y0 to Y7 are demultiplexer outputs, E1&E2 are active-low data enable and active-high data enable pins respectively, LE is the latch enable input ,Vcc and GND terminals are positive supply voltage and ground terminals.

With a 3-bit storage latch, this IC combines the 3-to-8 decoder function.

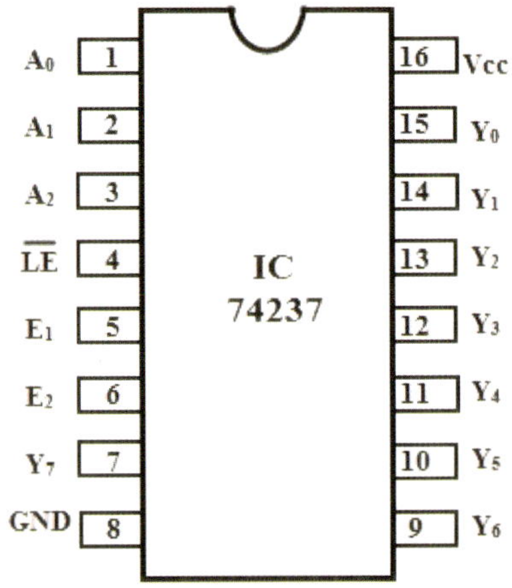

1-to-8 DEMUX using Two 1-to- 4 Demultiplexers

When the application requires a large demultiplexer with more number of output pins, then we cannot implement by a single integrated circuit. In case if more than 16 output pins are needed, then two or more demultiplexer ICs are cascaded to fulfill the requirement.

For example, if the application needs 32 output lines from a DEMUX, then we cascade two 1:16 demultiplexers or three 1:8 demultiplexers. Therefore, by cascading the two or more demultiplexers, a large demultiplexer can be implemented.

Consider the case that a 1-to-8 demultiplexer can be implemented by using two 1-to-4 demultiplexers with a proper cascading.

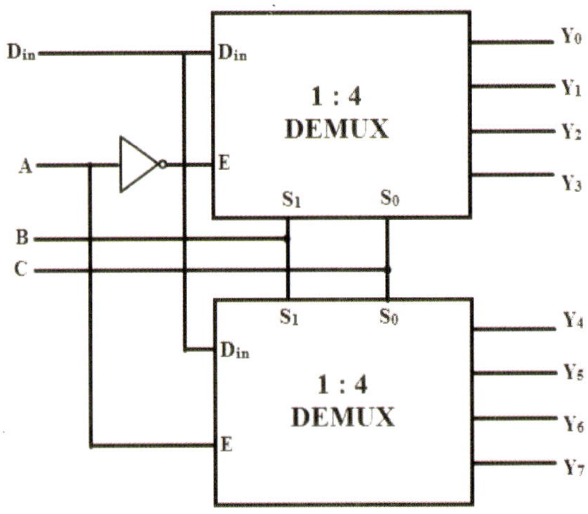

In the above figure, the highest significant bit A of the selection inputs are connected to the enable inputs such that it is complemented before connecting to one DEMUX and to the other it is directly connected.

By this configuration, when A is set to zero, one of the output lines from Y0 to Y3 is selected based on the combination of select lines B and C. Similarly, when A is set to one, based on the select lines one of the output lines from Y4 to Y7 will be selected.

Implementation of Full Subtractor using 1-to-8 DEMUX

As similar to the multiplexers, demultiplexers are also used for Boolean function implementation as well as combinational circuit design. We can design the demultiplexer to produce any truth table output by correspondingly controlling the select lines.

Consider the case for implementing a demultiplexer circuit in order to produce the full subtractor output. The truth table below shows the output of a full subtractor.

A	B	B_{in}	D	B_{out}
0	0	0	0	0
0	0	1	1	1
0	1	0	1	1
0	1	1	0	1
1	0	0	1	0
1	0	1	0	0
1	1	0	0	0
1	1	1	1	1

From the above table, the full subtractor output D can be written as

$$D = f(A, B, C)$$

$$= \sum m(1, 2, 4, 7)$$

And the borrow output can be expressed as

$$Bout = F(A, B, C) = \sum m(1, 2, 3, 7)$$

From these Boolean functions, a demultiplexer for producing full subtractor output can be built by properly configuring the 1-to-8 DEMUX such that with input D=1 it gives the minterms at the output.

And by logically ORing these minterms, the outputs of difference and borrow can be obtained as shown in figure.

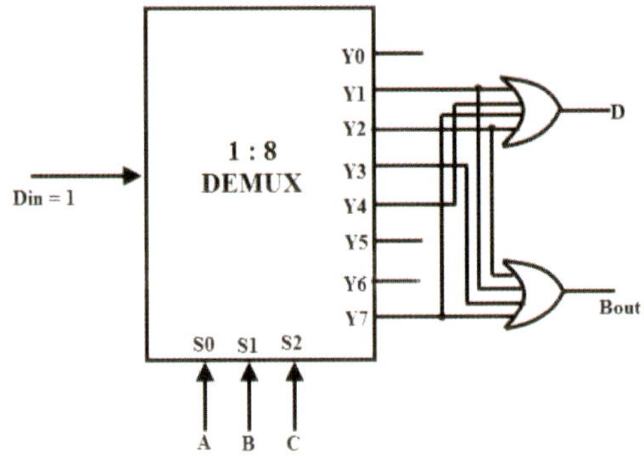

1 to 16 Demultiplexer

1 to 16 demultiplexer has one input data, four select lines A, B, C and D and 16 output lines Y0 to Y15. This is implemented using AND and NOT gate. 1 to 16 demultiplexer is implemented using the logic circuit below.

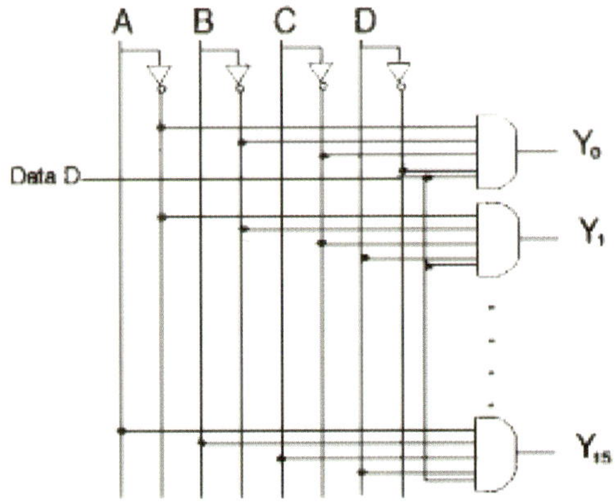

1 to 16 Demux

This can be implemented using 1 to 8 demultiplexer, 1 to 4 demultiplexer and 1 to 2 demultiplexer.

Truth Table

The below truth table shows the operation of 1 to 16 demultiplexer.

INPUTS						OUTPUTS															
G1	G2	D	C	B	A	0	1	2	3	4	5	6	7	8	9	10	11	12	13	14	15
L	L	L	L	L	L	L	H	H	H	H	H	H	H	H	H	H	H	H	H	H	H
L	L	L	L	L	H	H	L	H	H	H	H	H	H	H	H	H	H	H	H	H	H
L	L	L	L	H	L	H	H	L	H	H	H	H	H	H	H	H	H	H	H	H	H
L	L	L	L	H	H	H	H	H	L	H	H	H	H	H	H	H	H	H	H	H	H
L	L	L	H	L	L	H	H	H	H	L	H	H	H	H	H	H	H	H	H	H	H
L	L	L	H	L	H	H	H	H	H	H	L	H	H	H	H	H	H	H	H	H	H
L	L	L	H	H	L	H	H	H	H	H	H	L	H	H	H	H	H	H	H	H	H
L	L	L	H	H	H	H	H	H	H	H	H	H	L	H	H	H	H	H	H	H	H
L	L	H	L	L	L	H	H	H	H	H	H	H	H	L	H	H	H	H	H	H	H
L	L	H	L	L	H	H	H	H	H	H	H	H	H	H	L	H	H	H	H	H	H
L	L	H	L	H	L	H	H	H	H	H	H	H	H	H	H	L	H	H	H	H	H
L	L	H	L	H	H	H	H	H	H	H	H	H	H	H	H	H	L	H	H	H	H
L	L	H	H	L	L	H	H	H	H	H	H	H	H	H	H	H	H	L	H	H	H
L	L	H	H	L	H	H	H	H	H	H	H	H	H	H	H	H	H	H	L	H	H
L	L	H	H	H	L	H	H	H	H	H	H	H	H	H	H	H	H	H	H	L	H
L	L	H	H	H	H	H	H	H	H	H	H	H	H	H	H	H	H	H	H	H	L
L	H	X	X	X	X	H	H	H	H	H	H	H	H	H	H	H	H	H	H	H	H
H	L	X	X	X	X	H	H	H	H	H	H	H	H	H	H	H	H	H	H	H	H
H	H	X	X	X	X	H	H	H	H	H	H	H	H	H	H	H	H	H	H	H	H

H = high level, L = low level, X = irrelevant

1 to 16 Demux Truth Table

Applications of Demultiplexer

- A demultiplexer is used to connect a single source to multiple destinations. Demultiplexers are mainly used in the field of the communication system.

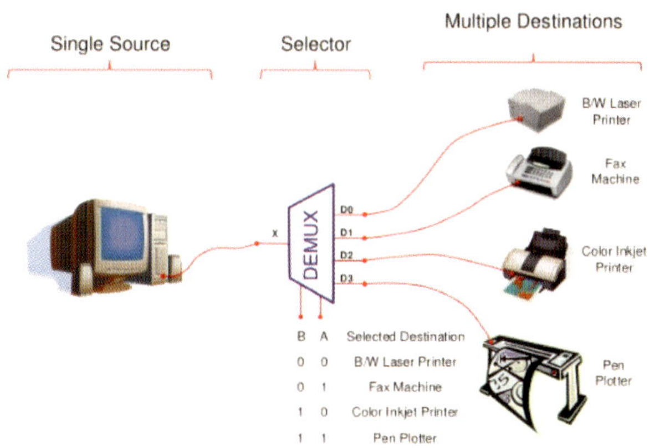

DEMUX-Applications

- Demultiplexers are used for reconstruction of parallel data and ALU circuits.
- The demultiplexer receives the output signals of the multiplexer and converts back to the original form of the data at the receiving end. MUX and DEMUX work together to carry out the process communication.
- Demultiplexer helps to store the output of the ALU in multiple registers and storage units in an ALU circuit. The output of the data the ALU is fed as data input to the DEMUX. Each output of the DEMUX is connected to the multiple registers which can be stored in the register.
- A serial to parallel converter is used for reconstructing the parallel data from the incoming serial data stream. In this technique, serial data from the incoming serial data stream is given as the input to the DEMUX at the regular intervals. A counter is attached to the control input of the demux. This counter directs the data signal output of the demux where these data signals are stored. When all data signals have been stored. The output of the demux can be retrieved and read out in parallel.

Sequential Logic Circuit

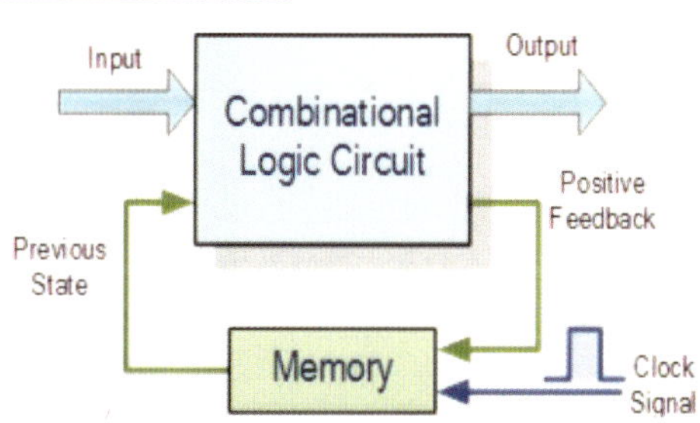

Sequential digital logic circuit

A Sequential digital logic circuit is different from combinational logic circuits. In sequential circuit the output of the logic device is not only dependent on the present inputs to the device, but also on past inputs.

In other words output of a sequential logic circuit depends on present input as well as present state of the circuit.

Unlike combinational circuits, the sequential circuits have memory devices in order to store the past outputs. In fact sequential digital logic circuits are nothing but combinational circuit with memory.These types of digital logic circuits are designed using finite state machine.

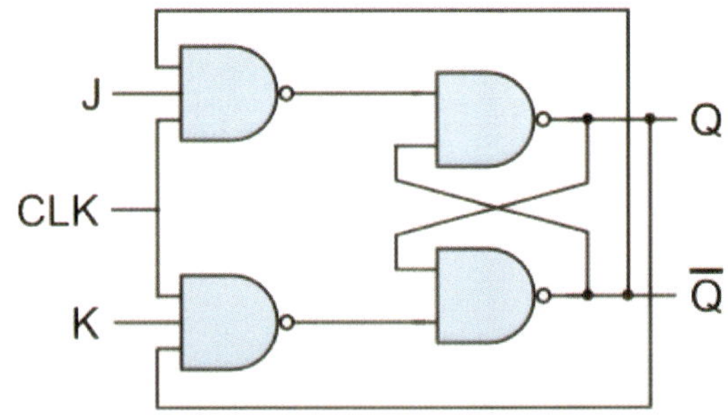

Sequential circuits: J/K Flip flop

Examples of sequential logic circuits are counters, flip flops, constructed using digital logic gates and memory.

There are two inputs which are operated by combinational logic circuits in order to produce various outputs.

The output from the memory devices are fed to the combinational logic circuit. The internal inputs and outputs form part of the secondary devices

Secondary inputs devices are state variables produced by the storage elements, where as secondary outputs devices are excitations for the storage elements.

Types of Sequential Circuits

The sequential circuits are classified into two types

- Synchronous Circuit
- Asynchronous Circuit

In synchronous sequential circuits, the state of device changes at discrete times in response to a clock signal. In asynchronous circuits, the state of the device changes in response to changing inputs.

Synchronous Circuits

In synchronous circuits, the inputs are pulses with certain restrictions on pulse width and propagation delay. Thus synchronous circuits can be divided into clocked and un-clocked or pulsed sequential circuits.

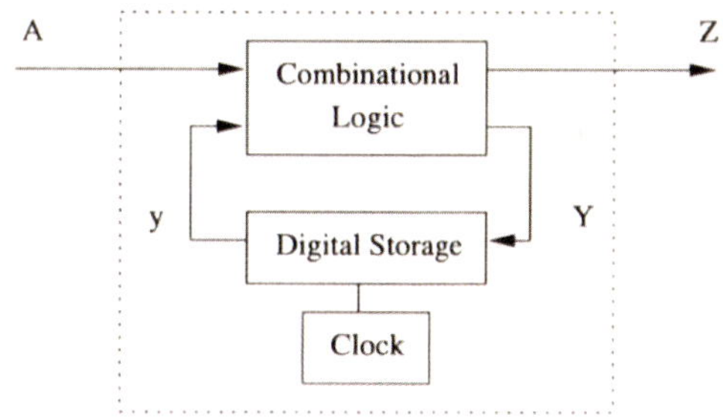

Synchronous Circuit

Clocked Sequential Circuit

The clocked sequential circuits have flip-flops or gated latches for its memory elements. There is a periodic clock connected to the clock inputs of all the memory elements of the circuit to synchronize all the internal changes of state. Hence the operation of the circuit is controlled and synchronized by the periodic pulse of the clock.

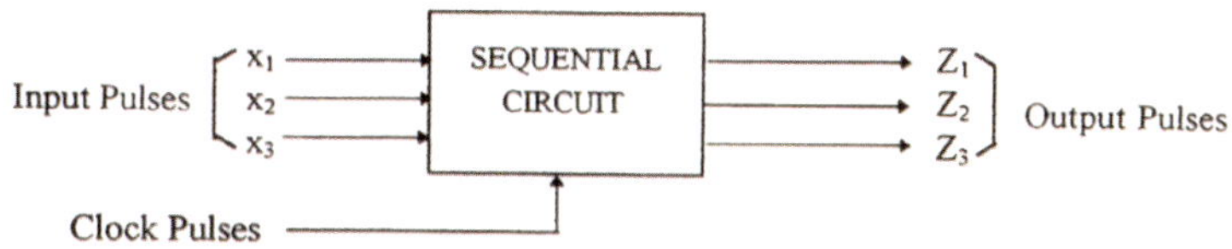

Clocked Sequential

Unclocked Sequential Circuit

In an unclocked sequential circuit requires two consecutive transitions between 0 and 1 to alternate the state of the circuit. An unclocked mode circuit is designed to respond to pulses of certain durations which do not affect the circuit's behaviour.

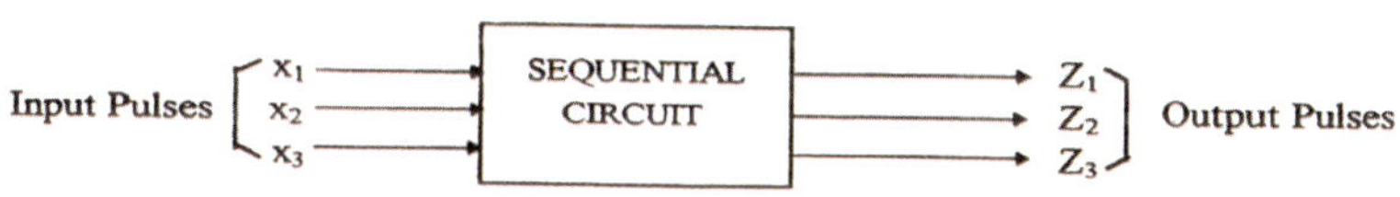

UnClocked Sequential

The synchronous logic circuit is very simple. The logic gates which perform the operations on the data, require a finite amount of time to respond to the changes in the input.

Asynchronous Circuits

An asynchronous circuit does not have a clock signal to synchronize its internal changes of the state. Hence the state change occurs in direct response to changes that occur in primary input lines. An asynchronous circuit does not require the precise timing control from flip-flops.

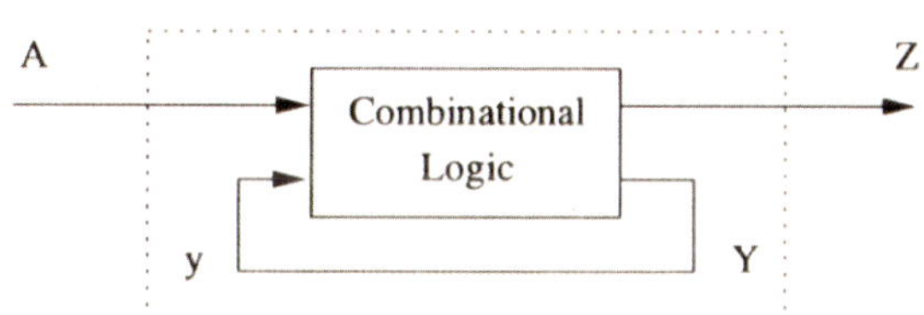

Asynchronous Circuit

Asynchronous logic is more difficult to design and it has some problems compared to synchronous logic. The main problem is that the digital memory is sensitive to the order that their input signals arrive them, like, if two signals arrive at a flip-flop at the same time, which state the circuit goes into can depend on which signal gets to the logic gate first.

Asynchronous circuits are used in critical parts of synchronous systems where the speed of the system is a priority, like as in microprocessors and digital signal processing circuits.

Clock Signal

Clocks and Timing Signals

Most sequential logic circuits are driven by a clock oscillator. This usually consists of an astable circuit producing regular pulses that should ideally:

- Be constant in frequency - Many clock oscillators use a crystal to control the frequency. Because crystal oscillators generate normally high frequencies, where lower frequencies are required the original oscillator frequency is divided down from a very high frequency to a lower one using counter circuits.
- Have fast rising and falling edges to its pulses - It is the edges of the pulses that are important in timing the operation of many sequential circuits, the rise and fall times are usually be less than 100ns. The outputs of clock circuits will typically have to drive more gates than any other output in a given system. To prevent this load distorting the clock signal, it is usual for clock oscillator outputs to be fed via a buffer amplifier.
- Have the correct logic levels - The signals produced by the clock circuits must have appropriate the logic levels for the circuits being supplied.

Simple Clock Oscillator

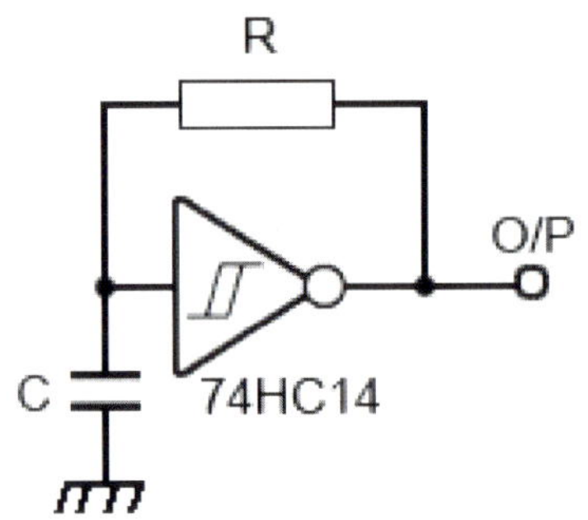

Fig. Simple Schmitt Inverter Clock Oscillator

Fig. is probably the simplest oscillator possible, having only three components. Notice that the gate is a Schmitt inverter. This device has an extremely fast change over between logic states. Also the level at which it responds to an input change from 0 to 1 (Vt+) is higher than the level at which it changes from 1 to 0 (Vt-). The operation of the circuit is as follows.

Suppose the gate input is at logic 0, because the gate is an inverter, the output must be at logic 1, and C will therefore charge up via R from the output. This will happen with the normal CR charging curve. Once Vt+ is reached at the gate input, the gate output will rapidly switch to 0. The resistor is now connected effectively between the positive plate of C and zero volts. Thus the capacitor now discharges via R until the gate input voltage reduces to Vt- when the output will change to logic 1 once more, starting the charging and discharging cycle over again.

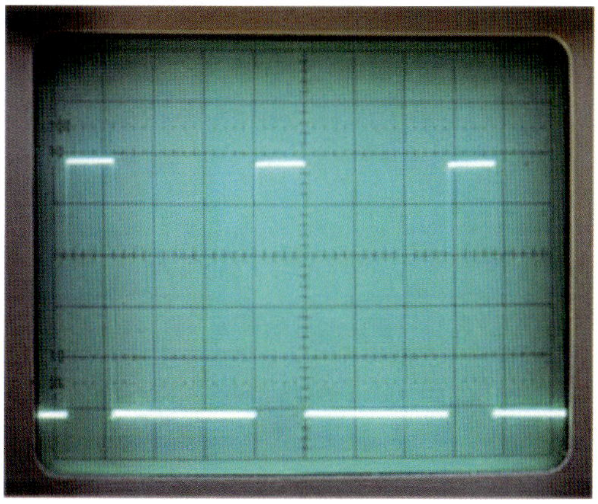

Fig. Typical Basic Schmitt Oscillator Output

This Schmitt RC oscillator can produce a pulse waveform with an excellent wave shape and very fast rise and fall times. The mark to space ratio, as shown in Fig. is approximately 1:3.

The frequency of oscillation depends on the time constant of R and C, but is also affected by the characteristics of the logic family used. For the 74HC14 the frequency (f)is calculated by:

$$f = \frac{1}{0.8RC}$$

When using the 74HCT14 the 0.8 correction factor is replaced by 0.67, however either of these formulae will give an approximate frequency. Whichever logic family is used, the frequency will vary with changes in supply voltage. Although this basic oscillator gives an excellent performance in many simple applications, if a stable frequency is an important factor in the choice of clock oscillator, there are of course better options.

Crystal Controlled Clock Oscillator

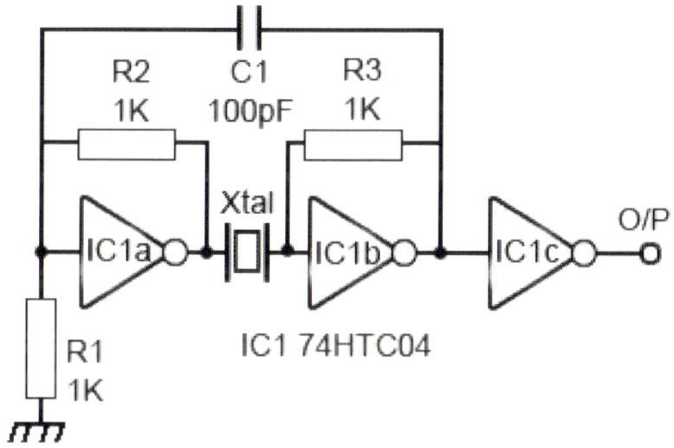

Fig. Crystal Controlled Clock Oscillator

Fig. uses three gates from a 74HCT04 IC, and a crystal to provide an accurate frequency of oscillation. Here, the oscillator is running at 3.276MHz but this can be reduced by dividing the output frequency down to a lower value by dividing it by 2 a number of times using a series of flip-flops.

The top waveform in figure below shows the clock signal generated by figure below, and beneath it is the clock signal frequency divided by 4 after passing it through two flip-flops. Notice that after passing the signal through flip-flops, as well as being reduced in frequency, the wave shape is considerably squarer and now has a 1:1 mark to space ratio.

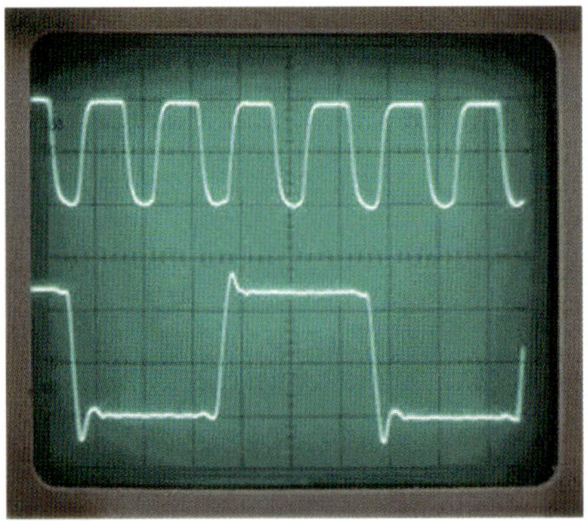

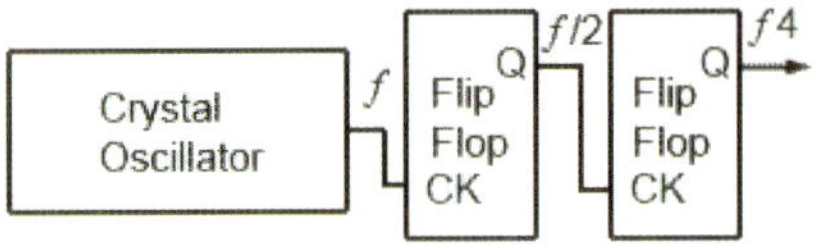

Clock Frequency Divided by 4

Two Phase Clock Signals

Some older microprocessor systems required two-phase clock signals which, provided that the source clock signal operated at twice the frequency required by the microprocessor, saved processing time as the microprocessor was able to carry out two actions per clock cycle instead of one.

Producing a Two-phase Clock Signal

If a clock signal with a 1:1 mark space ratio is used, two non-overlapping clock pulses can be created, using the circuit shown in Fig. These signals are usually called Φ01 and Φ02.

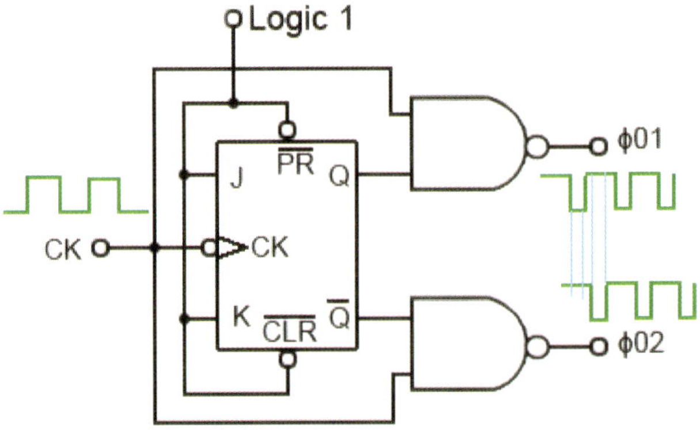

Fig. Two-Phase Clock Generator

In Fig. a single clock signal having a 1:1 mark to space ratio is fed into a JK flip-flop working in toggle mode. This is achieved by making both J and K logic 1. The active low $\overline{PR}$ and $\overline{CLR}$ inputs take no part in the operation of this circuit so are also tied to logic 1. In toggle mode the Q output of the JK flip-flop inverts the logic levels at Q and Q at every falling edge of the clock(CK) input, also Q and Q output always reaming at opposite logic states.

Each of the NAND gates will then produce a logic 0 output whenever both its inputs are at logic 1. The NAND gate producing Φ01 therefore creates a logic 0 pulse whenever CK and Q are at logic 1, and the NAND gate producing Φ02 creates a logic 0 pulse whenever CK and $\overline{Q}$ are at logic 1.

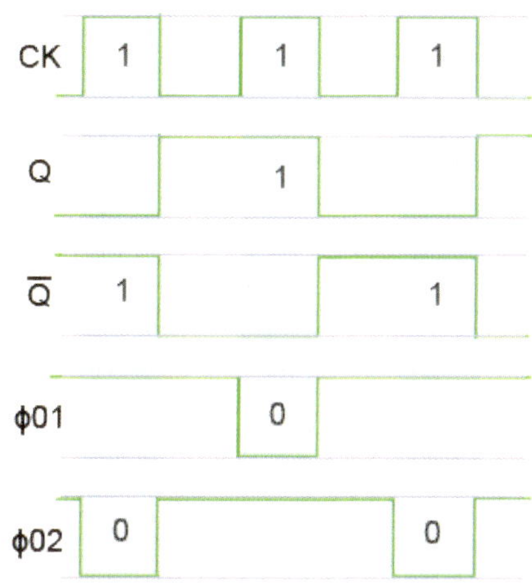

Fig. Two-Phase Clock Signal

Figure illustrates the operation of figure above. Each of the NAND gates will produce a logic 0 output whenever both its inputs are at logic 1. The NAND gate producing Φ01 therefore creates a logic 0 pulse whenever CK and Q are at logic 1, and the NAND gate producing Φ02 creates a logic 0 pulse whenever CK and Q are at logic 1. Typical output waveforms are illustrated.

If positive going clock pulses are required, the outputs from the NAND gates may be inverted using Schmitt inverters, which will also help to sharpen the rise and fall times of the clock waveforms.

Distributing Clock Signals

For more demanding applications there are very many specialised clock oscillator ICs available that are typically optimised for a particular range of applications, such as computer hardware, wireless communications, automotive or medical applications etc.

Clock Fan-out

Whatever circuit is used to generate a clock signal, it is important that its output has sufficient fan-out capability to drive the necessary number of ICs requiring a clock input, and that the clock signal is not degraded in amplitude, speed of its rise and fall times or accuracy of its frequency. Also, by maintaining fast rise and fall times, ringing on the waveform can become a problem. The waveform should be kept as close as possible to a perfect square wave shape.

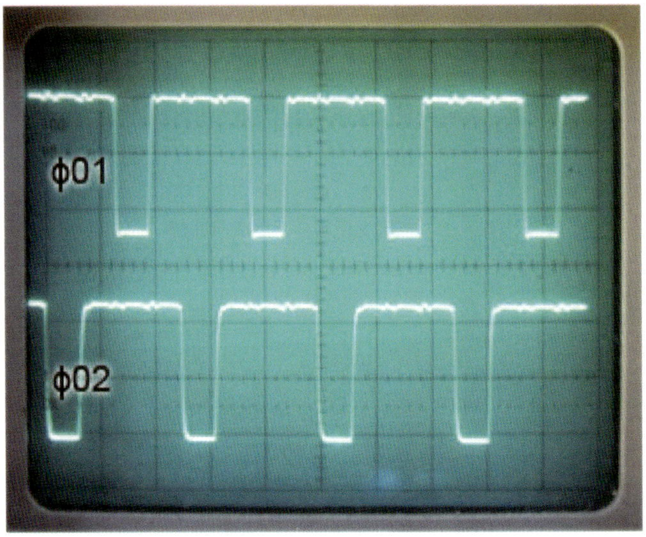

Fig. Two Phase Clock Waveforms

Circuit Capacitance

Because the clock must feed many gates, the small capacitance of each of these gates will add, to become an appreciable capacitance, which loads the clock output tending to slow the rise and fall time of the clock signal. To avoid this, the clock output must have a low enough impedance to rapidly charge and discharge any natural capacitance in the circuit. The usual way to achieve this is to feed the clock signal via a special clock buffer gate, which will have the necessary low output impedance and a large fan out factor. Schmitt trigger gates may also be used to restore the shape and integrity of clock signals before they are applied to gates in different parts of the circuit.

Cross-talk

Where the clock signal has to be distributed around large circuits, there is a greater chance of introducing noise, and possible 'cross-talk' where data in one conductor is radiated into another nearby conductor. Problems such as this will increase the likelihood of 'skew' errors, i.e. clock signals arriving at different parts of the circuit at slightly different times, due to small changes in the phase of some of the distributed clock signals. Miniaturisation brought about by surface mount technology can help minimise these problems. Also when clock signals need to be sent from one system to another over an external wired or wireless link it is common to use one of the several ECL or LVDS logic families with their differential outputs to minimise interference, and there are many application specific ICs (ASICS) using these technologies for high frequency clock distribution.

Types of Triggering

Following are the two possible types of triggering that are used in sequential circuits.

- Level triggering
- Edge triggering

Level Triggering

There are two levels, namely logic High and logic Low in clock signal. Following are the two types of level triggering.

- Positive level triggering
- Negative level triggering

If the sequential circuit is operated with the clock signal when it is in Logic High, then that type of triggering is known as Positive level triggering. It is highlighted in below figure.

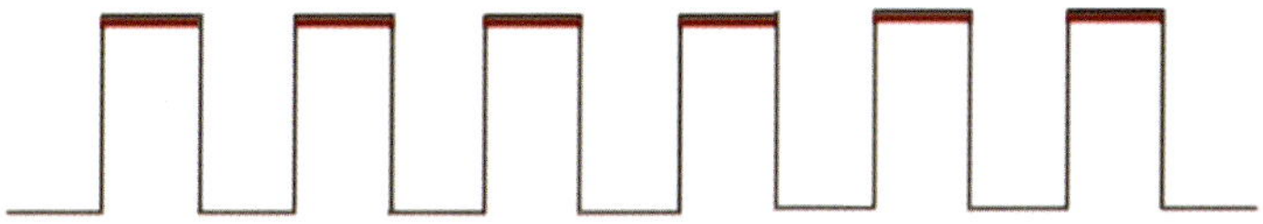

If the sequential circuit is operated with the clock signal when it is in Logic Low, then that type of triggering is known as Negative level triggering. It is highlighted in the following figure.

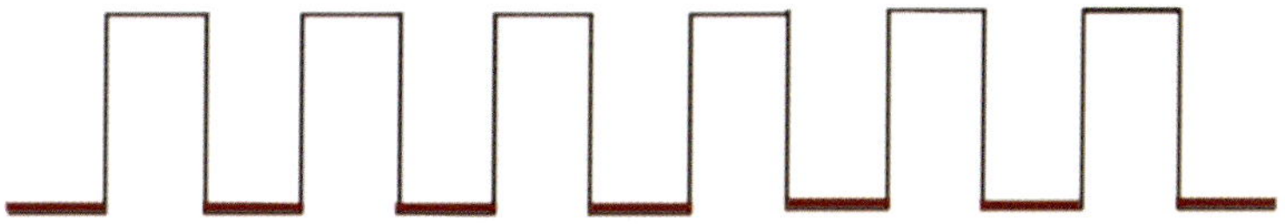

Edge Triggering

There are two types of transitions that occur in clock signal. That means, the clock signal transitions either from Logic Low to Logic High or Logic High to Logic Low.

Following are the two types of edge triggering based on the transitions of clock signal.

- Positive edge triggering
- Negative edge triggering

If the sequential circuit is operated with the clock signal that is transitioning from Logic Low to Logic High, then that type of triggering is known as Positive edge triggering. It is also called as rising edge triggering. It is shown in the following figure.

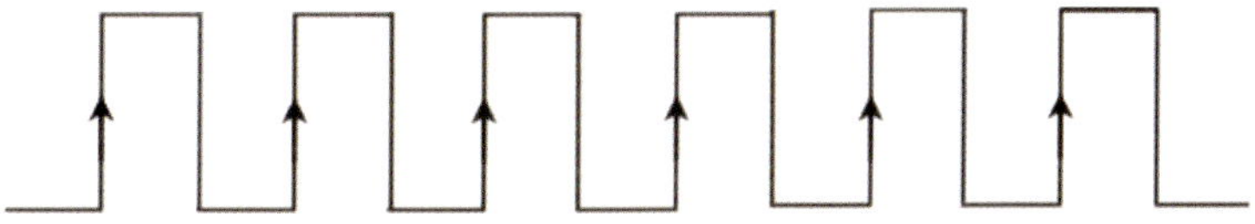

If the sequential circuit is operated with the clock signal that is transitioning from Logic High to Logic Low, then that type of triggering is known as Negative edge triggering. It is also called as falling edge triggering. It is shown in the following figure.

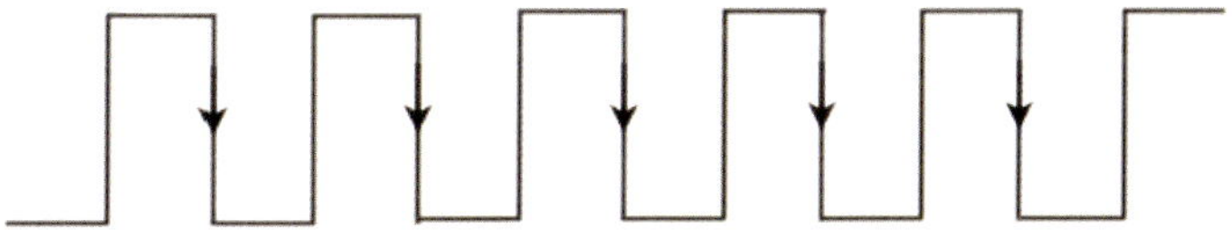

Synchronous Sequential Circuits

Latches

Latch is an electronic logic circuit with two stable states i.e. it is a bistable multivibrator. Latch has a feedback path to retain the information. Hence a latch can be a memory device. Latch can store one bit of information as long as the device is powered on. When enable is asserted, latch immediately changes the stored information when the input is changed i.e. they are level triggered devices. It continuously samples the inputs when the enable signal is on.

Latch circuits can work in two states depending on the triggering signal being high or low: Active – High or Active – Low.

- In case of Active – High latch circuits, normally both the inputs are low. The circuit is triggered by a momentary high on either of the inputs.
- In case of Active – Low latch circuits, normally both the inputs are high. The circuit is triggered by a momentary low on either of the inputs.

SR Latch

We can use static gates as basic building blocks in order to construct a simple latch and it can be constructed with two NOR gates by introducing feedback to a NOR gate circuit.

A simple NOR gate logic with feedback is shown below.

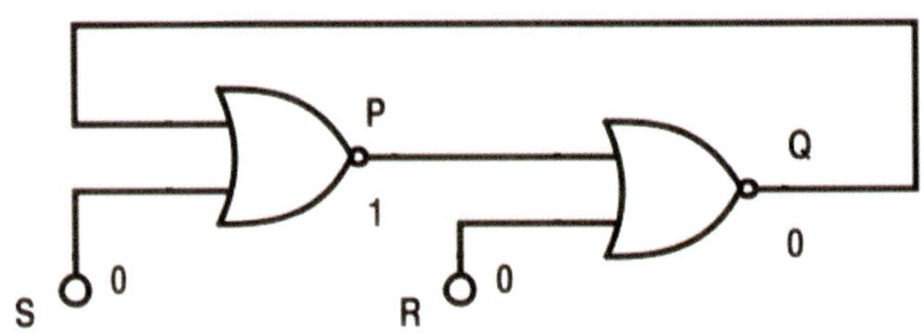

Here, both the inputs S and R are 0 (S = R = 0). The output of first NOR gate is P = 1. This is fed to the second NOR gate along with R = 0. Hence the output of the second gate is Q = 0. The circuit is said to be in stable state with P = 1 and Q = 0. If we make S = 1, then P = 0. This will make Q =1 as shown below.

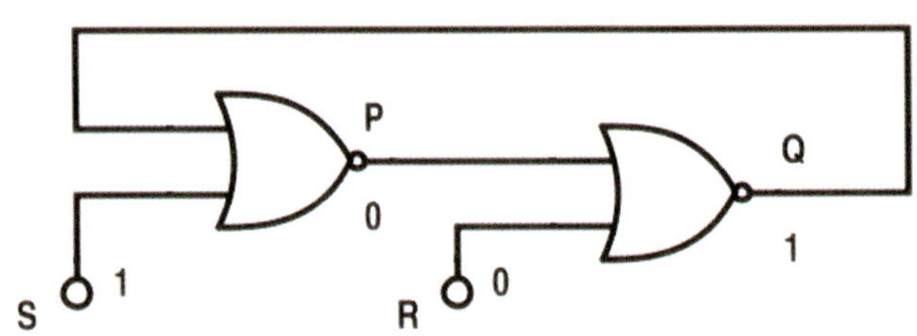

This is also a stable state. If S is made 0, there is no change as Q = 1 is fed back to first NOR and P still remains 0. It is shown in the below figure.

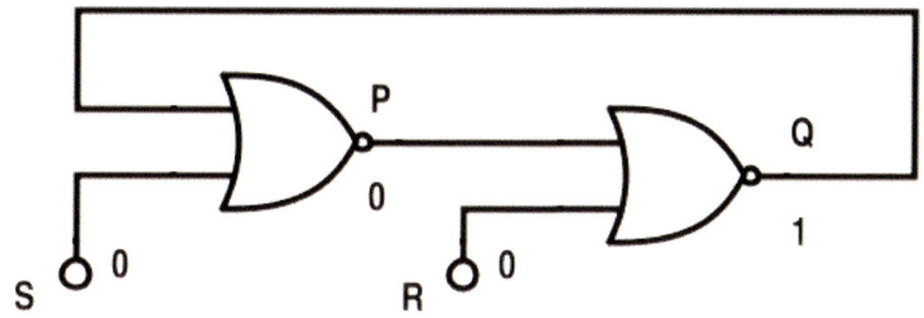

If R is made 1, then Q becomes 0 which will change P back to 1.

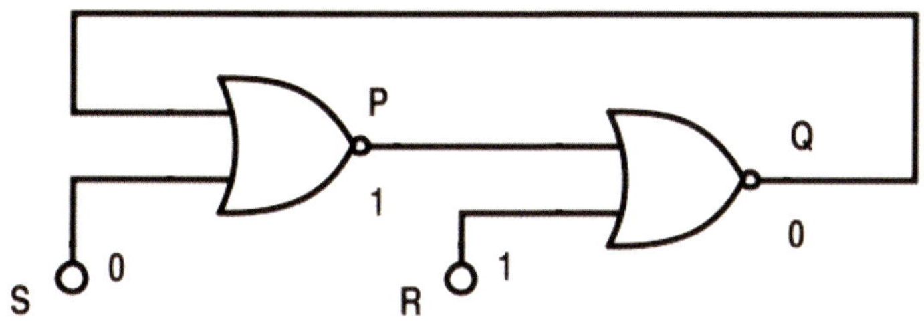

If R is made 0, then there is no change and we arrive at where we started.

As the output depends not only on present inputs but also on past sequence of inputs, the circuit is said to have memory. If the input condition S = R = 1 is not allowed, the stable state outputs are always complementary. When both S and R are equal to 1, P = 0 and Q = 0 which contradicts the complementary condition. Hence the input condition S = R = 1 is said to be not allowed. The latch circuit is always drawn as a cross coupled form to emphasize the symmetry between the gates.

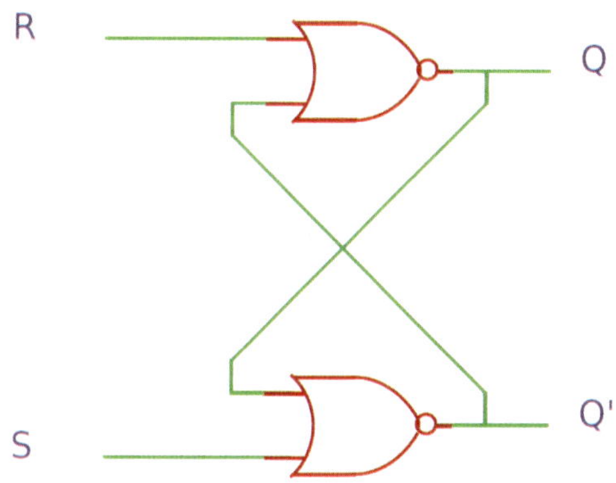

In this circuit, when S = 1, it 'sets' the output Q to 1 and when the input R = 1, it 'resets' the output Q to 0. With the restriction of S = R = 1, this circuit is called a Set – Reset Latch (SR Latch).

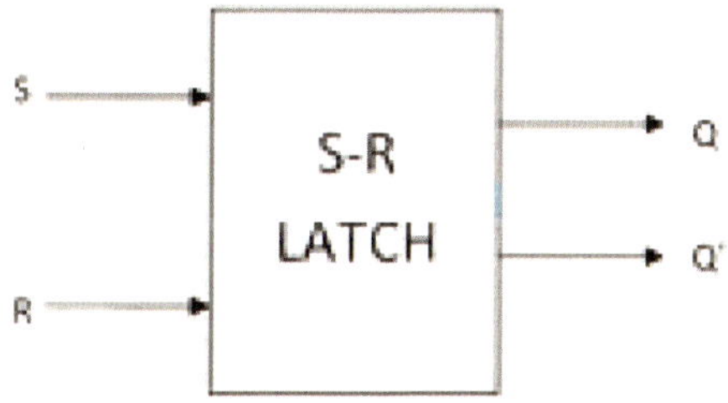

Race Around Condition

In logic circuits, Race condition means "The situation at which the two inputs of a logic circuit change at the same time and that will make the output tentative". The inputs are in competition to change the output. It generally happens in the devices which have the output as the feed-back input of the circuit. It is an unwanted situation that occurs when a device attempts to perform two operations at the same time (i.e. changing the state of two inputs simultaneously). There are few ways in which we can avoid race around condition like using Edge Triggering or by using Master Slave Flip – flop.

State Transition Table or Truth Table of SR Latch

State table is similar to truth table in combinational circuits that gives the information about the states of a circuit. As the outputs of sequential circuits depend on present and previous states, these are represented in the form of table called state table and it shows the next state based on the present state and other inputs.

The state table for SR latch is shown below.

S	R	Q_N	Q_{N+1}
0	0	0	0
0	0	1	1
0	1	0	0
0	1	1	0
1	0	0	1
1	0	1	1
1	1	0	-
1	1	1	-

State Diagram

In addition to tables and equations, a state machine (or a system) can be represented by a state diagram. In this state diagram, a state is represented by a circle, and the transition between states is represented by lines or arcs that connect the circles. State diagram for a simple SR latch is shown below.

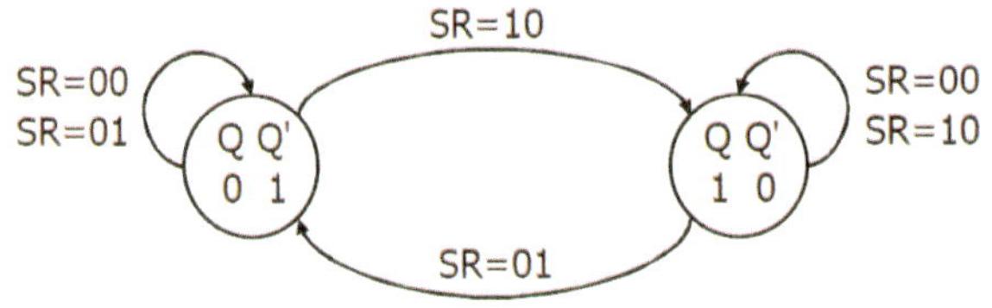

The state diagram provides all the information that a state table can have. This is obtained from the state table directly.

Gated S-R Latch

Generally, latches are transparent i.e. the output changes immediately when there is a change in the input. But for many applications, it is desirable to have an isolated period where the output doesn't change even when there is a change in the input. During this period, the outputs are said to be truly 'latched'. This can be achieved with the use of an extra input (enable or clock or gate). If the enable (or clock or gate) signal is not asserted, the inputs are ignored and the outputs are latched to the previous values. In order to use this extra signal, additional logic should be added. These circuits are called Gated or Clocked Latches. Gated SR latch can be made in two ways: by adding second level of AND gates to SR latch or by adding second level of NAND gates to $\bar{S}\bar{R}$ latch (Inverted SR latch). The circuit diagram of Gated SR latch constructed from NOR gates is shown below.

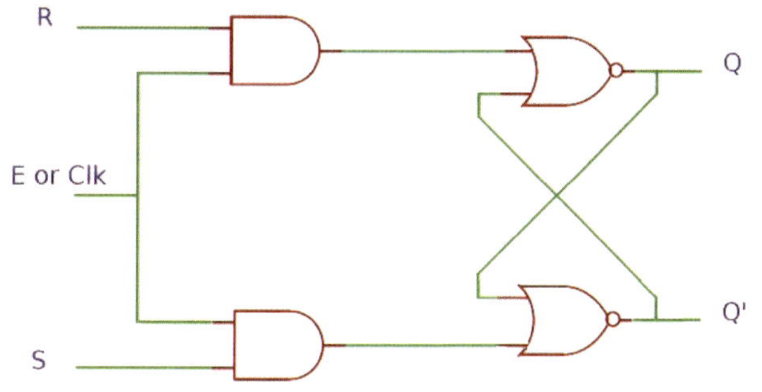

The circuit diagram of Gated SR latch constructed from NAND gates is shown below.

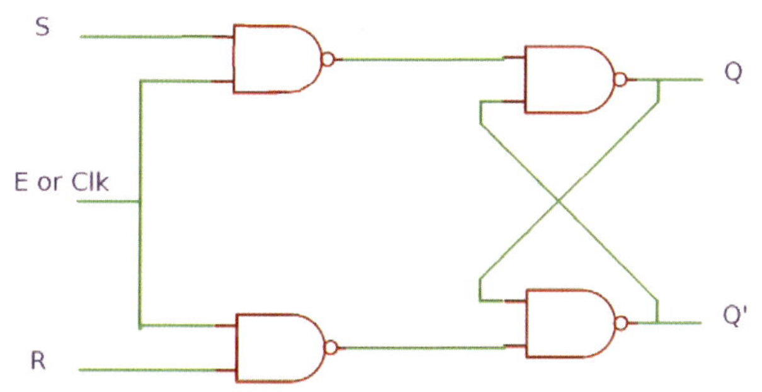

As the NAND gate inverts the inputs, $\bar{S}\bar{R}$ latch becomes a gated SR latch.

When enable (or clock) is high, the latch is said to be enabled i.e. the output responds to the inputs.

When enable (or clock) is low, the latch is disabled and remains in that state until enable is asserted.

The symbol for gated SR latch is shown below

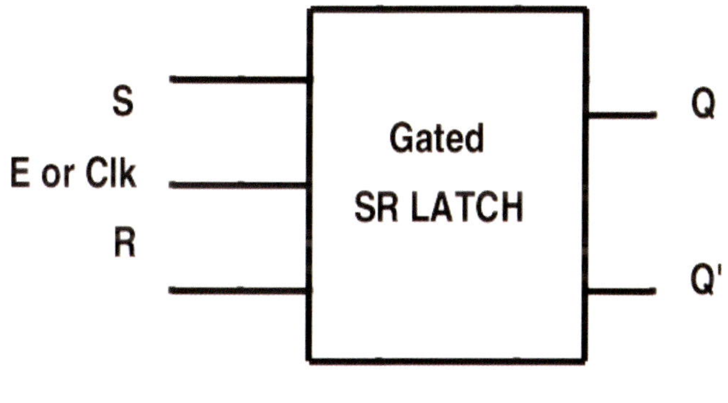

The truth table for gated SR latch is tabulated below.

E or Clk	S	R	Q	Q'
0	0	0	Latch	
0	0	1	Latch	
0	1	0	Latch	
0	1	1	Latch	
1	0	0	Latch	
1	0	1	0	1
1	1	0	1	0
1	1	1	Not allowed	

The state diagram of gated SR latch is shown below.

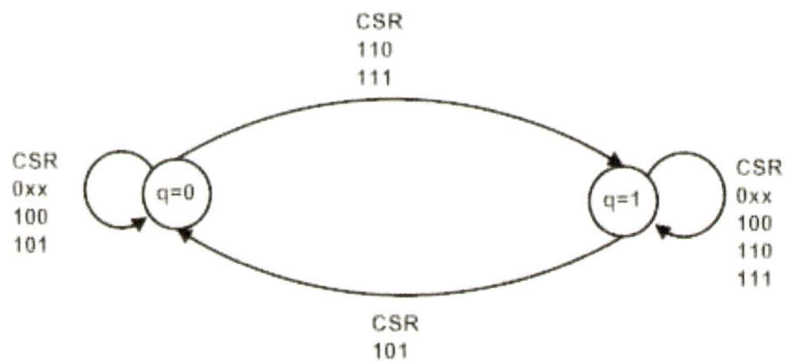

D Latch

Data latch or Delay latch (D latch) is one of the simple latches to store data. It is also called transparent latch. A simple D latch can be constructed with two NAND gates.

The race around condition in SR latch that occurs when S = R = 1 can be avoided in D latch as the R input is replaced with inverted S which is renamed to D. hence there are no illegal or forbidden inputs. In a D latch, Q always D.

The symbol for D latch is shown below.

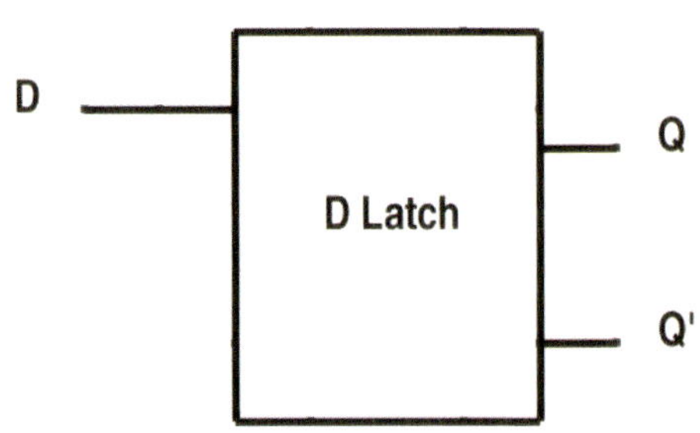

These simple D latches are not frequently used but Gated D latches are very common. The truth table of a simple D Latch is shown below.

D	Q	Q'
0	0	1
1	1	0

Gated D Latch

A gated D latch can be easily constructed by modifying a gated SR latch. The only modification to the gated SR latch is that the R input has to be changed to inverted S. A gated latch formed from NOR SR latch is shown below.

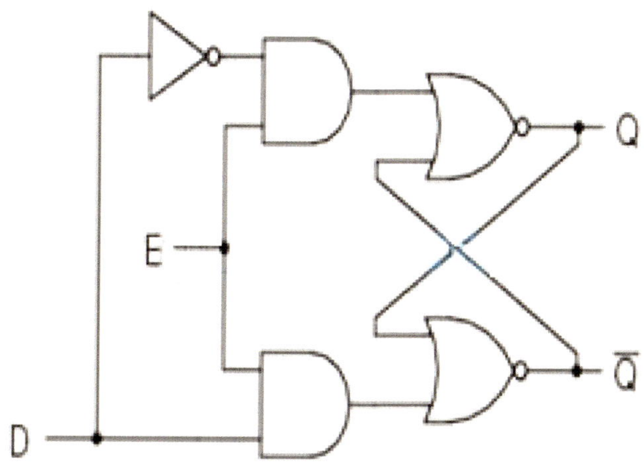

When the clock or enable is high (logic 1), the output latches whatever is on the D input. When the enable or clock is low (logic0), the D input for the last enable high will be the output.

This latch circuit will never experience a "Race" condition because the single D input is inverted to provide to both the inputs. Hence, there is no chance for same input condition. So the D latch circuit can be safely used in any circuit.

The symbol of gated D latch is shown below.

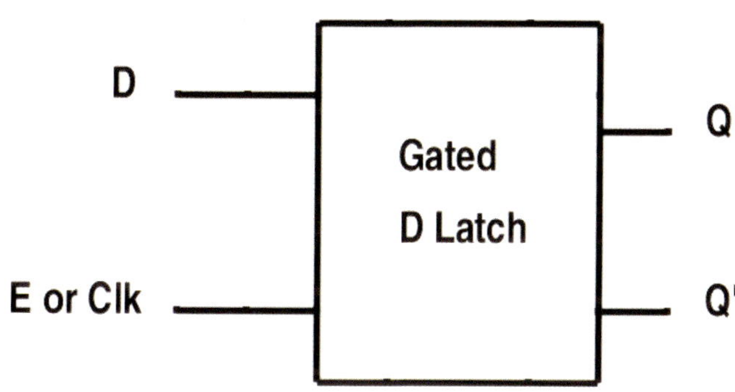

Similar to gated NOR SR latch, a gated D latch can also be constructed from gated NAND SR latch. The circuit for gated D latch from gated NAND SR larch is shown below.

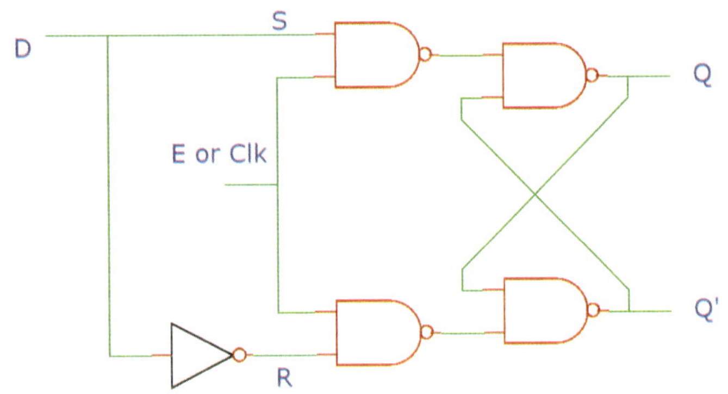

The usage of inverter can be avoided as the NAND gate can be used to obtain the inverted value. Some modification is required in the above circuit and the resultant circuit is shown below.

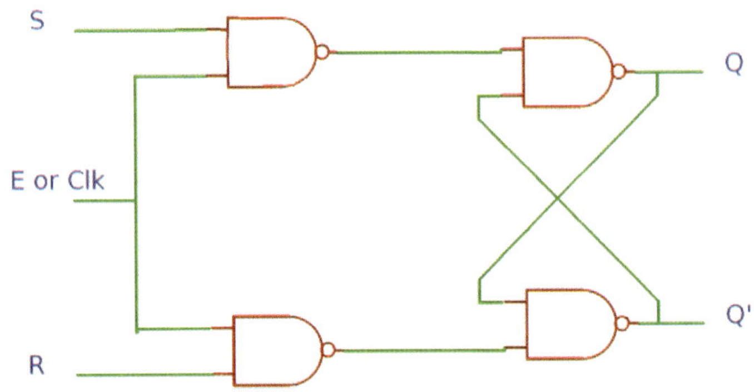

The truth table (or state table) of a gated D latch is shown below.

E or Clk	D	Q	Q'
0	0	Latch	
0	1	Latch	
1	0	0	1
1	1	1	0

The state diagram of s gated D latch is shown below.

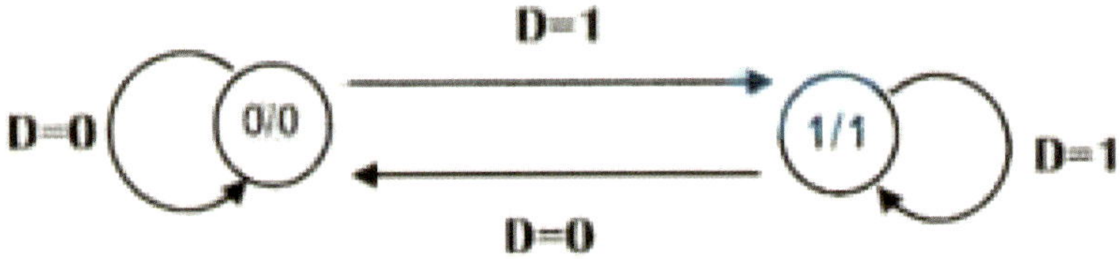

Applications

- They are basic 1 – bit memory devices.
- D latches are typically used as I/O ports in asynchronous systems.

- Data latches are sometimes used in synchronous two phase systems to reduce the transistor count.

Advantages of Latches

- Latch is faster because it has no need to wait for a clock signal so they are most used in high speed designs.
- They require less power.
- Latch based designs have small die-size.
- The major advantage of the latches is "Time-Borrowing". In which if an operation is not completed within time, the required time for executing the operation is borrowed from the other operational time.

Disadvantages of latches

- Latch less predictable because there is more chance to affect to race conditions.
- Level sensitive devices and hence more chance of metastability.
- Analysing of Latch circuits is difficult because of its level sensitive property.

Flip-flops

Flip flop is formed using logic gates, which are in turn made of transistors. Flip flop are basic building blocks in the memory of electronic devices. Each flip flop can store one bit of data.

Flip – flops have two stable states and hence they are bistable multivibrators. The two stable states are High (logic 1) and Low (logic 0).

The term flip – flop is used as they can switch between the states under the influence of a control signal (clock or enable) i.e. they can 'flip' to one state and 'flop' back to other state.

- Flip – flops are a binary storage device because they can store binary data (0 or 1).
- Flip – flops are edge sensitive or edge triggered devices i.e. they are sensitive to the transition rather than the duration or width of the clock signal.
- They are also known as signal change sensitive devices which mean that the change in the level of clock signal will bring change in output of the flip flop.
- A Flip – flop works depending on clock pulses.
- Flip flops are also used to control the digital circuit's functionality. They can change the operation of a digital circuit depending on the state.

Some of the most common flip – flops are SR Flip – flop (Set – Reset), D Flip – flop (Data or Delay), JK Flip – flop and T Flip – flop.

Latches vs Flip-Flops

Latches and flip – flops are both 1 – bit binary data storage devices. The main difference between a latch and a flip – flop is the triggering mechanism. Latches are transparent when enabled ,whereas flip – flops are dependent on the transition of the clock signal i.e. either positive edge or negative edge.

The modern usage of the term flip – flop is reserved to clocked devices and term latch is to describe much simpler devices. Some of the other differences between latches and flip – flops are listed in below table.

LATCH	FLIP – FLOP
Latches do not require clock signal.	Flip – flops have clock signals
A latch is an asynchronous device.	A flip – flop is a synchronous device.
Latches are transparent devices i.e. when they are enabled, the output changes immediately if the input changes.	A transition from low to high or high to low of the clock signal will cause the flip – flop to either change its output or retain it depending on the input signal.
A latch is a Level Sensitive device (Level Triggering is involved).	A flip – flop is an edge sensitive device (Edge Triggering is involved).
Latches are simpler to design as there is no clock signal (no careful routing of clock signal is required).	When compare to latches, flip – flops are more complex to design as they have clock signal and it has to be carefully routed. This is because all the flip – flops in a design should have a clock signal and the delay in the clock reaching each flip – flop must be minimum or negligible.
The operation of a latch is faster as they do not have to wait for any clock signal.	Flip - flops are comparatively slower than latches due to clock signal.
The power requirement of a latch is less.	Power requirement of a flip – flop is more.
A latch works based on the enable signal.	A flip – flop works based on the clock signal.

Types of Flip Flops

Based on their operations, flip flops are basically 4 types:

1. R-S flip flop
2. D flip flop
3. J-K flip flop
4. T flip flop.

S-R Flip Flop

The S-R flip-flop is basic flip-flop among all the flip-flops. All the other flip flops are developed after SR-flip-flop.

SR flip flop is represented as shown below.

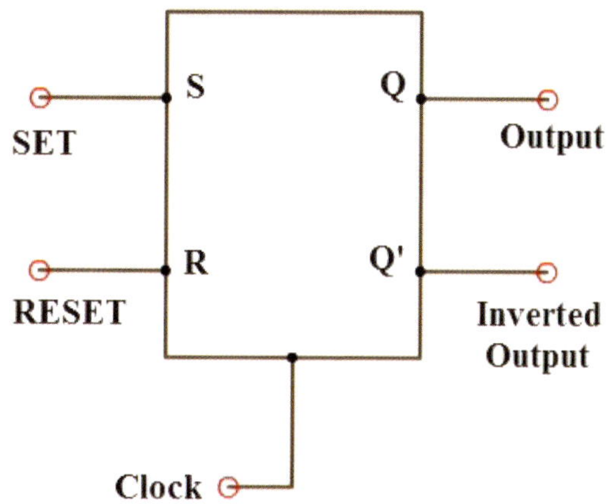

S-R stands for SET and RESET. This can also be called RS flip-flop. Difference is RS is inverted SR flip-flop.

Any flip flop can be build using logic gates. NAND and NOR gates were used as they are universal gates.

Here is the SR flip-flop using NAND gates.

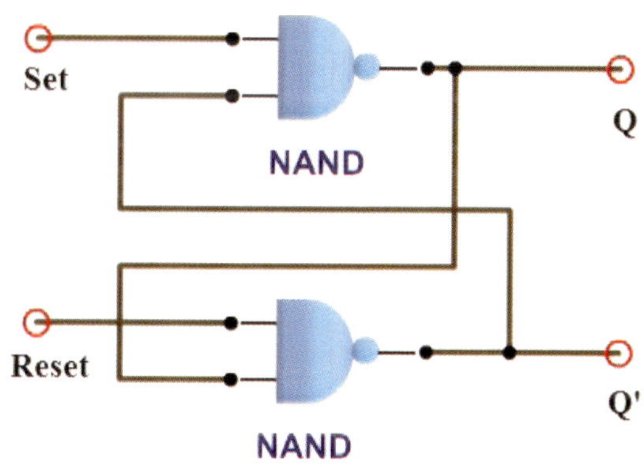

Truth Table of SR Flip Flop

Sno	S	R	Q	Q'	State
1	1	0	1	0	Q is set to 1
2	1	1	1	0	No change
3	0	1	0	1	Q' is set to 1
4	1	1	0	1	No change
5	0	0	1	1	Invalid

Working

From the above truth table it is clear that SR flip flop will be set or reset for four conditions.

1. For last condition it will be in invalid state.
2. SR Flip-flop will be set when S=1 and R=0, if S=1 and R=1 then previous state is remembered by the flip flop.
3. Flip-flop will be reset when S=0 and R=1, if S=1 and R=1, then it will remember the previous state.
4. But when both the inputs are zeros, SR Flip flop will be in an uncertain state where both Q and Q' will be same. This is not same allowed.

This is indeterminate state is avoided by adding gates extra gates to the existing flip flop. This is called clocked or gated SR Flip flop. This produces the output only for the High clock pulse.

The circuit of a clocked SR flip – flop using NAND gates is shown below.

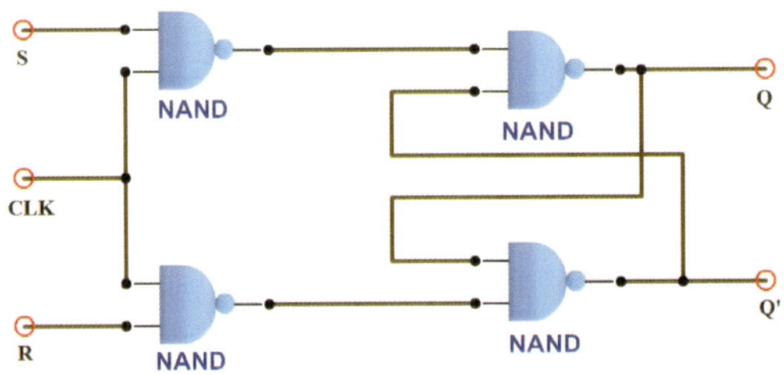

D Flip Flop

In the SR flip flop an uncertain state occurred. This can be avoided by using D flip flop. Here D stands for "Data".

It is constructed from SR flip flop. The two inputs (S &R) of the clocked SR flip flop are connected to an inverter.

It is one of the most widely used flip – flops. It has a clock signal (Clk) as one input and Data (D) as other. There are two outputs and these outputs are complement to each other. The symbol of D flip – flop is shown below.

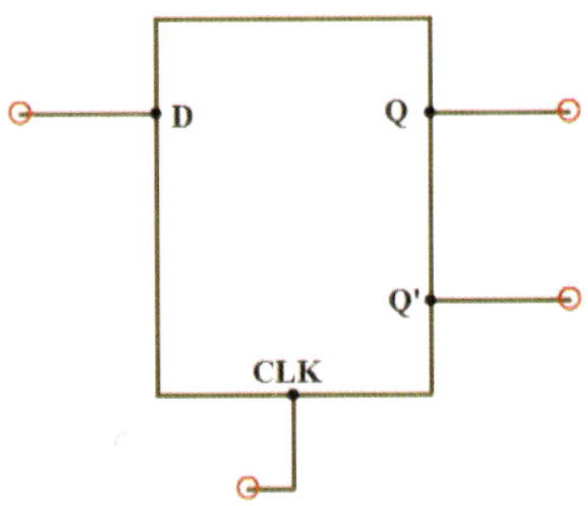

Truth Table

Clk	D	Q	Q'	State
0	0	Q	Q'	No change in state
1	0	0	1	Resets Q to 0
1	1	1	1	Sets Q to 1

D flip – flop using NAND gates is shown below.

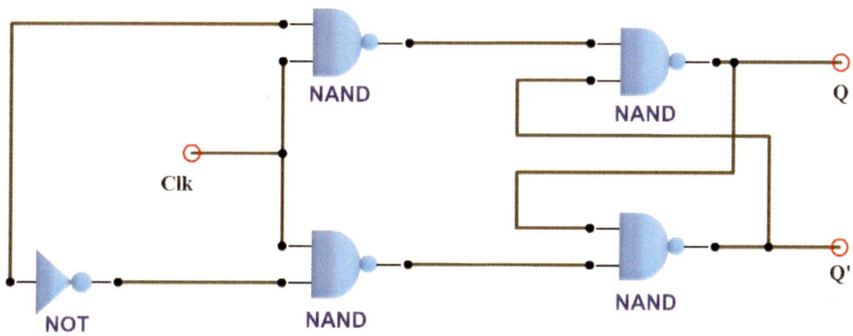

Working

- D flip flop will work depending on the clock signal.
- When the clock is low there will be no change in the output of the flip flop i.e. it remembers the previous state.
- When the clock signal is high and if it receives any data on its data pin, it Changes the state of output.
- When data is high Q reset to 0,while Q is set to 0 if data is low.

A master slave D flip flop can be constructed using D-flip flop.

J-K Flip Flop

JK flip – flop is named after Jack Kilby, an electrical engineer who invented IC.

A JK flip – flop is a modification of SR flip – flop. In this the J input is similar to the set input of SR flip – flop and the K input is similar to the reset input of SR flip – flop. The condition J = K = 1 which is not allowed in SR flip – flop (S = R = 1) is interpreted as a toggle command.

The JK flip flop has

- Two data inputs J and K.
- One clock signal input (CLK).
- Two outputs Q and Q'.

The symbol of a JK flip – flop is shown below.

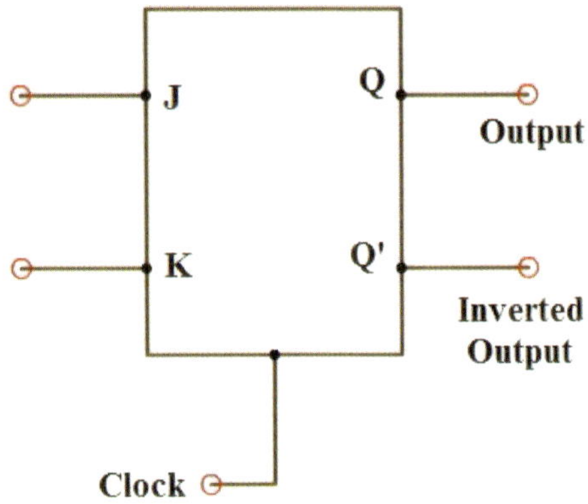

Truth Table

Clk	J	K	Q	Q'	State
1	0	0	Q	Q'	No change in state
1	0	1	0	1	Resets Q to 0
1	1	0	1	0	Sets Q to 1
1	1	1	-	-	Toggles

The circuit of a JK flip – flop using gates is shown below. It is similar to a modified NAND SR flip – flop.

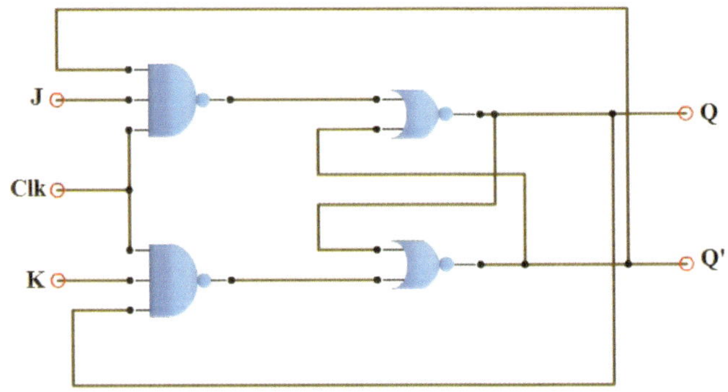

Working

- When J is low and K is low, then Q returns its previous state value i.e. it holds the current state.
- When J is low and K is high, then flip – flop will be in reset state i.e. Q = 0, Q' =1.
- When J is high and K is low then flip – flop will be in set state i.e. Q = 1, Q' =0.
- When J is high and K is high then flip – flop will be in Toggle state or flip state. This means that the output will complement to the previous state value.

T Flip Flop

T flip flop is also known as "Toggle Flip – flop". Toggle is to change the output to complement of the previous state in the presence of clock input signal.

The T flip flop has

- T input.
- One clock signal input (CLK).
- Two outputs Q and Q'.

The symbol of a T flip – flop is shown below.

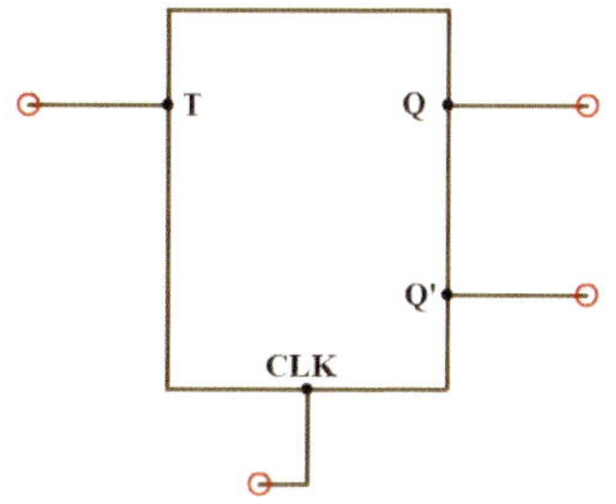

We can construct a T flip – flop by using any other flip – flops.

- SR flip – flop: By connecting the feedback of outputs of SR flip – flop to the inputs (S & R).
- D flip – flop: Connecting the Q' to its Data input of D flip – flop as feedback path.
- J K flip – flop: By combing the J & K inputs of JK flip – flop, to make as single input, we can design the T flip – flop.

Truth Table

T	Q	Q'
0	0	0
1	0	1
0	1	0
1	1	0

The circuit of a T flip – flop made from NAND JK flip – flop is shown below.

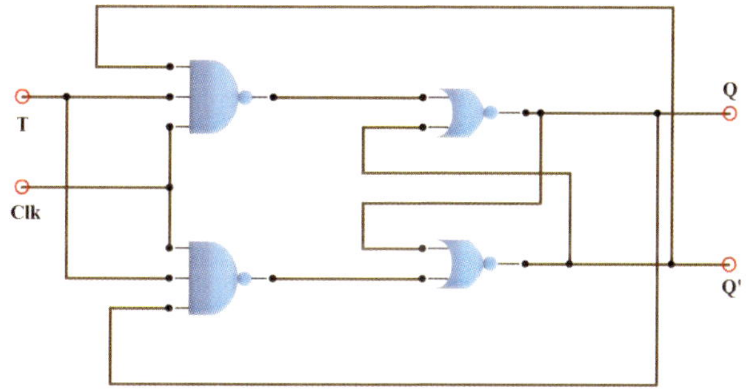

Working

The operation of the T flip flop is explained below.

When the T input is low, then the next sate of the T flip – flop is same as the present state i.e. it holds the current state.

- T = 0 and present state = 0 then the next state = 0.
- T = 0 and present state = 1 then the next state = 1.

When the T input is high, then the next sate of the T flip – flop is toggled i.e. it is same as the complement of present state on clock transition.

- T = 1 and present state = 0 then the next state = 1.
- T = 1 and present state = 1 then the next state = 0.

Master Slave Flip Flop

Shows the schematic diagram of master sloave J-K flip flop:

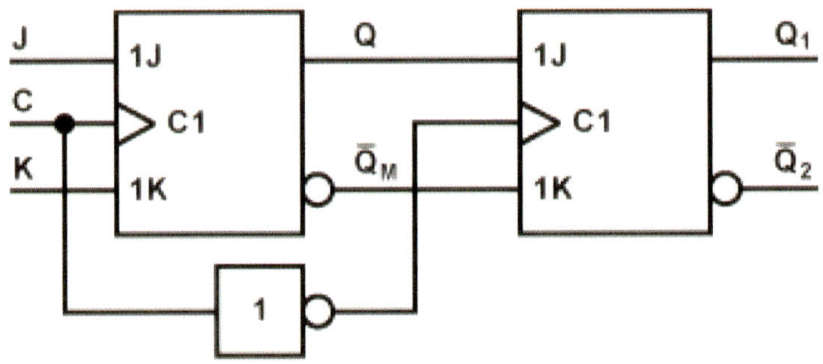

Master Slave JK Flip Flop

A master slave flip flop contains two clocked flip flops. The first is called master and the second slave. When the clock is high the master is active. The master slave flip flop output of the master is set or reset according to the state of the input. As the slave is incative during this period its output remains in the previous state. When clock becomes low the output of the slave flip flop changes because it become active during low clock period. The final output of master slave flip master slave flip flop is the output of the slave flip flop. So the output of master slave flip flop is available at the end of a clock pulse.

Uses of Flip-flops

Flip flops are widely used in:

- Registers: As the flip flops have two stable states, we use them in memory elements like registers, for data storage. Generally we use registers in electronic devices like computers.
- Counters: The groups of interconnected flip flops are uses as counters, to count the increment or decrement of an event occurrence.

- Frequency division: Flip flops are used as frequency division circuits, which divide the input frequency to exactly to its half. Frequency division circuits are used to regularize the frequency of electronic circuits.

- Data transfer: We use shift registers (A special-type of registers) to transfer the data from one flip flop to another, which are connected in a specific order.

Shift Register

The Shift Register is another type of sequential logic circuit that can be used for the storage or the transfer of binary data.

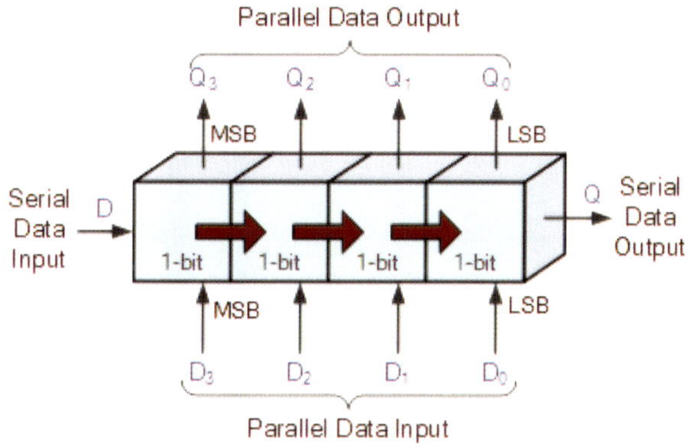

This sequential device loads the data present on its inputs and then moves or "shifts" it to its output once every clock cycle, hence the name Shift Register.

A shift register basically consists of several single bit "D-Type Data Latches", one for each data bit, either a logic "0" or a "1", connected together in a serial type daisy-chain arrangement so that the output from one data latch becomes the input of the next latch and so on.

Data bits may be fed in or out of a shift register serially, that is one after the other from either the left or the right direction, or all together at the same time in a parallel configuration.

The number of individual data latches required to make up a single Shift Register device is usually determined by the number of bits to be stored with the most common being 8-bits (one byte) wide constructed from eight individual data latches.

Shift Registers are used for data storage or for the movement of data and are therefore commonly used inside calculators or computers to store data such as two binary numbers before they are added together, or to convert the data from either a serial to parallel or parallel to serial format. The individual data latches that make up a single shift register are all driven by a common clock (Clk) signal making them synchronous devices.

Shift register IC's are generally provided with a clear or reset connection so that they can be "SET" or "RESET" as required. Generally, shift registers operate in one of four different modes with the basic movement of data through a shift register being:

- Serial-in to Parallel-out (SIPO) - the register is loaded with serial data, one bit at a time, with the stored data being available at the output in parallel form.

- Serial-in to Serial-out (SISO) - the data is shifted serially "IN" and "OUT" of the register, one bit at a time in either a left or right direction under clock control.
- Parallel-in to Serial-out (PISO) - the parallel data is loaded into the register simultaneously and is shifted out of the register serially one bit at a time under clock control.
- Parallel-in to Parallel-out (PIPO) - the parallel data is loaded simultaneously into the register, and transferred together to their respective outputs by the same clock pulse.

The effect of data movement from left to right through a shift register can be presented graphically as:

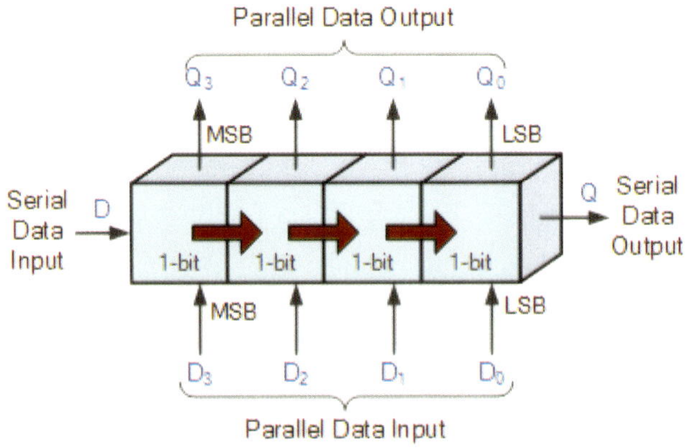

Also, the directional movement of the data through a shift register can be either to the left, (left shifting) to the right, (right shifting) left-in but right-out, (rotation) or both left and right shifting within the same register thereby making it bidirectional. In this topic it is assumed that all the data shifts to the right, (right shifting).

Serial-in to Parallel-out (SIPO) Shift Register

4-bit Serial-in to Parallel-out Shift Register

The operation is as follows. Lets assume that all the flip-flops (FFA to FFD) have just been RESET (CLEAR input) and that all the outputs Q_A to Q_D are at logic level "0" ie, no parallel data output.

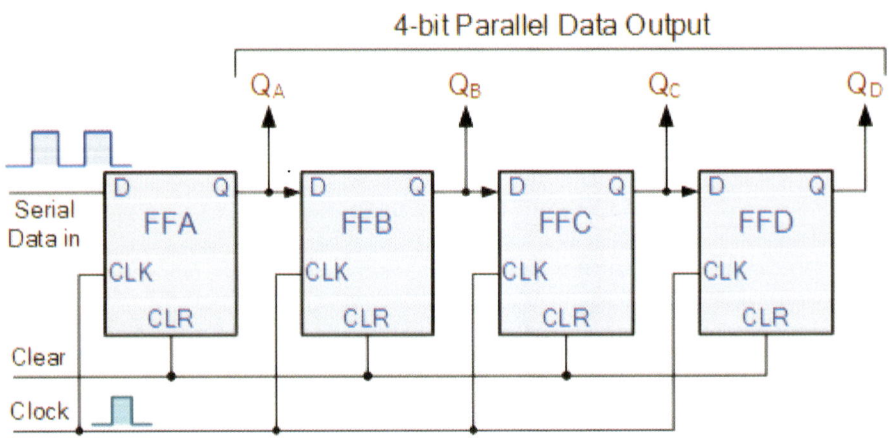

If a logic "1" is connected to the DATA input pin of FFA then on the first clock pulse the output of FFA and therefore the resulting Q_A will be set HIGH to logic "1" with all the other outputs still remaining LOW at logic "0". Assume now that the DATA input pin of FFA has returned LOW again to logic "0" giving us one data pulse or 0-1-0.

The second clock pulse will change the output of FFA to logic "0" and the output of FFB and Q_B HIGH to logic "1" as its input D has the logic "1" level on it from Q_A. The logic "1" has now moved or been "shifted" one place along the register to the right as it is now at Q_A.

When the third clock pulse arrives this logic "1" value moves to the output of FFC (Q_C) and so on until the arrival of the fifth clock pulse which sets all the outputs Q_A to Q_D back again to logic level "0" because the input to FFA has remained constant at logic level "0".

The effect of each clock pulse is to shift the data contents of each stage one place to the right, and this is shown in the following table until the complete data value of 0-0-0-1 is stored in the register. This data value can now be read directly from the outputs of Q_A to Q_D.

Then the data has been converted from a serial data input signal to a parallel data output. The truth table and following waveforms show the propagation of the logic "1" through the register from left to right as follows.

Basic Data Movement Through A Shift Register

Clock Pulse No	QA	QB	QC	QD
0	0	0	0	0
1	1	0	0	0
2	0	1	0	0
3	0	0	1	0
4	0	0	0	1
5	0	0	0	0

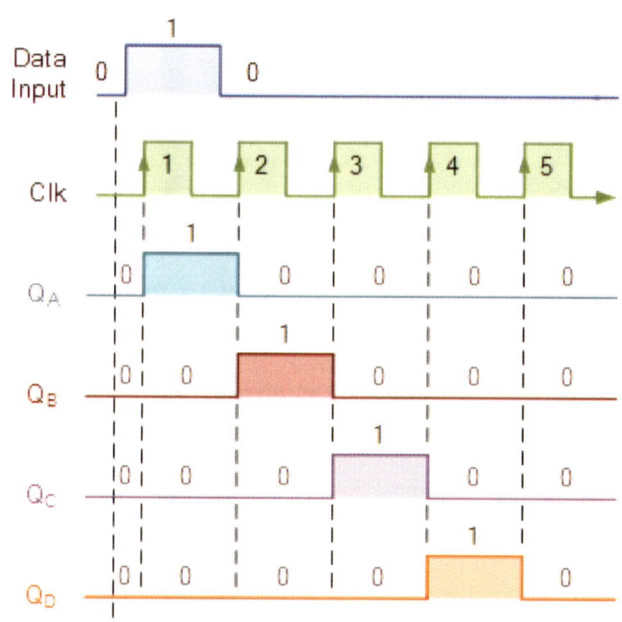

Note that after the fourth clock pulse has ended the 4-bits of data (0-0-0-1) are stored in the register and will remain there provided clocking of the register has stopped. In practice the input data to the register may consist of various combinations of logic "1" and "0". Commonly available SIPO IC's include the standard 8-bit 74LS164 or the 74LS594.

Serial-in to Serial-out (SISO) Shift Register

This shift register is very similar to the SIPO above, except were before the data was read directly in a parallel form from the outputs Q_A to Q_D, this time the data is allowed to flow straight through the register and out of the other end. Since there is only one output, the DATA leaves the shift register one bit at a time in a serial pattern, hence the name Serial-in to Serial-Out Shift Register or SISO.

The SISO shift register is one of the simplest of the four configurations as it has only three connections, the serial input (SI) which determines what enters the left hand flip-flop, the serial output (SO) which is taken from the output of the right hand flip-flop and the sequencing clock signal (Clk). The logic circuit diagram below shows a generalized serial-in serial-out shift register.

4-bit Serial-in to Serial-out Shift Register

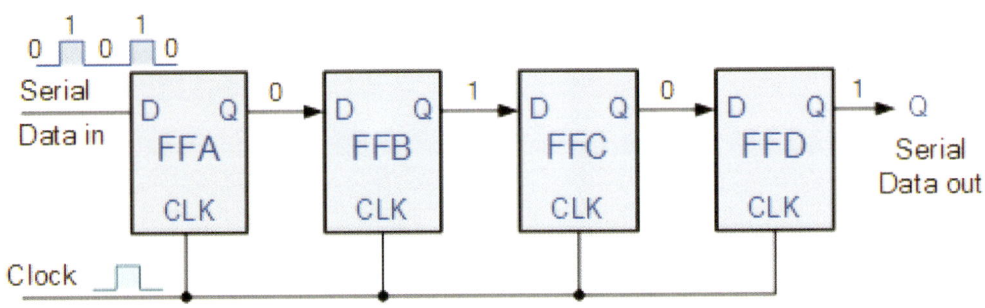

You may think what's the point of a SISO shift register if the output data is exactly the same as the input data. Well this type of Shift Register also acts as a temporary storage device or it can act as a time delay device for the data, with the amount of time delay being controlled by the number of stages in the register, 4, 8, 16 etc or by varying the application of the clock pulses. Commonly available IC's include the 74HC595 8-bit Serial-in to Serial-out Shift Register all with 3-state outputs.

Parallel-in to Serial-out (PISO) Shift Register

The Parallel-in to Serial-out shift register acts in the opposite way to the serial-in to parallel-out one above. The data is loaded into the register in a parallel format in which all the data bits enter their inputs simultaneously, to the parallel input pins P_A to P_D of the register. The data is then read out sequentially in the normal shift-right mode from the register at Q representing the data present at P_A to P_D.

This data is outputted one bit at a time on each clock cycle in a serial format. It is important to note that with this type of data register a clock pulse is not required to parallel load the register as it is already present, but four clock pulses are required to unload the data.

4-bit Parallel-in to Serial-out Shift Register

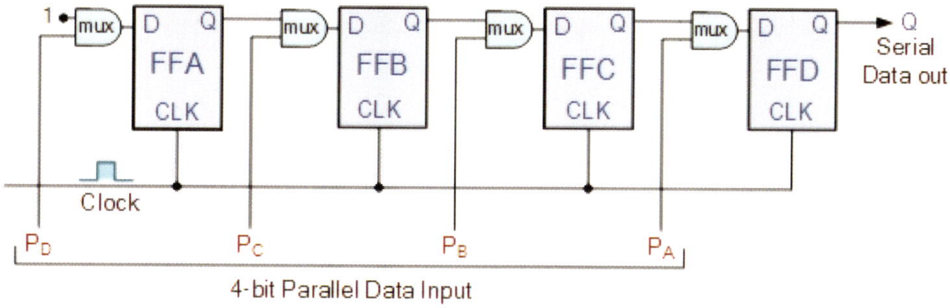

As this type of shift register converts parallel data, such as an 8-bit data word into serial format, it can be used to multiplex many different input lines into a single serial DATA stream which can be sent directly to a computer or transmitted over a communications line. Commonly available IC's include the 74HC166 8-bit Parallel-in/Serial-out Shift Registers.

Parallel-in to Parallel-out (PIPO) Shift Register

The final mode of operation is the Parallel-in to Parallel-out Shift Register. This type of shift register also acts as a temporary storage device or as a time delay device similar to the SISO configuration above. The data is presented in a parallel format to the parallel input pins P_A to P_D and then transferred together directly to their respective output pins Q_A to Q_A by the same clock pulse. Then one clock pulse loads and unloads the register. This arrangement for parallel loading and unloading is shown below.

4-bit Parallel-in to Parallel-out Shift Register

The PIPO shift register is the simplest of the four configurations as it has only three connections, the parallel input (PI) which determines what enters the flip-flop, the parallel output (PO) and the sequencing clock signal (Clk).

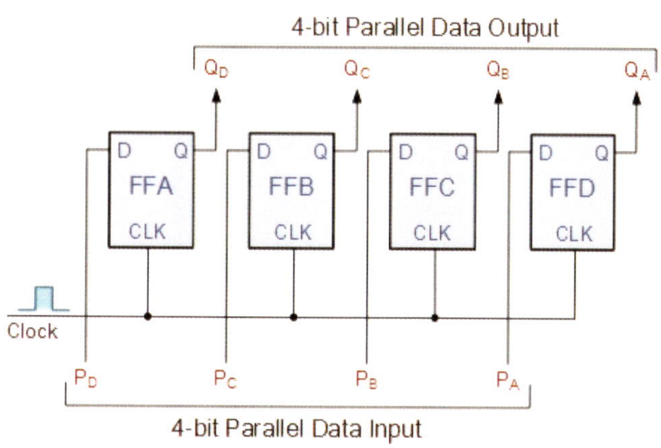

Similar to the Serial-in to Serial-out shift register, this type of register also acts as a temporary storage device or as a time delay device, with the amount of time delay being varied by the frequency of the clock pulses. Also, in this type of register there are no interconnections between the individual flip-flops since no serial shifting of the data is required.

Universal Shift Register

Today, there are many high speed bi-directional "universal" type Shift Registers available such as the TTL 74LS194, 74LS195 or the CMOS 4035 which are available as 4-bit multi-function devices that can be used in either serial-to-serial, left shifting, right shifting, serial-to-parallel, parallel-to-serial, or as a parallel-to-parallel multifunction data register, hence their name "Universal".

These universal shift registers can perform any combination of parallel and serial input to output operations but require additional inputs to specify desired function and to pre-load and reset the device. A commonly used universal shift register is the TTL 74LS194 as shown below.

4-bit Universal Shift Register 74LS194

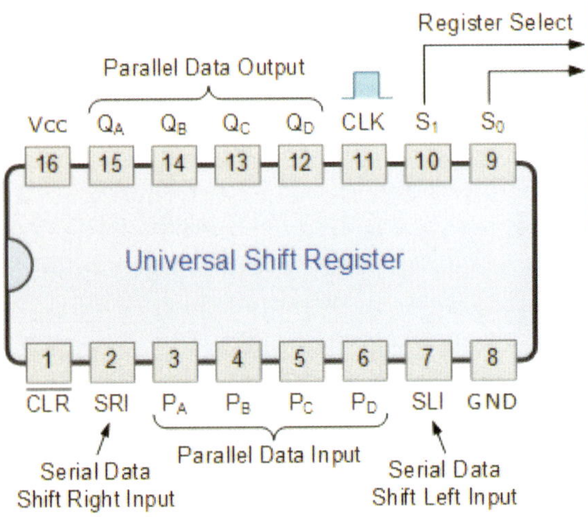

Mode Control		Register Type
S_1	S_0	
0	0	Hold
0	1	Shift to right
1	0	Shift to left
1	1	Parallel Mode

Universal shift registers are very useful digital devices. They can be configured to respond to operations that require some form of temporary memory storage or for the delay of information such as the SISO or PIPO configuration modes or transfer data from one point to another in either a serial or parallel format. Universal shift registers are frequently used in arithmetic operations to shift data to the left or right for multiplication or division.

Asynchronous Sequential Circuits

Asynchronous sequential circuits have state that is not synchronized with a clock. Like the synchronous sequential circuits we have studied up to this point they are realized by adding state feedback to combinational logic that implements a next-state function. Unlike synchronous circuits, the state variables of an asynchronous sequential circuit may change at any point in time. This asynchronous state update – from next state to current state – complicates the design process. We must be concerned with hazards in the next state function, as a momentary glitch may result in an incorrect final state. We must also be concerned with races between state variables on transitions between states whose encodings differ in more than one variable.

Flow Table Analysis

An asynchronous sequential circuit is formed when a feedback path is placed around combinational logic as shown in figure below. To analyze such circuits, we break the feedback path as shown in figure.

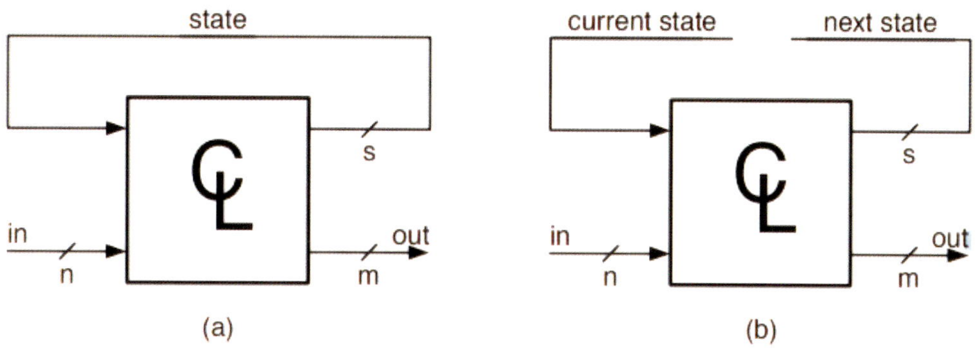

Figure: Asynchronous sequential circuit. (a) A sequential circuit is formed when a feedback path carrying state information is added to combinational logic. (b) To analyze an asynchronous sequential circuit, we break the feedback path and look at how the next state depends on the current state.

and write the equations for the next state variables as a function of the current state variables and the inputs. We can then reason about the dynamics of the circuit by exploring what happens when the current state variables are updated, in arbitrary order if multiple bits change, with their new values.

At first this may look just like the synchronous sequential circuits

In both cases we compute a next state based on current state and input. What's different is the dynamics of how the current state is updated with the next state. Without a clocked state register, the state of an asynchronous sequential circuit may change at any time (asynchronously). When multiple bits of state are changing at the same time (a condition called a race. The bits may change at different rates resulting in different end states. Also, a synchronous circuit will eventually reach a steady state where the next state and outputs will not change until the next clock cycle. An asynchronous circuit on the other hand may never reach a steady state. It is possible for it to oscillate indefinitely in the absence of input changes.

Consider the circuit shown in figure above. Each of the AND gates in the figure is labeled with the input state ab during which it is enabled. For example, the top gate, labeled 00, is enabled when a is high and b is low. To analyze the circuit we break the feedback loop as shown in Figure At this point we can write down the next-state function in terms of the inputs, a and b, and the current state. This function is shown in tabular form in the flow table of Figure.

Figure shows the next state for each of the eight combinations of inputs and current state. Input states are shown horizontally in Gray-code order. Current states are shown vertically. If the next state is the same as the current state, this state is stable since updating the current state with the next state doesn't change anything. If the next state is different than the current state, this state is transient since as soon as the current state is updated with the next state, the circuit will change states.

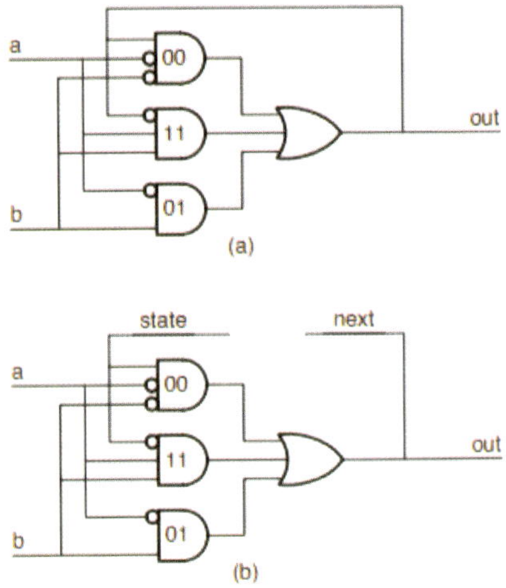

State	Next			
	00	01	11	10
0	(0)	1	1	(0)
1	(1)	(1)	0	0

(c)

Figure: An example asynchronous sequential circuit. (a) The original circuit. (b) With feedback loop broken. (c) Flow table showing next-state function. Circled entries in the flow table are stable states.

For example, suppose the circuit has inputs ab = 00 and the current state is 0. The next state is also 0, so this is a stable state - as shown by the circled 0 in the leftmost position of the top row of the table. If from this state input b goes high, making the input state ab = 01, we move one square to the right in the table. In this case, the 01 AND gate is enabled and the next-state is 1. This is an unstable or transient situation since the current state and next state are different. After some amount of time (for the change to propagate) the current state will become 1 and we move to the bottom row of the table. At this point we have reached a stable state since the current and next state are now both 1.

If there is a cycle of transient states with no stable states we have an oscillation. For example, if the inputs to the circuit of Figure are ab = 11, the next state is always the complement of the current state. With this input state, the circuit is never stable, but instead will oscillate indefinitely between the 0 and 1 states. This is almost never a desired behavior. An oscillation in an asynchronous circuit is almost always an error.

So, what does the circuit of Figure do? By this point the estute reader will have realized that its an RS flip-flop with an oscillation feature added. Input a is the reset input. When a is high and b is low, the state is made 0 when a is lowered the state remains 0. Similarly b is the set input. Making b high while a is low sets the state to 1 and it remains at 1 when b is lowered.

To simplify our analysis of asynchronous circuits we typically insist that the environment in which the circuits operate obey the fundamental mode restriction:

Fundamental-Mode: Only one input bit may be changed at a time and the circuit must reach a stable state before another input bit is changed.

A circuit operated in fundamental-mode need only worry about one input bit changing at a time.

Multiple-bit input changes are not allowed. Our setup and hold-time restrictions on flip-flops are an example of a fundamental-mode restriction. The clock and data inputs of the flip flop are not allowed to change at the same time. After the data input changes, the circuit must be allowed to reach a steady-state (setup time) before the clock input can change. Similarly, after the clock input changes, the circuit must be allowed to reach a steady-state (hold time) before the data input can change.

In looking at a flow-table, like the one in Figure, operating in the fundamental mode means that we need only consider input transitions to adjacent

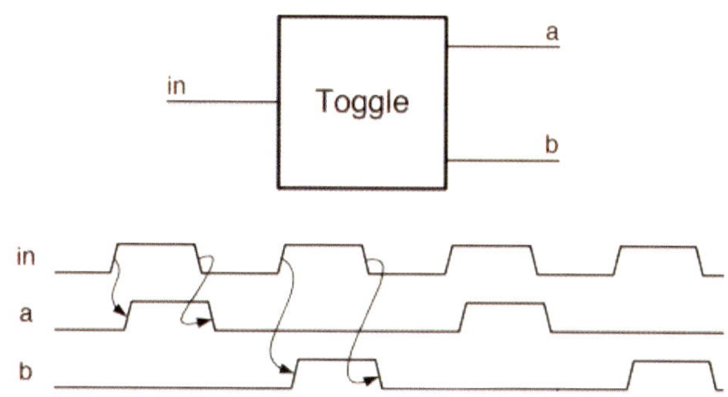

Figure: A toggle circuit alternates pulses on it input in between its two outputs a and b

squares (including wrapping from leftmost to rightmost). Thus, we don't have to worry about what happens when the input changes from 11 (oscillating) to 00 (storing). This can't happen. Since only one input can change at a time, we must first visit state 10 (reset) or 01 (set) before getting to 00.

In some real world situations, it is not possible to restrict the inputs to operate in fundamental mode. In these cases we need consider multiple input changes.

Flow-Table Synthesis: The Toggle Circuit

We now understand how to use a flow-table to analyze the behavior of an asynchronus circuit. That is, given a schematic, we can draw a flow table and understand the function of the circuit.

Consider the specification of a toggle circuit - shown graphically in Figure The toggle circuit has a single input in and two outputs a and b. 1 Whenever in is low, both outputs are low. The first time in goes high, output a goes high. On the next rising transition of in, output b goes high. On the third rising input, a goes high again. The circuit continues steering pulses on in alternately between a and b.

The first step in synthesizing a toggle circuit is to write down its flow table. We can do this directly from the waveforms of Figure Each transition of the input potentially takes us to a new state. Thus, we can partition the waveform into potential states as shown in Figure We start in state A. When in rises we go to state B where output a is high. when in falls again we go to state C. Even though C has the same output as A, we know its a different state because the next transition on in will cause a different output. The second rising edge on in takes us to state D with output b high. When in falls for the second time we go back to state A. We know that this state is the same as state A since the behavior of the circuit at this point under all possible inputs is indistinguishable from where we started.

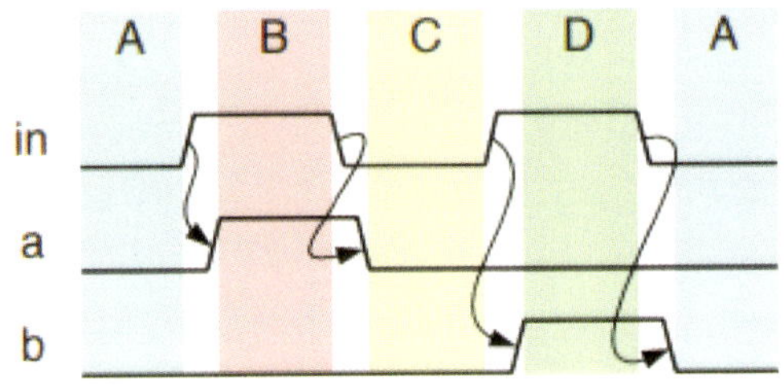

State	Next (in) 0	Next (in) 1	Out (a,b)
A	(A)	B	00
B	C	(B)	10
C	(C)	D	00
D	A	(D)	01

Figure: A flow table is created from the specification of the toggle circuit by creating a new state for every input transition until the circuit is obviously back to the same state.

Once we have a flow table for the toggle circuit, the next step is to assign binary codes to each of the states. This state assignment is more critical than with synchronous machines. If two states X and Y differ in more than one state bit, a transition from X to Y requires first visiting a transient state with one state bit changed before arriving at Y. In some cases, a race between the two state bits may result. For now, we pick a state assignment (shown in Figure where each state transition switches only a single bit).

With the state assignment, realizing the logic for the toggle circuit is a simple matter of combinational logic synthesis. We redraw the flow table as a Karnaugh map in Figure. The Karnaugh map shows the symbolic next state function - i.e., each square shows the next state name (A through D) for that input and current state. The arrows show the path through the states followed during operation of the circuit. Understanding this path is important for avoiding races and hazards. We refer to this Karnaugh map showing the state transitions as a trajectory map since it shows the trajectory of the state variables.

We redraw the Karnaugh map with state names replaced by their binary codes in Figure, and separate maps for the two state variables s_0 and s_1

State	Code	Next (in) 0	Next (in) 1	Out (a,b)
A	00	(A)	B	00
B	01	C	(B)	10
C	11	(C)	D	00
D	10	A	(D)	01

(a)

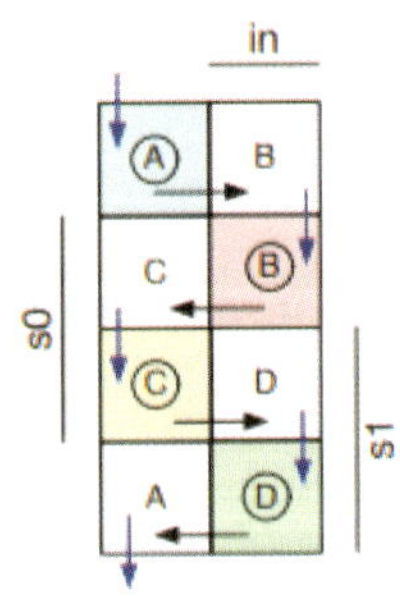

(b)

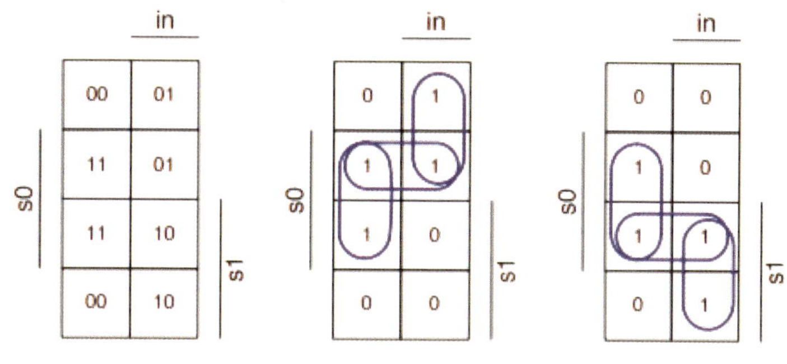

(c) (d) (e)

Figure: Implementing the toggle circuit from its flow table. (a) Flow table with state assignment. (b) Flow table mapped to Karnaugh map. (c) Next state codes mapped to Karnaugh map. (d) Karnaugh map for so. (e) Karnaugh map for s1.

are shown in Figure (d) and (e) respectively. From these Karnaugh maps we write down the equations for s_0 and s_1:

$$so = (\overline{s1} \wedge in) \vee (so \wedge \overline{in}) \vee (so \wedge \overline{s1}),$$
$$s1 = (s1 \wedge in) \vee (so \wedge \overline{in}) \vee (so \wedge s1).$$

The last implicant in each expression is required to avoid a hazard that would otherwise occur. Asynchronous circuits must be hazard free along their path through the input/state space. Because the current state is being constantly fed back a glitch during a state transition can result in the circuit switching to a different state - and hence not implementing the desired function. For example, suppose we left the $s_0 \wedge \overline{s1}$ term out of (22.2). When in goes low in state B, s_0 might go momentarily low before s_1 comes high. At this point the middle term of both equations becomes false and s1 never goes high - the circuit goes to state A rather than C.

All that remains to complete our synthesis is to write the output equations. Output a is true in state 01 and output b is true in state 10. The equations are thus:

$$a = \overline{s_1} \wedge s_0,$$
$$b = s_1 \wedge \overline{s_0}$$

Races and State Assignment

To illustrate the problem of multiple state variables changing simultaneously, consider an alternate state assignment for the toggle circuit shown in Figure. Here we observe that the two outputs, a and b can also serve as state variables, so we can add to the outputs just one additional state variable c to distinguish between states A and C giving the codes shown in the figure.

With this state assignment, the transition from state A (cab = 000) to state B (110) changes both c and a. If the logic is designed so that in going high in state A makes both c and a go high, they could change in an arbitrary order. Variable a could change first, variable c could change first, or they could change simultaneously. If they change simultaneously, we go from state A directly to state B with no intermediate stops. If a changes first, we go first to state 010 which is not assigned and then, if the logic in state 010 does the right thing, to state 110. If c changes first, the machine will go to state C (100) where the high input will then drive it to state D. Clearly, we cannot allow c to change first. This situation where multiple state variables can change at the same time is called a race. The state variables are racing to see which one can change first.

State	Code (c,a,b)	Next (in) 0	Next (in) 1	Out (a,b)
A	000	(A)	B	00
B	110	C	(B)	10
C	100	(C)	D	00
D	001	A	(D)	01

(a)

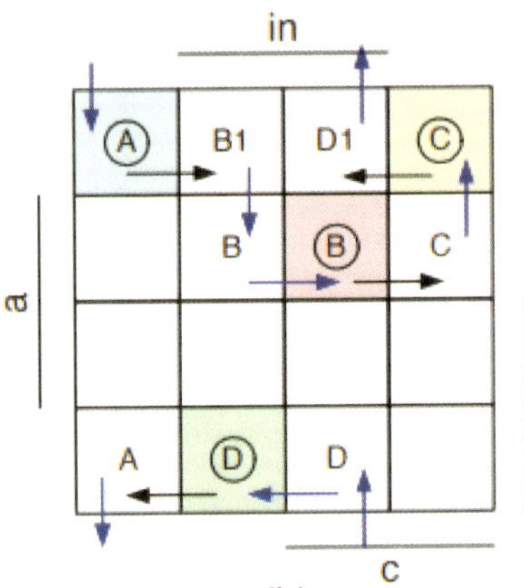

(b)

Figure: An alternate state assignment for the toggle circuit requires multiple state variables to change on a single transition. (a) The flow table with the revised state assignment. (b) A trajectory map showing the introduction of transient states B1 = 010 and D1 = 101.

When the outcome of the race affects the end state - as in this case - we call the race a critical race.

To avoid the critical race that could occur if we allow both a and c to change at the same time, we specify the next-state function so that only a can change in state A. This takes us to a transient state 010, which we will call B1. When the machine reaches state B1, the next state logic then enables c to change.

The introduction of this transient state is illustrated in the trajectory map of Figure. When the input goes high in state A, the next state function specifies B1 rather than B. This enables only a single transition, downward as shown by the blue arrow - which corresponds to a rising, to state B1. A change in c is not enabled in this state to avoid a horizontal transition into the square marked D1. Once the machine reaches state B1, the next state function becomes B which enables the change in c, a horizontal transition, to stable state B.

A transient state is also required for the transition from state C 100 to state D 001. Both variables c and b change between these two states. An uncontrolled race in which variable c changes first could wind up in state A 000 which is not correct. To prevent this race, we enable only b to change when in rises in state C. This takes us to a transient state D1 (101). Once in state D1, c is allowed to fall, taking us to state D (001).

Figure illustrates the process of implementing the revised toggle circuit. Figure shows a Karnaugh map of the next-state function. Each square of the Karnaugh map shows the code for the next state for that present state and input. Note that where the next state equals the present state the state is stable. Transient state B1 (at 010 - the second square along the diagonal) is not stable since it has a next state of 110.

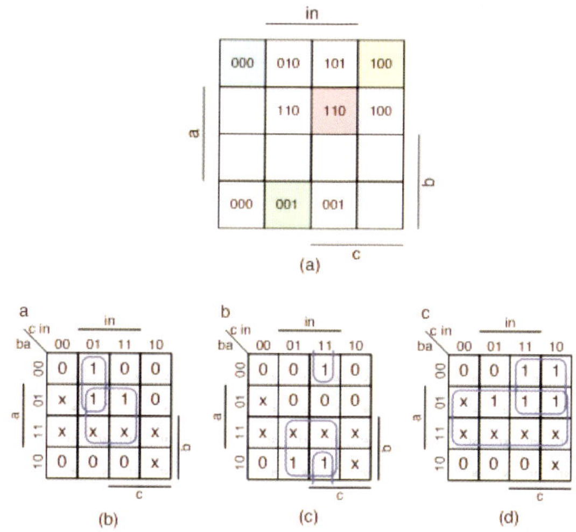

Figure: Implementation of the toggle circuit with the alternate state assignment of Figure. (a) Karnaugh map showing 3-bit next state function c,a,b. (b) Karnaugh map for a. (c) Karnaugh map for b. (d) Karnaugh map for c.

From the next-state Karnaugh map we can write the individual Karnaugh maps for each state variable. Figure (b) through (d) show the Karnaugh maps for the individual variables - a, b, and c respectively. Note that the states that are not visited along the state trajectory (the blank squares in Figure are don't cares. The machine will never be in these states, thus we don't care what the next state function is in an unvisited state.

From these Karnaugh maps we write the state variable equations as:

$$a = (in \wedge \bar{b} \wedge \bar{c}) \vee (in \wedge a),$$

$$b = (in \wedge \bar{a} \wedge c) \vee (in \wedge b),$$

$$c = a \vee (\bar{b} \wedge c).$$

Note that we don't require separate equations for output variables since a and b are both state variables and output variables.

Arithmetic Logic Unit

An arithmetic logic unit (ALU) is a major component of the central processing unit of a computer system. It does all processes related to arithmetic and logic operations that need to be done on instruction words. In some microprocessor architectures, the ALU is divided into the arithmetic unit (AU) and the logic unit (LU).

An ALU can be designed by engineers to calculate any operation. As the operations become more complex, the ALU also becomes more expensive, takes up more space in the CPU and dissipates more heat. That is why engineers make the ALU powerful enough to ensure that the CPU is also powerful and fast, but not so complex as to become prohibitive in terms of cost and other disadvantages.

An arithmetic logic unit is also known as an integer unit (IU).

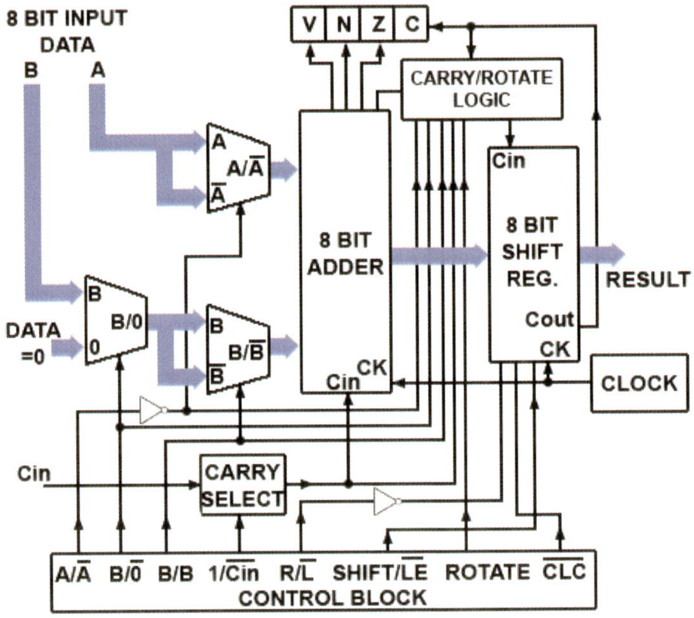

Fig. ALU Block Diagram

A simplified ALU is illustrated in Fig., which uses an arrangement of both combinational and sequential circuits. Their purpose is to perform the basic (though still complex) binary arithmetic.

Data passing through the ALU circuit does so on a system of buses, shown by the broad arrows in Fig. 5.8.1. These buses consist of groups of wires (usually as 8 parallel bits in simple systems) each carrying a single byte of binary data. In this system, data word A is the primary data source, and data word B is the secondary data source that may be added to, or subtracted from word A.

The ALU can also perform other operations. It can increment, add 1 to word A, or decrement, subtract 1 from it. By complementing (inverting) the logic value of individual bits of the data word A and adding 1 to the result, it is possible to use twos complement arithmetic to perform subtractions.

The shift register at the ALU output can also perform a 'logical shift-left' on word A by shifting the 8 bits consecutively into the carry bit, alternatively the shift register can create a rotating pattern of bits, rotating left, and using the carry bit as a ninth bit in the sequence, or rotate the 8 bits right ignoring the carry bit. Any of these functions can be selected by the control block, using various combinations of the eight control lines shown in Fig.

Putting the correct pattern of 1s and 0s (the control word) on the control lines will cause the ALU to perform the required arithmetic or logical operation on the data being input at A and B. With a control word of 8-bits, this could potentially allow up to 256 different combinations, or control words, which would be more than ample, even for very complex microprocessors or micro controllers. However this basic ALU needs only eight control words to control the different operations available.

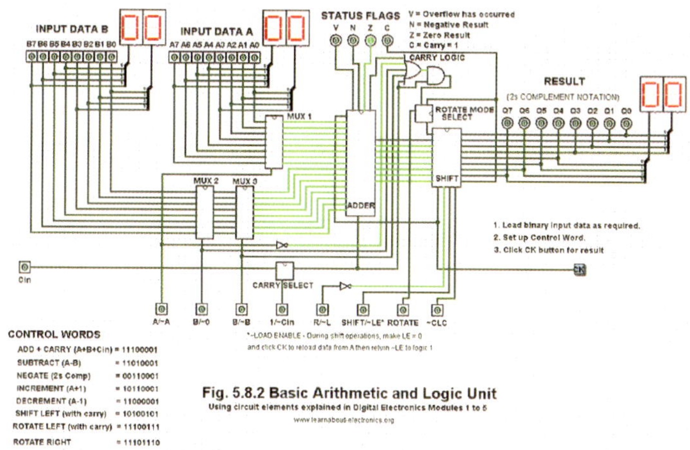

Fig. Basic Arithmetic and Logic Unit - Logisim Simulation

Component Parts

Any of the component parts of the Logisim design can be examined in detail by double clicking on the component (in simulation mode). To return to the main document, click 'main' in the component menu at the left of the screen.

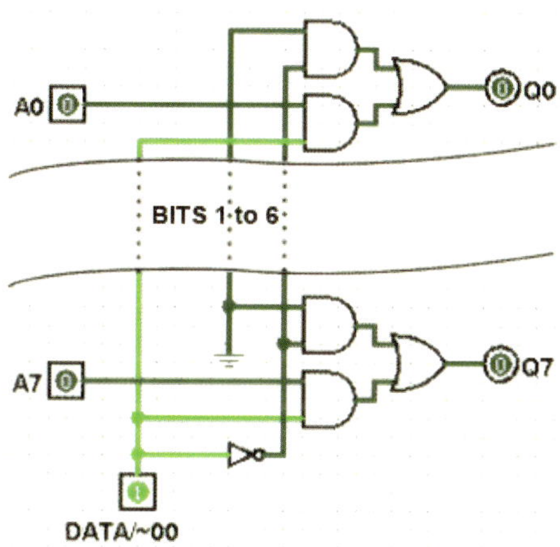

Fig. MUX 2

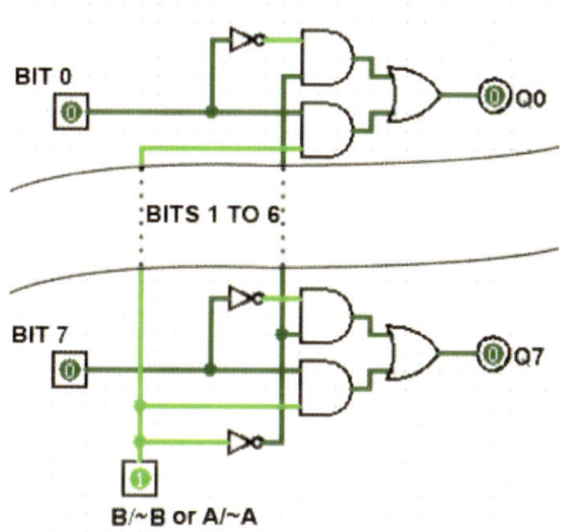

Fig. MUX 1 and MUX 3

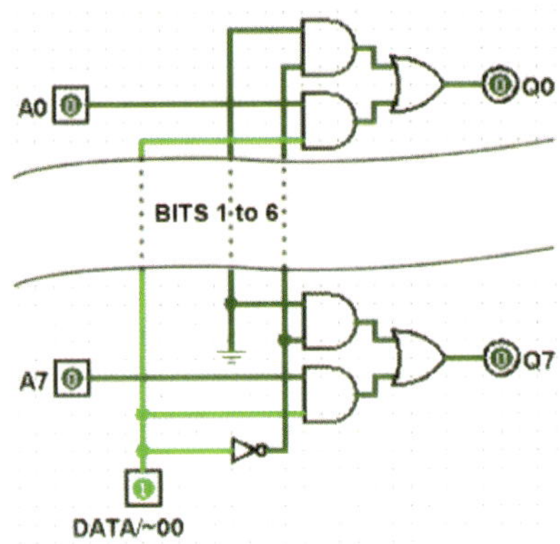

Fig. MUX 2

Multiplexers

MUX 1 and MUX 3 are identical 8 bit multiplexers that select either the input data word A (MUX 1) or data word B (MUX 3) or their internally generated complement, as shown in Fig.

MUX 2 is a similar design but selects either the data word B or the zero value 00_{HEX}, as shown in Figure.

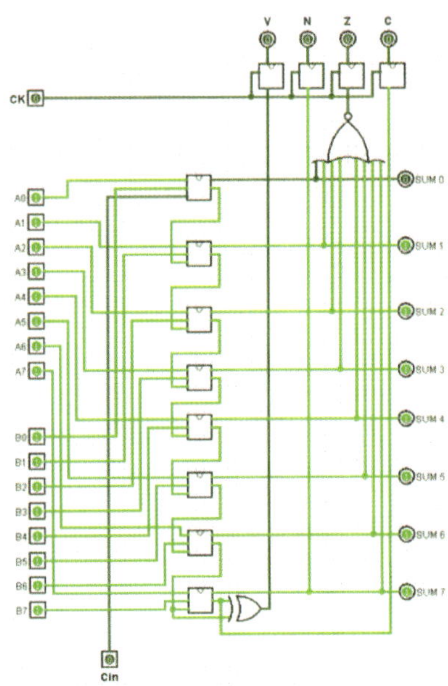

Fig. The ALU Adder Component

8-Bit Adder

The adder component is an 8-bit ripple carry adder; real ALUs would normally feature a 'carry-look-ahead' adder, allowing for high-speed operation. However for this example the much simpler ripple carry adder is adequate, as the operation is totally manual.

The adder component is illustrated in Figure and consists of eight full-adder circuits with additional logic consisting of an XOR gate to detect overflow errors, and an 8-input NOR gate to detect a zero result.

Negative results are indicated by sampling the most significant bit of the 'sum' output, and a 'carry' is indicated by sampling the carry output of the most significant full adder.

Four D type flip-flops are used as 'flag' outputs to indicate the current state of the ALU after each operation.

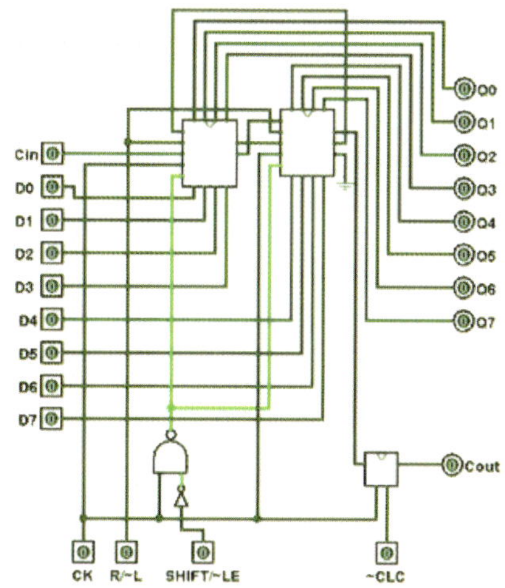

Fig. The ALU Shift Register Component

Shift Register

This component uses two 4-bit shift registers connected in cascade as shown in figure. Inputs are provided for clock pulses, (CK), a right/left shift control (R/~L) and an input to control whether the shift register is in shift, or load-enable modes (SHIFT/~LE).

If ~LE is chosen temporarily during shift operations, the shift register can be reloaded from the data placed on the 8-bit 'Data A' and 'carry-in' (CIN) inputs. This action is synchronised to the CK pulse by the external NAND and NOT gates connecting the SHIFT/~LE input to the two ~LOAD inputs of the 4-bit shift registers.

An additional JK flip-flop (mimicking a D type flip-flop) is placed between the 'serial-right' output of the shift register and COUT to allow the 'clear carry' input (~CLC) to clear the carry flag.

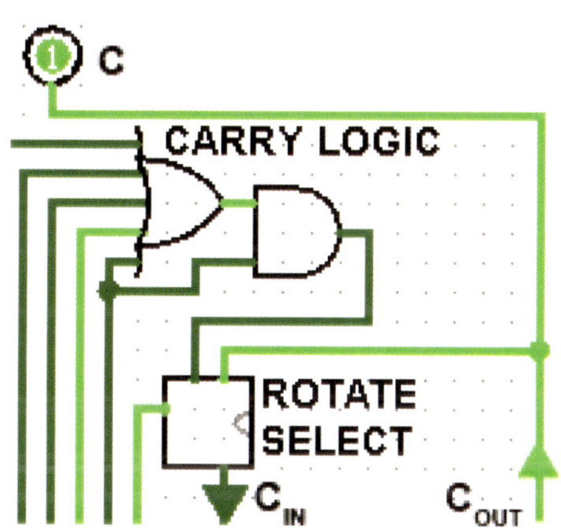

Fig. ALU Carry Logic

Carry Logic and Rotate Select

The carry logic circuit shown in Fig. prevents the carry flag being set in rotate right mode, as bits rotate from bit 0 and re-enter the shift register at bit 7, therefore allowing correct carry flag operation in both left and right rotate modes.

When the ROTATE input is at logic 1, the Rotate Select circuit in Figure allows COUT from the shift register to be fed back to the shift register CIN input for continuous bit rotation.

ALU Operation

Addition

To perform an addition, input data B is added to A. This is achieved by putting logic 1 on the control inputs of multiplexers 1, 2 and 3. This causes data A and B to be applied to the adder inputs. Also, to allow any carry bit from the CIN input to be included in the addition, the 1 bit carry multiplexer must have logic 0 on its control input. The shift register is only used as a PIPO register in addition mode, so its input lines R/~L and ROTATE must be at logic 0. SHIFT/~LE must also be at logic 0 to enable parallel loading of the shift register, which will hold the result of the addition (A plus B) after the application of a single CK pulse.

Status Flags

The Flag flip-flops are special outputs from the adder circuit. They consist of four separate D type flip-flops, each of which can be set to 1 or cleared to 0. They are set or cleared by the result in the adder. They signal, or 'flag' to the user, that a particular event has occurred.

Carry Flag (C)

The carry flag will be set if the result of any arithmetic or logic event causes a logic 1 to be carried over from bit 7 into the 'carry bit', (which is the carry flag). The carry flag can be cleared at any time by making the 'clear carry' input (~CLC) logic 0.

Overflow Flag (V)

When carrying out twos complement arithmetic, errors can occur if large numbers are involved. For example if two positive numbers less than 12710 are added and produce a negative result (any value greater than 12710). This would cause the sign of the result (indicated by bit 7) to be wrong. The overflow flag gives an indication that an error has occurred by being set to 1 to indicate an 'overflow error'. An error is sensed and the overflow flag is set when either of two conditions occurs.

There is a carry of logic 1 from bit 6 to bit 7 of the result, but the carry flip-flop is not at logic 1.

There is no carry from bit 6 to bit 7 of the result, but the carry flip-flop is at logic 1.

By using the carry-out from bit 6 and the carry-out from bit 7 of the result as inputs to an XOR gate, the output of the gate will be set to logic 1 for either of the above error conditions, signalling an overflow error at the overflow (V) flag.

Zero Flag (Z)

This flip-flop is set when every bit of the result is zero.

Negative Flag (N)

A negative result, i.e. bit 7 = 1 sets this flip-flop to logic 1.

Flag Register

The status flags are individual bits of a register called the Flag Register, and are operative not only when the ALU is in addition mode, but also in all other arithmetic modes, the C flag is also operative in shift and rotate left modes. In microprocessors the flag register not only indicates ALU results, but can also be used in decision-making. For example the ALU can be used to compare (by subtracting) two values and take various actions depending on the state of particular flags; e.g. after comparing two values, A and B, an action may be taken if A = B, indicated by the zero flag being set to 1, otherwise (if the zero flag is set to 0) take no action.

Subtraction

Subtraction is performed using twos complement arithmetic. That is, to subtract B from A, input B is complemented and 1 added to the complemented value to form the twos complement. Then the twos complement of B is added to A in the adder to find the result. To achieve this action with data A and data B present at the inputs, logic 1 is applied to the control inputs of MUX 1 and MUX 2. MUX 3 has logic 0 applied to its control input to complement data B, while the CARRY MUX has a logic 1 applied to its control line so that the carry-in (CIN) to the adder is forced to logic 1. This adds 1 to the result so that the twos complement of data B is added to data A. The result at the adder output is a twos complement number representing A - B. The flags are again set by the result as in the addition operation.

Counting with the ALU

Although the ALU does not include a binary counter circuit, it can also be used to count, by INCREMENTING or DECREMENTING, i.e. to add 1 to data A (incrementing), or subtract 1 from data A (decrementing). To count using this method would normally be carried out using (machine code or assembly language)software. A typical use could be to initiate a time delay by loading the ALU with some number, and then execute a looping routine to count down to zero by repeatedly decrementing data A. The zero result would be detected from the zero flag being set. However this would not be a common method, as the ALU (and therefore the CPU) would be occupied during the delay, and therefore not usable for other purposes. Most computer systems would also have dedicated counters for implementing similar time delays.

Incrementing

Data A can be incremented if logic 1 is applied to the control inputs of MUX 1 and MUX 3. This will add B to A, with data B made zero by applying logic 0 to the control input of MUX 2. The 1 that must be added to data A is supplied by making the control input of the CARRY SELECT block logic

1, causing the carry input to the adder to be logic 1. The result at the adder output is therefore A + 1, again the flags are set by the result.

Decrementing

To decrement data A, 1 must be subtracted from A. Because the ALU uses twos complement arithmetic, the twos complement of 1 added to A will in effect subtract 1 from A.

The twos complement of 1 is minus 1, which in 8-bit twos complement notation is 111111112. Therefore to subtract 1 from data A, data B must equal minus 1 (all bits = 1). To do this, and to make sure that the correct result is not changed by any data appearing on the data B input, logic 0 is applied to the control input of MUX 2 to make sure all data B bits = 0.

Logic 0 is also applied to the control input of MUX 3. This inverts data B, (which is 000000002) to give 111111112 at the adder input.

MUX 1 must have logic 1 on its control line, to apply data A to the other adder input. The adder's carry input is set to 1 by applying logic 0 to the control line of the CARRY MUX. This ensures that, provided there is no carry-in on the CIN input, the correct result at the adder output will be A – 1.

Negation

Negation is simply the inverse of a value; therefore any value and its inverse will add to produce zero. In binary arithmetic the additive inverse of a value is its twos complement. The ALU can be used to negate (find the twos complement of) data A by complementing data A and then adding 1. This involves a similar process to decrementing, except that data B is treated differently, as follows:

The control input of MUX 1 is set to logic 0, which complements data A, also data B is made zero by putting logic 0 on MUX 2 control, and logic 1 on MUX 3. The Carry Select control input is set at logic 1, to add 1 to data A in the adder.

The shift register is used as a simple PIPO register by applying logic 0 to the three shift controls and logic 1 to the ~CLC input to make sure the carry is not cleared. This gives a final result of A+1, which is the twos complement of A.

The Shift Operations

Shift operations are controlled by the four lower order control lines, R/~L controls the direction of shift or rotation, SHIFT/~LE has the dual purpose of enabling the shift operations if logic 1 is applied, or acting as a LOAD ENABLE when at logic 0, allowing the shift register to be loaded or reloaded with appropriate data. Each action of the shift register (shift, rotate or load) is actuated by a single CK pulse. Also note that the shift register in this design does not affect the V, N or Z flags.

Shift Left (with Carry)

In this mode (with control word 10100101) input data B is kept at zero and, after the shift register is loaded by temporarily making SHIFT/~LE logic 0 to move data from input A into the shift register, shift is enabled by returning SHIFT/~LE to logic 1, and both ROTATE and ~CLC are disabled. The data in the shift register will now shift one bit to the left with each CK pulse applied. This appears to multiply the value of the data by two for each shift left, but it is a very limited multiplication operation, because the result is reduced each time the left most bit is lost as it passes through the carry bit. This action is therefore considered a logical, rather than an arithmetic shift.

Rotate Left (with Carry)

If rotate is activated by applying logic 1 to the ROTATE control input with SHIFT/~LE and ~CLC also at logic 1, the data being shifted left from bit 7 and through the carry flag, is returned via the CIN input of the shift register to re-enter at bit 0 by the action of the ROTATE MODE SELECT data selector.

Rotate Right

When data in the shift register is rotated right, it leaves the register via bit 0 and is returned directly to bit 7 via an internal link, without passing through the carry flag.

There are a number of other operations, such as performing 8 bit logic functions, commonly found on microprocessors that this ALU is not designed to do. The purpose of this design is to illustrate how the circuits described in Digital Electronics they can be inter-connected in many ways to make many different circuits.

Counter

Counter is a digital device and the output of the counter includes a predefined state based on the clock pulse applications. The output of the counter can be used to count the number of pulses. Generally, counters consist of a flip-flop arrangement which can be synchronous counter or asynchronous counter. In synchronous counter, only one clock i/p is given to all flip-flops, whereas in asynchronous counter, the o/p of the flip flop is the clock signal from the nearby one. The applications of the microcontroller need counting of exterior events such as exact internal time delay generation and the frequency of the pulse trains. These events are frequently used in digital systems & computers. Both these events can be executed by software techniques, but software loops for counting will not give the exact result slightly more important functions are not done. These problems can be rectified by timers and counters in the microcontrollers which are used as interrupts.

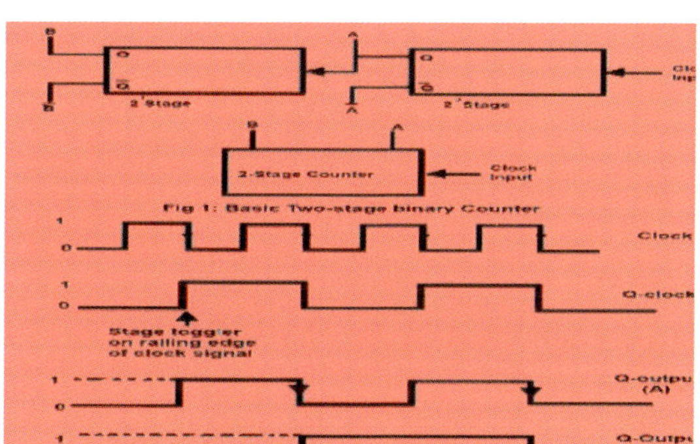

Counters

Types of Counters

Counters can be categorized into different types according to the way they are clocked. They are

- Asynchronous Counters
- Synchronous Counters
- Asynchronous Decade Counters
- Synchronous Decade Counters
- Asynchronous Up-Down Counters
- Synchronous Up-Down Counters

For better understanding of this type of counters, here we are discussing some of the counters.

Asynchronous Counters

The diagram of a 2-bit asynchronous counter is shown below. The exterior clock is connected to the clock i/p of the FFo (first flip-flop) only. So, this FF changes the state at the decreasing edge of every clock pulse, but FF1 changes only when activated by the decreasing edge of the Q o/p of FFo. Because of the integral propagation delay through a FF, the change of the i/p clock pulse and a change of the Q o/p of FFo can never occur at precisely the same time. So, the FF's cannot be activated concurrently, generating an asynchronous operation.

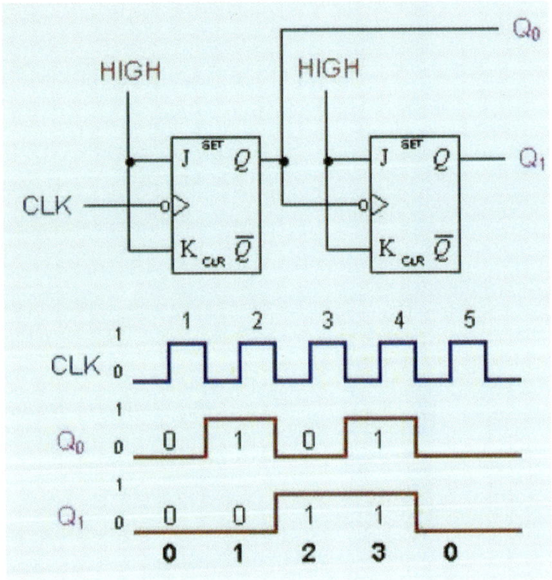

Asynchronous Counters

Note that for ease, the changes of Qo,Q1 & CLK in the above diagram are shown as concurrent, even though this is an asynchronous counter. Actually, there is a small delay b/n the Qo, Q1 and CLK changes.

Generally, all the CLEAR i/ps are connected together, so before counting starts then that a single pulse can clear all the FFs. The clock pulse fed into FFo is rippled through the new counters after propagation delays, such as a ripple on the water, hence the term Ripple Counter.

The circuit diagram of the two bit ripple counter includes four different states ,each one consisting with a count value. Likewise, a counter with n FFs can have 2N states. The number of states in a counter is called as its mod number. Therefore a two-bit counter is a mod-4 counter.

Asynchronous Decade Counters

In the previous counter have 2n states. But, counters with states less than 2n is also possible. These are designed to have the no. of states in their series.These are called shortened sequences which are accomplished by driving the counter to recycle before going through all of its states. A common modulus for counters with shortened sequence is 10. A counter with 10-states in its series is called a decade counter.The implemented decade counter circuit is given below.

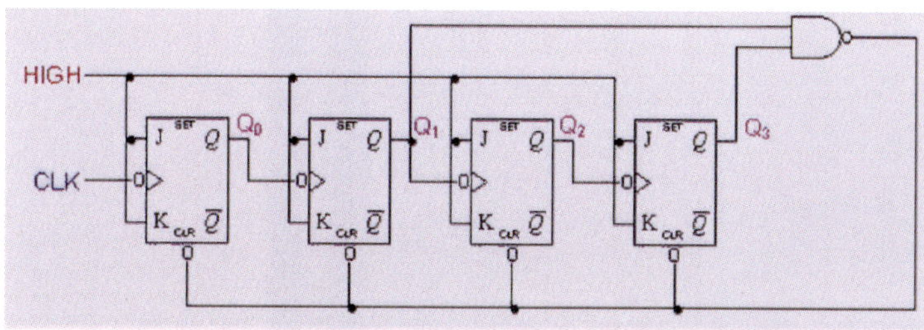

Asynchronous Decade Counter Circuit Diagram

When the counter counts to ten, then all the FFs will be cleared. Notice that only Q1&Q3 both are used to decode the count of 10, that is called partial decoding. At the same time one of the other states from 0-9 have both Q1&Q3 will be high. The series of the decade counter table is given below.

Clock Pulse	Q3	Q2	Q1	Q0
0	0	0	0	0
1	0	0	0	1
2	0	0	1	0
3	0	0	1	1
4	0	1	0	0
5	0	1	0	1
6	0	1	1	0
7	0	1	1	1
8	1	0	0	0
9	1	0	0	1

Sequence of the Decade Counter

Asynchronous Up-Down Counters

In particular applications, a counter must be capable to count both up & down. The below circuit is a three bit up & down counter, that counts UP or DOWN based on the control signal status. When the UP i/p is at 1 & the DOWN i/p is at 0, the NAND gate between FF0 & FF1 will gate the non-inverted o/p (Q) of flip flop (FF0) into the clock i/p of flip flop (FF1). Likewise, the non-inverted o/p of Flip Flop1 will be gated through the other NAND gate into the clock i/p of flip-flop2. Therefore the counter will count up.

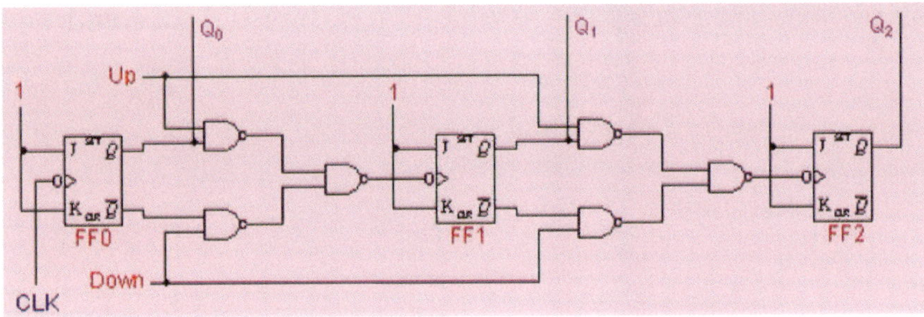

Asynchronous Up-Down Counter Circuit Diagram

Once the control i/p (UP) is at 0 & DOWN is at 1, the inverted o/ps of flip-flop0 (FF0) and flip-flop1 (FF) are gated into the clock i/ps of FF1 & FF2 separately. If the FFs are initially changed to 0's, then the counter will go through the below series as i/p pulses are applied. Notice that an asynchronous up-down counter is slower than an UP counter/down counter because of an extra propagation delay introduced by the NAND gates.

FF2	FF1	FF0
0	0	0
1	1	1
1	1	0
1	0	1
1	0	0
0	1	1
0	1	0
0	0	1

Sequence of the Asynchronous Up-Down Counter

Synchronous Counters

In this type of counters, the CLK i/ps of all the FFs are connected together and are activated by the i/p pulses. So, all the FFs change states instantaneously. The circuit diagram below is a three bit synchronous counter. The inputs J and K of flip-flop0 are connected to HIGH. Flip-flop 1 has its J &K i/ps connected to the o/p of flip-flop0 (FF0), and the inputs J & K of flip-flop2 (FF2) are connected to the o/p of an AND gate that is fed by the o/ps of flip-flop0 and flip-flop1. When the both the outputs of FF0 & FF1 are HIGH. The positive edge of the fourth CLK pulse will cause FF2 to alter its state because of the AND gate.

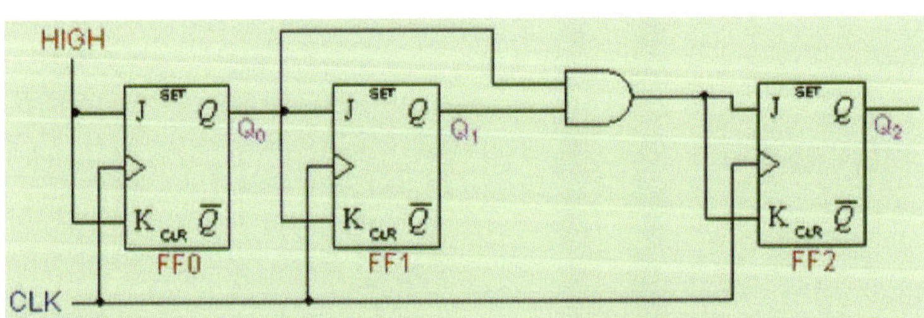

Synchronous Counter Circuit Diagram

The series of the three bit counter table is given below.The major advantage of these counters is that there is no increasing time delay due to all FFs are activated in parallel. Thus, the max operating frequency of this synchronous counter will be considerably higher than for the equivalent ripple counter.

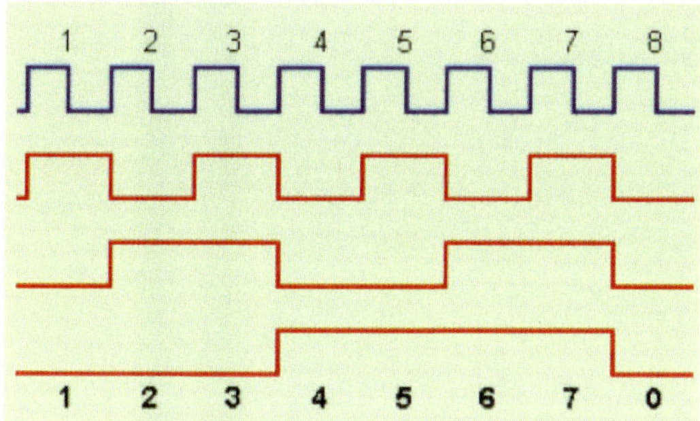

CLK Pulses of the Synchronous Counters

Synchronous Decade Counters

Synchronous counter counts from 0-9 similar to asynchronous counter and then again recycles zero. This process is done by driving the 1010 states back to the 0000 state. This is termed as truncated sequence, that can be designed by the below circuit.

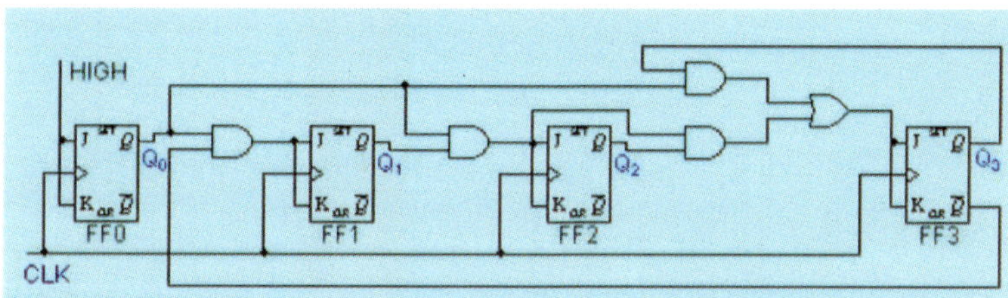

Synchronous Decade Counter Circuit Diagram

From the series on the left table, we can observe that

- Q0 ties on each and every CLK pulse
- Q1 alters on the next clock pulse every time when Q0=1 & Q3=0.
- Q2 alters on the next clock pulse every time when Q0=Q1=1.
- Q3 alters on the next CLK pulse each and every time when Q0=1, Q1=1 & Q2=1 (count 7), or when Q0=1 & Q3=1 (count 9).

Clock Pulse	Q3	Q2	Q1	Q0
0	0	0	0	0
1	0	0	0	1
2	0	0	1	0
3	0	0	1	1
4	0	1	0	0
5	0	1	0	1
6	0	1	1	0
7	0	1	1	1
8	1	0	0	0
9	1	0	0	1

Sequence of the Synchronous Decade Counter

The above characteristics are employed with the AND gate or OR gate. The logic diagram of this is shown in the above diagram.

Synchronous Up-Down Counters

A three bit synchronous Up-Down counter, tabular form and series are given below. This type of counter has an up-down control i/p similar to asynchronous up-down counter, that is used to control the counter's direction through a certain series.

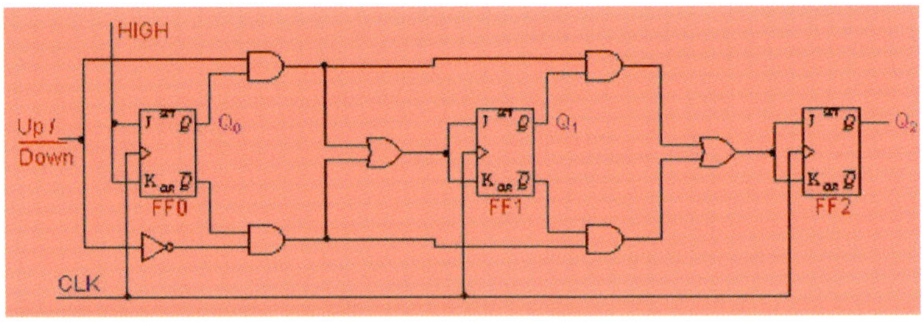

Synchronous Up-Down Counters Circuit Diagram

The series of the table shows

- Q0 ties on each CLK pulse for both up & down series
- When Q0=1 for the up series, then the state of the Q1 changes on the next CLK pulse.
- When Q0=0 for the down series, then the state of the Q1 changes on the next CLK pulse.
- When Q0=Q1=1 for the up series, then the state of the Q2 changes on the next CLK pulse.
- When Q0=Q1=0 for the down series, then the state of the Q2 changes on the next CLK pulse.

Clock Pulse	Q3	Q2	Q1	Q0
0	0	0	0	0
1	0	0	0	1
2	0	0	1	0
3	0	0	1	1
4	0	1	0	0
5	0	1	0	1
6	0	1	1	0
7	0	1	1	1
8	1	0	0	0
9	1	0	0	1

Sequence of the Synchronous Decade Counters

The above characteristics are employed with the AND gate, OR gate and NOT gate. The logic diagram of this is shown in the above diagram.

Applications of Counters

The applications of the counters mainly involve in digital clocks and in multiplexing. The best example of the counter is parallel to serial data conversion logic discussed below.

A set of bits, performing concurrently on parallel lines is called parallel data. A set of bits, performing on a single line in a time series is called serial data. The Parallel-to-serial data conversion is normally is done by using a counter to afford a binary series of the data, select i/ps of a MUX, as explained in the circuit below.

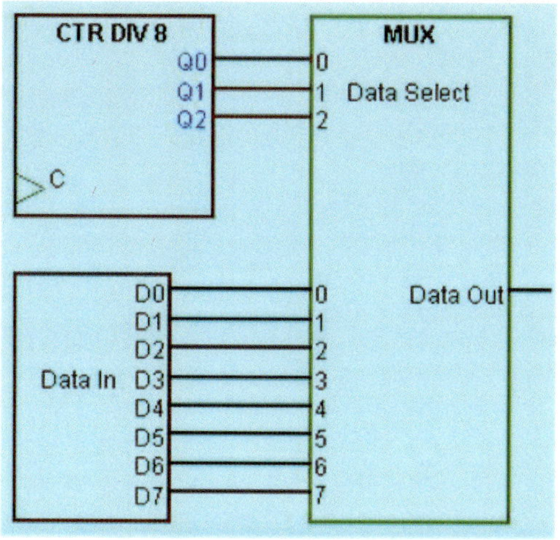

Parallel-to-Serial Data Conversion

In the above circuit, modulo-8 counter consist of Q o/ps, that are connected to the data, select i/ps of an 8-bit MUX. The first 8-bit group of parallel data is applied to the inputs of the MUX. As the counter goes through a binary series from 0-7, each bit starts with D0, is serially selected & passed through the MUX to the o/p line. After 8-CLK pulses, the data byte has been changed to a serial format & sent out through the transmission line. Then, the counter reprocesses back to 0 and changes another parallel byte serially again in the similar process.

References

- Brent, Richard Peirce; Kung, Hsiang Te (March 1982). "A Regular Layout for Parallel Adders". IEEE Transactions on Computers. C–31 (3): 260–264. doi:10.1109/TC.1982.1675982. ISSN 0018-9340
- What-is-a-digital-circuit: wisegeek.com, Retrieved 29 March 2018
- Han, Tackdon; Carlson, David A.; Levitan, Steven P. (October 1982). "VLSI design of high-speed, low-area addition circuitry". Proceedings 1981 IEEE International Conference on Computer Design: VLSI in Computers & Processors. IEEE: 418–422. ISBN 0-81860802-1
- Difference-between-analog-circuit-and-digital-circuit: elprocus.com, Retrieved 29 May 2018
- Lynch, Thomas Walker (May 1996). "Binary Adders" (Thesis). University of Texas. Archived (PDF) from the original on 2018-04-14. Retrieved 2018-04-14
- Half-adder-and-full-adder: elprocus.com, Retrieved 19 July 2018
- Kogge, Peter Michael; Stone, Harold S. (August 1973). "A Parallel Algorithm for the Efficient Solution of a General Class of Recurrence Equations". IEEE Transactions on Computers. C–22 (8): 786–793. doi:10.1109/TC.1973.5009159
- Carry-look-ahead-adder: electronicshub.org, Retrieved 30 June 2018
- *Kochanski, Martin (2003-08-19).* "A New Method of Serial Modular Multiplication" *(PDF).* Archived *(PDF) from the original on 2018-07-16.* Retrieved 2018-07-16
- Binary-multiplication: electronicshub.org, Retrieved 3 July 2018

Permissions

All chapters in this book are published with permission under the Creative Commons Attribution Share Alike License or equivalent. Every chapter published in this book has been scrutinized by our experts. Their significance has been extensively debated. The topics covered herein carry significant information for a comprehensive understanding. They may even be implemented as practical applications or may be referred to as a beginning point for further studies.

We would like to thank the editorial team for lending their expertise to make the book truly unique. They have played a crucial role in the development of this book. Without their invaluable contributions this book wouldn't have been possible. They have made vital efforts to compile up to date information on the varied aspects of this subject to make this book a valuable addition to the collection of many professionals and students.

This book was conceptualized with the vision of imparting up-to-date and integrated information in this field. To ensure the same, a matchless editorial board was set up. Every individual on the board went through rigorous rounds of assessment to prove their worth. After which they invested a large part of their time researching and compiling the most relevant data for our readers.

The editorial board has been involved in producing this book since its inception. They have spent rigorous hours researching and exploring the diverse topics which have resulted in the successful publishing of this book. They have passed on their knowledge of decades through this book. To expedite this challenging task, the publisher supported the team at every step. A small team of assistant editors was also appointed to further simplify the editing procedure and attain best results for the readers.

Apart from the editorial board, the designing team has also invested a significant amount of their time in understanding the subject and creating the most relevant covers. They scrutinized every image to scout for the most suitable representation of the subject and create an appropriate cover for the book.

The publishing team has been an ardent support to the editorial, designing and production team. Their endless efforts to recruit the best for this project, has resulted in the accomplishment of this book. They are a veteran in the field of academics and their pool of knowledge is as vast as their experience in printing. Their expertise and guidance has proved useful at every step. Their uncompromising quality standards have made this book an exceptional effort. Their encouragement from time to time has been an inspiration for everyone.

The publisher and the editorial board hope that this book will prove to be a valuable piece of knowledge for students, practitioners and scholars across the globe.

Index